Nikolaos Halidias
**Calculus of One Variable**

## Also of Interest

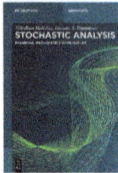

*Stochastic Analysis*
*Financial Mathematics with Matlab®*
Nikolaos Halidias, Ioannis S. Stamatiou, 2025
ISBN 978-3-11-144285-3, e-ISBN 978-3-11-144373-7

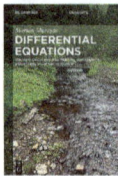

*Differential Equations*
*Solving Ordinary and Partial Differential Equations with Mathematica®*
Marian Mureşan, 2024
ISBN 978-3-11-141109-5, e-ISBN (PDF) 978-3-11-141139-2

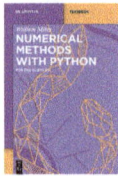

*Numerical Methods with Python for the Sciences*
William Miles, 2023
ISBN 978-3-11-077645-4, e-ISBN (PDF) 978-3-11-077664-5

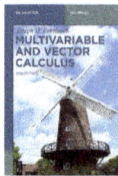

*Multivariable and Vector Calculus*
Joseph D. Fehribach, 2024
ISBN 978-3-11-139238-7, e-ISBN (PDF) 978-3-11-139348-3

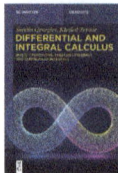

*Differential and Integral Calculus*
*Implicit Functions, Stieltjes Integrals and Curvilinear Integrals*
Svetlin G. Georgiev, Khaled Zennir, 2026
ISBN 978-3-11-914462-9, e-ISBN (PDF) 978-3-11221808-2

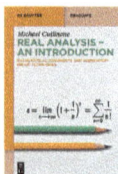

*Real Analysis - An Introduction*
*Mathematical Arguments and Elementary Proof Techniques*
Michael Cullinane, 2025
ISBN 978-3-11-142928-1, e-ISBN (PDF) 978-3-11-142956-4

Nikolaos Halidias

# Calculus of One Variable

——

with Python and AI

**DE GRUYTER**

**Mathematics Subject Classification 2020**
Primary: 26-01, 26-04, 24A06; Secondary: 97I10, 97I40, 97I50

**Author**
Prof. Dr. Nikolaos Halidias
Department of Statistics and
Actuarial-Financial Mathematics
University of the Aegean
Samos
Greece
nick@aegean.gr

ISBN 978-3-11-914171-0
e-ISBN (PDF) 978-3-11-222828-9
e-ISBN (EPUB) 978-3-11-222829-6

**Library of Congress Cataloging-in-Publication Data**
A CIP catalog record for this book has been applied for at the Library of Congress.

**Bibliographic information published by the Deutsche Nationalbibliothek**
The Deutsche Nationalbibliothek lists this publication in the Deutsche Nationalbibliografie;
detailed bibliographic data are available on the Internet at http://dnb.dnb.de.

© 2026 Walter de Gruyter GmbH, Berlin/Boston, Genthiner Straße 13, 10785 Berlin
Cover image: selimaksan / iStock / Getty Images Plus
Typesetting: VTeX UAB, Lithuania

www.degruyterbrill.com
Questions about General Product Safety Regulation:
productsafety@degruyterbrill.com

# Preface

Calculus forms the foundation of modern science, engineering, and quantitative analysis. This book was written to provide students and educators with a rigorous yet accessible introduction to the calculus of one variable. Our aim is to blend mathematical theory with practical computation, offering clear explanations, abundant examples, and exercises that foster both conceptual understanding and computational skill.

A distinguishing feature of this book is the integration of Python and Artificial Intelligence tools throughout the text. These resources empower readers to visualize concepts, verify computations, and explore mathematical ideas interactively. By combining traditional mathematical rigor with the power of computation, we hope to enrich the learning experience and inspire further exploration. Of cource one can find more on [6, 7, 9–13].

This volume begins with a careful development of the real number system and its fundamental properties, proceeds through functions, limits, and continuity, and covers the essential results of differential and integral calculus. Each chapter contains numerous solved problems, exercises of varying difficulty, and sections dedicated to computational approaches using Python.

We observed that a good way to communicate with AI for the purpose of mathematical calculations is through the LaTeX text editor. In the link below, you will find 10 lessons on LaTeX for mathematics. The text was entirely created by AI. Similarly, you will also find another 10 lessons on Python, also written entirely by AI. Finally, you will find some Python codes for various mathematical problems discussed in this book. However, it should be noted that these were also generated by AI and have not been checked or corrected thoroughly. That is your task. To reach this level, though, you need to fully understand the corresponding mathematical concepts, and of course, learn the Python language as well. Ask AI to inform you on how you can use LaTeX (for example with Overleaf) and how to use Python (for example with Colab).

http://github.com/nikoshalidias/Calculus-of-One-Variable-Python

AI is not only useful for mathematical calculations regarding this particular book. It can also assist you in getting informed about historical facts related to the corresponding material. For example, you can ask the AI:

---

**AI Request**

Provide me with an in-depth historical overview of Pythagoras of Samos.

---

Nikolaos Halidias

https://doi.org/10.1515/9783112228289-202

# Contents

# List of Figures

https://doi.org/10.1515/9783112228289-204

# List of Tables

https://doi.org/10.1515/9783112228289-205

# 1 Real numbers

## Contents

The set of real numbers, denoted by $\mathbb{R}$, forms the foundation of mathematical analysis. This chapter introduces the fundamental properties of real numbers, including their algebraic structure and order properties.

We begin with an axiomatic approach, defining the basic operations and inequalities that govern $\mathbb{R}$. A key feature of the real number system is its **completeness**, which ensures that every bounded set has a supremum (least upper bound) and infimum (greatest lower bound).

We also explore important techniques, such as **mathematical induction**, and discuss concepts like the **absolute value**, **integer part**, and the **topological structure** of $\mathbb{R}$.

These ideas provide the essential tools needed for the study of limits, continuity, and calculus.

## 1.1 Introduction

The system of **real numbers** lies at the heart of modern mathematical analysis. It represents a culmination of centuries of mathematical thought, from the geometric intuition of the ancient Greeks to the rigorous constructions of 19th-century mathematicians.

The idea that not all lengths could be expressed as ratios of whole numbers was first recognized by the Pythagoreans, who discovered that $\sqrt{2}$ is irrational, a revelation so profound that it was initially kept secret. This realization marked the beginning of a long journey toward understanding the continuum of real numbers beyond the rationals.

In the 19th century, the need for rigor in calculus led to formal constructions of the real number system. Mathematicians such as *Richard Dedekind, George Cantor*, and *Karl Weierstrass* independently developed methods to define real numbers through cuts, sequences, and limits, respectively. These constructions provided the foundation for the completeness property, a key feature distinguishing the real numbers from the rationals.

https://doi.org/10.1515/9783112228289-001

This chapter introduces the essential properties of real numbers, including:
- The field axioms (addition, multiplication, distributivity).
- The order axioms and the trichotomy principle.
- The completeness axiom: existence of suprema and infima.
- Consequences such as the Archimedean property and density of rational and irrational numbers.

These ideas are not only fundamental to real analysis but also serve as the basis for advanced topics like continuity, convergence, and measure theory. Through precise definitions and illustrative examples, this chapter aims to build a deep understanding of the structure and behavior of the real number system, an indispensable tool for both pure and applied mathematics.

## 1.2 Axiomatic foundation of real numbers

In this section, we present some basic elements and properties of the set of real numbers, which we denote by $\mathbb{R}$. Beginning with the fundamentals, we note that $\mathbb{R}$ is equipped with two operations: addition (+) and multiplication (·). Subsequently, we will describe the axioms and fundamental theorems that provide an axiomatic foundation for the real numbers.

**Axiom 1.1** (Commutative laws). We assume that for all $x, y \in \mathbb{R}$,

$$x + y = y + x, \quad x \cdot y = y \cdot x.$$

**Axiom 1.2** (Associative laws). We assume that for all $x, y, z \in \mathbb{R}$,

$$x + (y + z) = (x + y) + z, \quad x \cdot (y \cdot z) = (x \cdot y) \cdot z.$$

**Axiom 1.3** (Distributive law). We assume that for all $x, y, z \in \mathbb{R}$,

$$x \cdot (y + z) = x \cdot y + x \cdot z.$$

**Axiom 1.4** (Existence of identity elements). There exist two distinct real numbers, denoted by 0 and 1, such that for every $x \in \mathbb{R}$, we have

$$x + 0 = x \quad \text{and} \quad 1 \cdot x = x.$$

**Axiom 1.5** (Existence of additive inverse). For every real number $x \in \mathbb{R}$, there exists a real number $y \in \mathbb{R}$ such that

$$x + y = 0.$$

**Axiom 1.6** (Existence of multiplicative inverse). For every nonzero real number $x \neq 0$, there exists a real number $y \in \mathbb{R}$ such that

$$x \cdot y = 1.$$

We now prove that the identity elements for addition and multiplication are unique. This ensures there is only one number 0 such that $x + 0 = x$, and only one number 1 such that $x \cdot 1 = x$.

**Theorem 1.1** (Uniqueness of identity elements). *The identity elements 0 and 1 are unique.*

*Proof.* Assume that there exists a number $z \neq 0$ such that for some $x \in \mathbb{R}$,

$$x + z = x.$$

Adding the additive inverse of $x$ to both sides yields

$$z = 0,$$

which is a contradiction.

Similarly, suppose there exists a number $d \neq 1$ such that for some nonzero $x \in \mathbb{R}$,

$$d \cdot x = x.$$

Multiplying both sides by the multiplicative inverse of $x$, we obtain

$$d = 1,$$

which is also a contradiction. □

We now establish the existence and uniqueness of additive inverses in $\mathbb{R}$. This ensures that for every real number $x$, there is exactly one number $y$ such that $x + y = 0$.

**Theorem 1.2** (Existence and uniqueness of additive inverse). *For every $x \in \mathbb{R}$, there exists a unique real number $y \in \mathbb{R}$ such that*

$$x + y = 0.$$

*This number is denoted by $-x$. Consequently, we have*

$$-(-x) = x.$$

*Proof.* Assume there exist two such numbers $y$ and $z$, with $y \neq z$, satisfying

$$x + y = 0 \quad \text{and} \quad x + z = 0.$$

Subtracting the two equations (or adding $-x$ to both sides), we get

$$y = z,$$

which is a contradiction. Therefore, the additive inverse is unique.

To prove $-(-x) = x$, note that by definition, $-x$ is the unique number such that

$$x + (-x) = 0.$$

But this also implies that $x$ is the unique number such that

$$(-x) + x = 0,$$

hence $x = -(-x)$. □

We now establish the existence and uniqueness of multiplicative inverses in $\mathbb{R}$ for nonzero real numbers. This ensures that for every $x \neq 0$, there is exactly one number $y$ such that $x \cdot y = 1$.

**Theorem 1.3** (Existence and uniqueness of multiplicative inverse). *For every nonzero real number $x \neq 0$, there exists a unique real number $y \in \mathbb{R}$ such that*

$$x \cdot y = 1.$$

*This number is denoted by $\frac{1}{x}$ or $x^{-1}$. Consequently, we have*

$$\left(x^{-1}\right)^{-1} = x.$$

*Proof.* Assume there exist two such numbers $y$ and $z$ satisfying

$$x \cdot y = 1 \quad \text{and} \quad x \cdot z = 1.$$

Multiplying both sides of the first equation by $z$, we obtain

$$x \cdot y \cdot z = z.$$

Using associativity and the fact that $x \cdot z = 1$, it follows that

$$y = z,$$

which contradicts our assumption that $y \neq z$. Hence, the multiplicative inverse is unique.
To show $(x^{-1})^{-1} = x$, observe that by definition

$$x^{-1} \cdot x = 1,$$

which means $x$ is the unique multiplicative inverse of $x^{-1}$, so

$$\left(x^{-1}\right)^{-1} = x.$$ □

## 1.2.1 Exercises

### True or false

Determine whether each statement is **true** or **false**. Justify your answer.

1. The real number 0 is unique.
   **Answer:** _____
2. For every $x \in \mathbb{R}$, there exists a unique $y \in \mathbb{R}$ such that $x + y = 0$.
   **Answer:** _____
3. The multiplicative inverse of $x \neq 0$ is not unique.
   **Answer:** _____
4. The distributive law states that $x \cdot (y + z) = x \cdot y + x \cdot z$.
   **Answer:** _____
5. There exists a real number $x$ such that $x + 1 = x$.
   **Answer:** _____

### Multiple choice questions

Choose the correct option (A, B, C, or D) for each question.

1. Which of the following is an axiom of addition in $\mathbb{R}$?
   A. $x \cdot y = y \cdot x$          B. $x + y = y + x$
   C. $x \cdot (y + z) = x \cdot y + x \cdot z$    D. $x \cdot 1 = x$
   **Answer:** _____
2. Which property guarantees that $x + (-x) = 0$?
   A. Existence of additive identity    B. Associative law
   C. Commutative law            D. Existence of additive inverse
   **Answer:** _____
3. What is the meaning of the statement "$-(-x) = x$"?
   A. $x$ has no inverse
   B. The additive inverse of the additive inverse is $x$
   C. The inverse of $x$ is 0
   D. $x + x = 0$
   **Answer:** _____
4. What is the multiplicative identity in $\mathbb{R}$?
   A. 0    B. –1
   C. 1    D. $x^{-1}$
   **Answer:** _____
5. Which of the following best defines the additive inverse of $x$?
   A. The number that satisfies $x \cdot y = 1$
   B. The number such that $x + y = x$
   C. The number such that $x + y = 0$
   D. The number that satisfies $x \cdot y = 0$
   **Answer:** _____

## Matching
Match each term in **Column A** with the most appropriate description in **Column B**.

| Column A | Column B |
| --- | --- |
| A. Commutative law | 1. For every $x \neq 0$ there exists $y$ with $x \cdot y = 1$. |
| B. Additive identity | 2. $x + (y + z) = (x + y) + z$ and $x \cdot (y \cdot z) = (x \cdot y) \cdot z$. |
| C. Distributive law | 3. $x + y = y + x$ and $x \cdot y = y \cdot x$. |
| D. Multiplicative inverse | 4. $x \cdot (y + z) = x \cdot y + x \cdot z$. |
| E. Associative law | 5. Unique real number 0 such that $x + 0 = x$. |

**Answers:** _____ \_ _____ \_ _____ \_ _____ \_ _____

## Fill in the blanks
1. The real number that satisfies $x + 0 = x$ for all $x \in \mathbb{R}$ is called the _____.
2. For every nonzero $x \in \mathbb{R}$, the number $x^{-1}$ is defined so that $x \cdot x^{-1} =$ _____.
3. The statement $x \cdot (y + z) = x \cdot y + x \cdot z$ exemplifies the _____ law.
4. If $x + (-x) = 0$, then $-x$ is the _____ of $x$.
5. The fact that $-(-x) = x$ shows that the additive inverse is _____.

## 1.3 Positive real numbers

We now introduce the set of positive real numbers and its defining properties. This will allow us to define an order on $\mathbb{R}$ and prove important algebraic facts.

**Definition 1.1** (Positive real numbers). We denote by $\mathbb{R}^+$ a specific subset of $\mathbb{R}$ that satisfies the following three axioms.

We now introduce the closure property of positive real numbers under addition and multiplication. This ensures that sums and products of positive numbers remain positive.

**Axiom 1.7** (Closure of $\mathbb{R}^+$ under addition and multiplication). If $x, y \in \mathbb{R}^+$, then both $x + y \in \mathbb{R}^+$ and $x \cdot y \in \mathbb{R}^+$.

The dichotomy property guarantees that every nonzero real number is either positive or negative, but never both.

**Axiom 1.8** (Dichotomy property). For every nonzero real number $x \neq 0$, either $x \in \mathbb{R}^+$ or $-x \in \mathbb{R}^+$, but not both.

This axiom explicitly states that zero is not included in the set of positive real numbers. It complements the dichotomy property.

**Axiom 1.9** (Zero is not positive). We have $0 \notin \mathbb{R}^+$.

The cancellation law for addition allows us to "cancel" equal terms on both sides of an equation. It is a direct consequence of the existence of additive inverses.

**Theorem 1.4** (Cancellation law for addition). *If $a + b = a + c$, then $b = c$.*

*Proof.* Using the axiom of existence of additive inverses, we add $-a$ to both sides:

$$-a + (a + b) = -a + (a + c).$$

By the associative property, this simplifies to

$$(-a + a) + b = (-a + a) + c \quad \Rightarrow \quad 0 + b = 0 + c,$$

which implies

$$b = c.$$

This completes the proof. □

Multiplication by zero always yields zero. This result will be used frequently in later proofs.

**Theorem 1.5** (Multiplication by zero). *For every real number $x \in \mathbb{R}$, it holds that*

$$0 \cdot x = 0.$$

*Proof.* We know from the axioms that $1 + (-1) = 0$. Then

$$0 \cdot x = (1 + (-1)) \cdot x.$$

Using the distributive law,

$$(1 + (-1)) \cdot x = 1 \cdot x + (-1) \cdot x = x + (-1) \cdot x.$$

Therefore

$$x + (-1) \cdot x = 0 \quad \Rightarrow \quad (-1) \cdot x = -x.$$

Hence

$$0 \cdot x = x + (-x) = 0.$$

This proves the claim. □

Multiplying any real number by –1 gives its additive inverse. This identity is crucial in many algebraic manipulations.

**Theorem 1.6** (Negation via multiplication by –1). *For every real number $a \in \mathbb{R}$, we have*

$$(-1) \cdot a = -a.$$

*Consequently*

$$(-1) \cdot (-1) \cdot a = a.$$

*Proof.* From the identity $1 + (-1) = 0$, multiplying both sides by $a$ gives

$$(1 + (-1)) \cdot a = 0 \cdot a.$$

Using the distributive law and Theorem 1.5, this becomes

$$1 \cdot a + (-1) \cdot a = 0 \quad \Rightarrow \quad a + (-1) \cdot a = 0.$$

Thus, $(-1) \cdot a$ is the additive inverse of $a$, and since the additive inverse is unique, we conclude

$$(-1) \cdot a = -a.$$

Now, applying this result twice

$$(-1) \cdot (-1) \cdot a = (-1) \cdot (-a) = -(-a) = a.$$

This completes the proof. □

One of the most important properties of multiplication is that if a product is zero, then at least one of the factors must be zero.

**Theorem 1.7** (Zero product property). *If $a \cdot b = 0$, then at least one of the numbers $a$ or $b$ must be zero.*

*Proof.* Suppose $b \neq 0$. Then we can multiply both sides of the equation by the multiplicative inverse of $b$, denoted by $b^{-1}$:

$$a = a \cdot (b \cdot b^{-1}) = (a \cdot b) \cdot b^{-1} = 0 \cdot b^{-1} = 0.$$

Therefore, $a = 0$.

Similarly, if we assume that $a \neq 0$, multiplying both sides by $a^{-1}$ yields

$$b = 0.$$

This completes the proof. □

The order on real numbers is defined using the set of positive real numbers. This definition introduces the basic notation and terminology used throughout this section.

**Definition 1.2** (Order on real numbers). We say that a real number $x$ is **less than** $y$ and write

$$x < y,$$

if and only if

$$y + (-x) \in \mathbb{R}^+.$$

For simplicity, we write $y - x$ instead of $y + (-x)$.
  The following notations are also used:
- $y > x$ means the same as $x < y$.
- $x \leq y$ means that either $x = y$ or $x < y$.
- A number $x$ is called **positive** if $x > 0$.
- A number $x$ is called **negative** if $x < 0$.

The trichotomy law ensures that any two real numbers can be uniquely compared. No two of the three possible relations can hold simultaneously.

**Theorem 1.8** (Trichotomy law). *For any real numbers $a, b \in \mathbb{R}$, exactly one of the following holds:*

$$a < b, \quad b < a, \quad or \quad a = b.$$

*Proof.* Let $x = b - a$. If $x = 0$, then since $0 \notin \mathbb{R}^+$, neither $a < b$ nor $b < a$ can hold. If $x \neq 0$, then by the dichotomy property of $\mathbb{R}^+$, either $x \in \mathbb{R}^+$ or $-x \in \mathbb{R}^+$. Thus, either $a < b$ or $b < a$, and the three cases are mutually exclusive. □

One of the fundamental properties of inequality is its transitivity. This allows us to compare multiple real numbers in sequence.

**Theorem 1.9** (Transitivity of inequality). *If $a < b$ and $b < c$, then $a < c$.*

*Proof.* Since $a < b$, we have $b - a \in \mathbb{R}^+$. Similarly, since $b < c$, we have $c - b \in \mathbb{R}^+$. Adding these two elements gives

$$(c - b) + (b - a) = c - a \in \mathbb{R}^+,$$

which means $a < c$, as required. □

Multiplication by a positive number preserves inequalities. This is crucial for solving inequalities algebraically.

**Theorem 1.10** (Multiplication preserves inequality for positive numbers). *If $a < b$ and $c > 0$, then $a \cdot c < b \cdot c$.*

*Proof.* Since $a < b$, we have $b - a \in \mathbb{R}^+$. Also, since $c > 0$, we have $c \in \mathbb{R}^+$. By closure of $\mathbb{R}^+$ under multiplication, it follows that

$$c \cdot (b - a) \in \mathbb{R}^+,$$

which implies

$$a \cdot c < b \cdot c.$$

This completes the proof. □

This important result shows that squares of nonzero real numbers are always positive. It will be used frequently in later proofs.

**Theorem 1.11** (Square of a nonzero real number is positive). *If $a \neq 0$, then $a^2 = a \cdot a > 0$.*

*Proof.* Since $a \neq 0$, either $a > 0$ or $a < 0$. If $a > 0$, then clearly $a^2 > 0$. If $a < 0$, then $(-a) > 0$, and by the closure property of $\mathbb{R}^+$, we have

$$(-a) \cdot (-a) \in \mathbb{R}^+ \quad \Rightarrow \quad (-1)^2 \cdot a^2 = a^2 \in \mathbb{R}^+.$$

Thus, in both cases, $a^2 > 0$. □

This theorem establishes that the multiplicative identity is a positive real number. This is essential for ordering arguments.

**Theorem 1.12** (The number one is positive). *We have $1 \in \mathbb{R}^+$.*

*Proof.* Since $1 \neq 0$, either $1 \in \mathbb{R}^+$ or $-1 \in \mathbb{R}^+$. Suppose $-1 \in \mathbb{R}^+$. Then, for any $x \in \mathbb{R}^+$, we would also have $-x = x \cdot (-1) \in \mathbb{R}^+$, which contradicts the dichotomy property. Therefore, $-1 \notin \mathbb{R}^+$, and so $1 \in \mathbb{R}^+$. □

Multiplying an inequality by a negative number reverses the direction of the inequality. This property must be carefully applied when solving inequalities.

**Theorem 1.13** (Multiplication reverses inequality with negative multiplier). *If $a < b$ and $c < 0$, then $a \cdot c > b \cdot c$.*

*Proof.* Since $a < b$, we have $b - a \in \mathbb{R}^+$. Also, since $c < 0$, we have $-c \in \mathbb{R}^+$. By closure under multiplication,

$$-c \cdot (b - a) \in \mathbb{R}^+ \quad \Rightarrow \quad a \cdot c - b \cdot c \in \mathbb{R}^+,$$

which implies

$$a \cdot c > b \cdot c.$$

This completes the proof. □

Adding two inequalities yields a new valid inequality. This is a useful tool in proving other results.

**Theorem 1.14** (Addition preserves inequality). *If $a < c$ and $b < d$, then $a + b < c + d$.*

*Proof.* Since $a < c$, we have $c - a \in \mathbb{R}^+$. Similarly, since $b < d$, we have $d - b \in \mathbb{R}^+$. Adding these two positive numbers gives

$$(c - a) + (d - b) = (c + d) - (a + b) \in \mathbb{R}^+,$$

which implies

$$a + b < c + d.$$

This proves the result. □

## 1.3.1 Exercises

### True or false
Determine whether each statement is **true** or **false**. Justify your answer.
1. If $x \in \mathbb{R}^+$ and $y \in \mathbb{R}^+$, then $x + y \in \mathbb{R}^+$.
   **Answer:** _____
2. Zero is a positive real number.
   **Answer:** _____
3. For any real number $x \neq 0$, either $x \in \mathbb{R}^+$ or $-x \in \mathbb{R}^+$, but not both.
   **Answer:** _____
4. If $a + b = a + c$, then $b = c$.
   **Answer:** _____
5. If $a < b$ and $c > 0$, then $a \cdot c < b \cdot c$.
   **Answer:** _____
6. For every $x \in \mathbb{R}$, we have $0 \cdot x = 0$.
   **Answer:** _____
7. The square of every nonzero real number is positive.
   **Answer:** _____

### Multiple choice questions
Choose the correct option (A, B, C, or D) for each question.

1. Which of the following is **true** about the set $\mathbb{R}^+$?
   A. It contains 0
   B. It is not closed under addition
   C. It is not closed under multiplication
   D. It satisfies the dichotomy property
   **Answer:** _____

2. Which of the following proves that $(-1) \cdot a = -a$ for all $a \in \mathbb{R}$?
   A. $1 + (-1) = 0$ and the distributive law
   B. $\mathbb{R}^+$ is closed under multiplication
   C. $a + a = 2a$
   D. $a^2 > 0$
   **Answer:** _____

3. Which statement correctly defines $x < y$?
   A. $x < y$ if and only if $x \cdot y \in \mathbb{R}^+$
   B. $x < y$ if and only if $y - x \in \mathbb{R}^+$
   C. $x < y$ if and only if $x + y = 0$
   D. $x < y$ if and only if $x = y$
   **Answer:** _____

4. If $a < b$ and $c < 0$, which of the following must be true?
   A. $a \cdot c < b \cdot c$    B. $a \cdot c = b \cdot c$
   C. $a \cdot c > b \cdot c$    D. $a + c > b + c$
   **Answer:** _____

5. According to the trichotomy law, for any real numbers $a$, $b$, which of the following holds?
   A. $a < b$ and $a = b$ may both be true
   B. Exactly one of $a < b$, $a = b$, $a > b$ is true
   C. If $a \neq b$, then $a = -b$
   D. $a > b$ implies $b > a$
   **Answer:** _____

## Matching

Match each item in **Column A** with the best description in **Column B**.

| Column A | Column B |
| --- | --- |
| A. Closure of $\mathbb{R}^+$ | 1. Exactly one of $a < b$, $a = b$, or $a > b$ holds for any $a, b \in \mathbb{R}$. |
| B. Zero product property | 2. $(-1) \cdot a = -a$ for all $a \in \mathbb{R}$. |
| C. Negation via $-1$ | 3. If $a < b$ and $c > 0$, then $ac < bc$. |
| D. Trichotomy law | 4. For $x, y \in \mathbb{R}^+$, both $x + y$ and $xy$ also lie in $\mathbb{R}^+$. |
| E. Multiplication preserves inequality (positive factor) | 5. If $a \cdot b = 0$, then $a = 0$ or $b = 0$. |

**Answers:** _____ – _____ – _____ – _____ – _____

**Fill in the blanks**

1. The number that *is not* contained in $\mathbb{R}^+$ is _____.
2. By definition, $a < b$ if and only if _____ belongs to $\mathbb{R}^+$.
3. If $a < b$ and $c < 0$, then $ac$ is _____ than $bc$.
4. The fact that every product of the form $(-1) \cdot a$ equals $-a$ relies on the identity $1 + (-1) =$ _____ and the _____ law.
5. The proof that $a^2 > 0$ for all $a \neq 0$ uses the closure of $\mathbb{R}^+$ under _____.

## 1.4 Positive integers, integers, and rational numbers

The set of positive integers is constructed inductively from the number 1. This construction forms the basis for defining the natural numbers within $\mathbb{R}$.

**Definition 1.3** (Positive integers). Starting from the number 1 and adding it to itself, we obtain a new real number:
$$1 + 1 \in \mathbb{R}^+,$$
since $1 \in \mathbb{R}^+$. This new number is denoted by 2. Next, we add 1 to 2, resulting in another positive real number, which we denote by 3.

Continuing this process, we construct the numbers:
$$1, 2, 3, 4, \ldots$$
We denote this set of numbers by $\mathbb{N}$ and call it as the set of positive integers.

The set of integers extends the positive integers to include negative numbers and zero. It allows for subtraction to be always defined within the set.

**Definition 1.4** (Integers). The set of **integers**, denoted by $\mathbb{Z}$, extends the positive integers by including their additive inverses and zero:
$$\mathbb{Z} = \{\ldots, -3, -2, -1, 0, 1, 2, 3, \ldots\}.$$
This set is closed under addition, subtraction, and multiplication.

Rational numbers are those that can be expressed as fractions of integers. They form a dense subset of the real line and are essential in many proofs. See Figure 1.1 to understand the relation between these sets.

**Definition 1.5** (Rational numbers). The set of **rational numbers**, denoted by $\mathbb{Q}$, consists of all numbers that can be expressed as a quotient of two integers
$$\mathbb{Q} = \left\{ \frac{a}{b} \,\middle|\, a, b \in \mathbb{Z},\, b \neq 0 \right\}.$$
This means that for every rational number $q \in \mathbb{Q}$, there exist integers $a$ and $b \neq 0$ such that $q = \frac{a}{b}$.

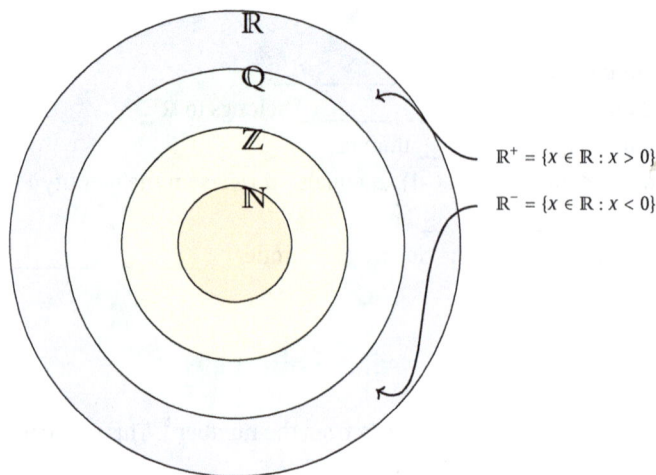

$R^+ = \{x \in \mathbb{R} : x > 0\}$

$R^- = \{x \in \mathbb{R} : x < 0\}$

**Figure 1.1:** Hierarchical inclusion of number sets: $\mathbb{N} \subset \mathbb{Z} \subset \mathbb{Q} \subset \mathbb{R}$, with $\mathbb{R}^+$ and $\mathbb{R}^-$ as subsets of $\mathbb{R}$.

### 1.4.1 Exercises

#### True or false
Determine whether each statement is **true** or **false**. Justify your answer.
1. Every natural number is a rational number.
   **Answer:** _____
2. Zero belongs to the set of positive integers.
   **Answer:** _____
3. The number $-3$ is an element of $\mathbb{Z}$ but not of $\mathbb{N}$.
   **Answer:** _____
4. The number $\frac{4}{5}$ is a rational number.
   **Answer:** _____
5. The set of rational numbers is a subset of the set of real numbers.
   **Answer:** _____

#### Multiple choice questions
Choose the correct option (A, B, C, or D) for each question.
1. Which of the following sets contains only positive integers?
   A. $\mathbb{R}^+$   B. $\mathbb{Q}$
   C. $\mathbb{N}$    D. $\mathbb{Z}$
   **Answer:** _____
2. Which of the following best describes the set $\mathbb{Z}$?
   A. All rational numbers
   B. All positive integers
   C. All integers, including negatives and zero
   D. All numbers greater than zero

**Answer:** _____

3. What does the notation $\frac{a}{b}$ with $a, b \in \mathbb{Z}$ and $b \neq 0$ define?

   A. A natural number

   B. A rational number

   C. An irrational number

   D. A real number not in $\mathbb{Q}$

   **Answer:** _____

4. Which of the following inclusions is correct?

   A. $\mathbb{Q} \subset \mathbb{Z} \subset \mathbb{N}$

   B. $\mathbb{R} \subset \mathbb{Q}$

   C. $\mathbb{N} \subset \mathbb{Z} \subset \mathbb{Q} \subset \mathbb{R}$

   D. $\mathbb{Z} \subset \mathbb{N} \subset \mathbb{Q}$

   **Answer:** _____

5. Which of the following numbers does **not** belong to the set $\mathbb{Q}$?

   A. $-7$   B. $0$

   C. $\frac{1}{2}$   D. $\sqrt{2}$

   **Answer:** _____

## Matching

Match each description in **Column B** with the set named in **Column A**.

| Column A | Column B |
|---|---|
| A. $\mathbb{N}$ | 1. All quotients $\frac{a}{b}$ with $a, b \in \mathbb{Z}, b \neq 0$. |
| B. $\mathbb{Z}$ | 2. Positive integers together with 0 and their negatives. |
| C. $\mathbb{Q}$ | 3. Numbers $1, 2, 3, \ldots$ obtained by repeated addition of 1. |
| D. $\mathbb{R}^{-}$ | 4. Real numbers strictly less than zero. |

**Answers:** _____ – _____ – _____ – _____

## Fill in the blanks

1. The first positive integer is _____.
2. Every integer is a real number, so we have the inclusion $\mathbb{Z} \subset$ _____.
3. The fraction $\frac{-7}{1}$ shows that $-7$ is a(n) _____ number.
4. A number of the form $\frac{a}{b}$ with $a, b \in \mathbb{Z}$ and $b \neq 0$ is called a _____.
5. The diagram demonstrates that $\mathbb{Q}$ is _____ inside $\mathbb{R}$.

## 1.5  Inductive set and mathematical induction

Inductive sets play a central role in defining the natural numbers within the real numbers. They ensure that the set of natural numbers is well-defined and closed under addition by one. Understanding inductive sets provides the foundation for the principle of mathematical induction and Peano arithmetic.

**Definition 1.6** (Inductive set). A set of real numbers is called **inductive** if it satisfies the following two properties:
(i)   The number 1 belongs to the set.
(ii)  For every element $x$ in the set, the number $x + 1$ also belongs to the set.

Clearly, every inductive set contains the numbers

$$\{1, 2, 3, \ldots\}.$$

The sets $\mathbb{R}$ and $\mathbb{R}^+$ are inductive sets, as are the sets

$$\{1, 2, 3, \ldots\} \quad \text{and} \quad \left\{\frac{1}{2}, 1, \frac{3}{2}, 2, \frac{5}{2}, \ldots\right\}.$$

The natural numbers can be characterized as the minimal set satisfying the properties of inductive sets.

We now provide a formal definition based on this idea.

**Definition 1.7** (Natural numbers). A number $x \in \mathbb{R}^+$ is called a **natural number** if it belongs to the intersection of all inductive sets.

**Corollary 1.1** (The set of natural numbers is $\mathbb{N}$). *The set of natural numbers equals $\mathbb{N}$.*

*Proof.* Every inductive set contains the number 1 by the first property of inductive sets. Furthermore, by the second property, it necessarily also contains the numbers $2, 3, \ldots$. Therefore, the set $\mathbb{N}$ is the smallest inductive set and thus contains all the natural numbers. $\square$

A central tool in proving statements about the natural numbers is the principle of mathematical induction.

We state this principle rigorously below.

**Theorem 1.15** (Principle of mathematical induction). *Let $P(n)$ be a statement about $n \in \mathbb{N}$ (with $1 \in \mathbb{N}$). Suppose that*
(i)    **Base step:** *$P(1)$ is true.*
(ii)   **Inductive step:** *For every $n \in \mathbb{N}$, if $P(n)$ is true, then $P(n + 1)$ is true.*

*Then $P(n)$ is true for all $n \in \mathbb{N}$.*

*Proof.* Let us define the set:

$$A = \{x \in \mathbb{N} : p(x) \text{ holds}\}.$$

From the assumptions of the theorem, it follows that $A$ is an inductive set and, moreover, a subset of $\mathbb{N}$. Since $\mathbb{N}$ is the smallest inductive set, we must have $A = \mathbb{N}$. This completes the proof. □

Using mathematical induction, we can verify some classical formulas involving sums of powers of integers.

We present three fundamental identities.

**Proposition 1.1** (Classical summation identities). *Using mathematical induction, the following formulas hold for every $n \in \mathbb{N}$:*

- $\sum_{k=0}^{n} k = \frac{n(n+1)}{2}$
- $\sum_{k=0}^{n} k^2 = \frac{n(n+1)(2n+1)}{6}$
- $\sum_{k=0}^{n} k^3 = (\frac{n(n+1)}{2})^2$

*Proof.*

- We prove the identity

$$\sum_{k=0}^{n} k = \frac{n(n+1)}{2}$$

by induction on $n$.

**Base case** ($n = 1$): $\sum_{k=0}^{1} k = 0 + 1 = 1 = \frac{1(1+1)}{2}$.

**Inductive step**: Assume it holds for $n = i$:

$$\sum_{k=0}^{i} k = \frac{i(i+1)}{2}.$$

Then, for $n = i + 1$,

$$\sum_{k=0}^{i+1} k = \sum_{k=0}^{i} k + (i+1) = \frac{i(i+1)}{2} + (i+1) = \frac{(i+1)(i+2)}{2}.$$

Thus, the identity holds for $n = i + 1$.

- We prove now

$$\sum_{k=0}^{n} k^2 = \frac{n(n+1)(2n+1)}{6}$$

by induction on $n$.

**Base case** ($n = 1$): $0^2 + 1^2 = 1 = \frac{1 \cdot 2 \cdot 3}{6}$.

**Inductive step**: Assume it holds for $n = i$:

$$\sum_{k=0}^{i} k^2 = \frac{i(i+1)(2i+1)}{6}.$$

Then, for $n = i+1$,

$$\sum_{k=0}^{i+1} k^2 = \sum_{k=0}^{i} k^2 + (i+1)^2$$
$$= \frac{i(i+1)(2i+1)}{6} + (i+1)^2$$
$$= \frac{i(i+1)(2i+1) + 6(i+1)^2}{6}$$
$$= \frac{(i+1)(i+2)(2i+3)}{6}.$$

Thus, the identity holds.

- Next we prove

$$\sum_{k=0}^{n} k^3 = \left( \frac{n(n+1)}{2} \right)^2$$

by induction on $n$.

**Base case** ($n = 1$): $0^3 + 1^3 = 1 = (\frac{1 \cdot 2}{2})^2$.

**Inductive step**: Assume the formula holds for $n = i$:

$$\sum_{k=0}^{i} k^3 = \left( \frac{i(i+1)}{2} \right)^2.$$

Then, for $n = i+1$,

$$\sum_{k=0}^{i+1} k^3 = \sum_{k=0}^{i} k^3 + (i+1)^3$$
$$= \left( \frac{i(i+1)}{2} \right)^2 + (i+1)^3$$
$$= \frac{i^2(i+1)^2 + 4(i+1)^3}{4}$$
$$= \frac{(i+1)^2(i+2)^2}{4}$$
$$= \left( \frac{(i+1)(i+2)}{2} \right)^2.$$

This completes the proof. $\square$

A crucial combinatorial quantity arising in algebra and analysis is the binomial coefficient.

Its formal definition is given below.

**Definition 1.8** (Binomial coefficient). The quantity

$$\frac{n!}{k!(n-k)!}$$

for integers $n, k \in \mathbb{N}$ with $0 \le k \le n$ is denoted by

$$\binom{n}{k}.$$

It is called the **binomial coefficient**.

Two important results involving binomial coefficients are Pascal's identity and the binomial theorem.

We state both of them here.

**Proposition 1.2** (Binomial identities). *Pascal's identity:*

$$\binom{n+1}{k} = \binom{n}{k} + \binom{n}{k-1}$$

*Binomial theorem:*

$$(a+b)^n = \sum_{k=0}^{n} \binom{n}{k} a^{n-k} b^k$$

*Proof.* We first prove Pascal's identity by expanding both terms on the right-hand side

$$\begin{aligned}
\binom{n}{k} + \binom{n}{k-1} &= \frac{n!}{k!(n-k)!} + \frac{n!}{(k-1)!(n+1-k)!} \\
&= \frac{n!(n+1-k)}{k!(n+1-k)!} + \frac{n!k}{k!(n+1-k)!} \\
&= \frac{n!(n+1)}{k!(n+1-k)!} = \binom{n+1}{k}.
\end{aligned}$$

Now, we prove the binomial theorem by induction on $n$.

**Base case** ($n = 2$):

$$(a+b)^2 = a^2 + 2ab + b^2 = \sum_{k=0}^{2} \binom{2}{k} a^{2-k} b^k.$$

**Inductive step**: Assume the formula holds for $n$:

$$(a+b)^n = \sum_{k=0}^{n} \binom{n}{k} a^{n-k} b^k.$$

We want to prove it for $n + 1$. Using Pascal's identity and separating boundary terms,

$$(a + b)^{n+1} = (a + b)(a + b)^n$$

$$= a \sum_{k=0}^{n} \binom{n}{k} a^{n-k} b^k + b \sum_{k=0}^{n} \binom{n}{k} a^{n-k} b^k$$

$$= \sum_{k=0}^{n} \binom{n}{k} a^{n+1-k} b^k + \sum_{k=0}^{n} \binom{n}{k} a^{n-k} b^{k+1}.$$

Now, re-index the second sum with $l = k + 1$, and combine

$$= a^{n+1} + b^{n+1} + \sum_{k=1}^{n} \left( \binom{n}{k} + \binom{n}{k-1} \right) a^{n+1-k} b^k = \sum_{k=0}^{n+1} \binom{n+1}{k} a^{n+1-k} b^k.$$

Thus, the binomial theorem holds for $n + 1$, completing the proof. □

The following inequality provides a useful lower bound for powers of numbers slightly greater than 1. It is known as Bernoulli's inequality.

**Proposition 1.3** (Bernoulli's inequality). *For every $n \in \mathbb{N}$ and every real number $x > -1$, the following inequality holds:*

$$(1 + x)^n \geq 1 + nx.$$

*Proof.* We prove the result by induction on $n$.

**Base case** ($n = 1$): Clearly,

$$(1 + x)^1 = 1 + x \geq 1 + 1 \cdot x.$$

**Inductive step**: Assume the inequality holds for $n = k$:

$$(1 + x)^k \geq 1 + kx.$$

We show it holds for $n = k + 1$:

$$(1 + x)^{k+1} = (1 + x)^k (1 + x)$$

$$\geq (1 + kx)(1 + x)$$

$$= 1 + (k + 1)x + kx^2$$

$$\geq 1 + (k + 1)x.$$

The last inequality follows from the fact that $kx^2 \geq 0$ since $k \in \mathbb{N}$ and $x^2 \geq 0$.
Therefore, the inequality holds for all $n \in \mathbb{N}$. □

One of the most elegant inequalities in mathematics relates the arithmetic and geometric means of nonnegative numbers. We now state and prove this inequality.

**Proposition 1.4** (Arithmetic-geometric mean inequality (AMGM)). *Let* $a_1, a_2, \ldots, a_n$ *be nonnegative real numbers. Then the following inequality holds:*

$$\frac{a_1 + a_2 + \cdots + a_n}{n} \geq \sqrt[n]{a_1 a_2 \cdots a_n},$$

*with equality if and only if* $a_1 = a_2 = \cdots = a_n$.

*Proof.* We first prove the following auxiliary inequality:

Let $b_1, \ldots, b_n > 0$ with $b_1 b_2 \cdots b_n = 1$, and suppose not all $b_j$'s are equal. Then

$$b_1 + b_2 + \cdots + b_n > n.$$

We prove this by induction on $n$.

**Base case** ($n = 2$): If $b_1 \neq b_2$, then

$$0 < \left( \sqrt{b_1} - \sqrt{b_2} \right)^2 = b_1 + b_2 - 2 \quad \Rightarrow \quad b_1 + b_2 > 2.$$

**Inductive step**: Assume the inequality holds for some $k \in \mathbb{N}$. We prove it for $k + 1$. Without loss of generality, suppose

$$b_1 \leq b_j \leq b_{k+1}, \quad \text{for all } j,$$

and $b_1 < b_{k+1}$. Then, necessarily $b_1 < 1 < b_{k+1}$, otherwise $b_1 b_2 \cdots b_{k+1} \neq 1$.
By the inductive hypothesis,

$$b_2 + \cdots + b_k + b_1 b_k \geq k.$$

Now

$$b_1 + \cdots + b_{k+1} = (b_1 b_{k+1} + b_2 + \cdots + b_k) + 1 + (b_{k+1} - 1)(1 - b_1)$$
$$\geq k + 1 + (b_{k+1} - 1)(1 - b_1) > k + 1.$$

So the inequality holds for $k + 1$ as well, and thus by induction

$$b_1 + \cdots + b_n > n.$$

Now we prove the desired inequality.

If $n = 1$, the inequality is trivial. If any $a_j = 0$, then the geometric mean is zero and the inequality becomes

$$(\text{positive sum}) \geq 0,$$

which is clearly true.

If all $a_j > 0$ and not all equal, set

$$G = \sqrt[n]{a_1 a_2 \cdots a_n}, \quad \text{and} \quad b_j = \frac{a_j}{G}.$$

Then $b_1 b_2 \cdots b_n = 1$, $b_j > 0$, and not all $b_j$ are equal.
By the inequality proven above,

$$\frac{a_1 + \cdots + a_n}{G} = b_1 + \cdots + b_n > n \quad \Rightarrow \quad \frac{a_1 + \cdots + a_n}{n} > G = \sqrt[n]{a_1 \cdots a_n}.$$

If all $a_j$ are equal, then equality holds.
Therefore, the inequality is proven. □

### 1.5.1 Exercises

#### True or false
Determine whether each statement is **true** or **false**. Justify your answer.
1. Every inductive set must contain the number 1.
   **Answer:** _____
2. The set $\{1, 3, 5, 7, \ldots\}$ is an inductive set.
   **Answer:** _____
3. The set of natural numbers $\mathbb{N}$ is the largest inductive set.
   **Answer:** _____
4. The set of positive real numbers $\mathbb{R}^+$ is an inductive set.
   **Answer:** _____
5. The principle of mathematical induction proves that a property holds for all integers.
   **Answer:** _____

#### Multiple choice questions
Choose the correct option (A, B, C, or D) for each question.
1. Which of the following best describes an inductive set?
   A. A set closed under subtraction
   B. A set that contains 1 and for every $x$ also contains $x + 1$
   C. A finite subset of $\mathbb{N}$
   D. Any set that contains only even numbers
   **Answer:** _____
2. Which of the following is the smallest inductive set?
   A. $\mathbb{Z}$    B. $\mathbb{R}$
   C. $\mathbb{Q}$    D. $\mathbb{N}$
   **Answer:** _____

3. The principle of mathematical induction requires
   A. a base case and a general formula
   B. a base case and a limit process
   C. a base case and an inductive step
   D. a counterexample and a hypothesis
   **Answer:** _____

4. Which statement best explains the goal of mathematical induction?
   A. To prove a property holds for some large values of $n$
   B. To prove a formula works for a fixed $n$
   C. To establish truth of a statement for all $n \in \mathbb{N}$
   D. To define a sequence recursively
   **Answer:** _____

5. Suppose a set $A \subseteq \mathbb{N}$ contains 1, and for every $n \in A$, it also contains $n + 1$. Then
   A. $A$ is finite    B. $A = \mathbb{N}$
   C. $A \subset \mathbb{N}$    D. $A$ is not inductive
   **Answer:** _____

## Matching

Match each item in **Column A** with the best description in **Column B**.

| Column A | Column B |
|---|---|
| A. Inductive set | 1. Provides the identity $\binom{n+1}{k} = \binom{n}{k} + \binom{n}{k-1}$. |
| B. Natural numbers $\mathbb{N}$ | 2. Inequality $(1 + x)^n \geq 1 + nx$ for $x > -1$. |
| C. Pascal's identity | 3. Smallest set containing 1 and closed under $x \mapsto x + 1$. |
| D. Bernoulli's inequality | 4. Intersection of all inductive sets. |
| E. AM-GM inequality | 5. $\frac{a_1 + \cdots + a_n}{n} \geq \sqrt[n]{a_1 \cdots a_n}$. |

**Answers:** _____ – _____ – _____ – _____ – _____

## Fill in the blanks

1. A set that contains 1 and is closed under the map $x \mapsto x+1$ is called _____.
2. The smallest inductive set is denoted by _____.
3. The induction proof structure has two parts: the _____ case and the _____ step.
4. Pascal's identity is used in the proof of the _____ theorem.
5. Bernoulli's inequality states that $(1 + x)^n \geq 1 + nx$ whenever $x > -1$ and $n \in$ _____.

# 1.6 Supremum and infimum

Before discussing suprema and infima, it is important to clarify when a set is considered bounded.

We begin with the definition of boundedness for subsets of the real numbers.

**Definition 1.9** (Bounded subsets of $\mathbb{R}$). We say that a subset $A$ of $\mathbb{R}$ is **bounded above** if there exists a real number $a$ such that
$$x \leq a \quad \text{for all } x \in A.$$
Similarly, $A$ is **bounded below** if there exists a real number $b$ such that
$$b \leq x \quad \text{for all } x \in A.$$
Finally, we say that $A$ is **bounded** if it is both bounded above and bounded below.

A fundamental concept in analysis is that of the least upper bound, or supremum, of a set.

The following definition formalizes this idea.

**Definition 1.10** (Supremum of a set). A real number $B$ is called the **least upper bound** of a nonempty set $S$ and is denoted by $\sup S$ if the following two properties hold:
(i)   $B$ is an upper bound of $S$, that is, $x \leq B$ for every $x \in S$.
(ii)  There does not exist a number less than $B$ that is also an upper bound of $S$.

The existence of least upper bounds for bounded sets is a crucial property of the real numbers.

This is encapsulated in the completeness axiom.

**Axiom 1.10** (Completeness axiom). Every nonempty set $S$ of real numbers that is bounded above has a least upper bound. That is, there exists a real number $B = \sup S$.

**Proposition 1.5** (Supremum and maximum). *Let $S \subseteq \mathbb{R}$ be nonempty and bounded above. Then*
$$\sup S \in S \quad \Longleftrightarrow \quad \sup S = \max S.$$
*In particular, $S$ contains its supremum exactly when it has a maximum.*

*Proof.* ($\Rightarrow$) If $\alpha = \sup S$ and $\alpha \in S$, then $\alpha$ is an upper bound of $S$, hence $x \leq \alpha$ for all $x \in S$; therefore $\alpha = \max S$.

($\Leftarrow$) If $M = \max S$, then $M$ is an upper bound. If $b < M$, then $b$ is not an upper bound because $M \in S$ and $M > b$. Thus $M$ satisfies the defining properties of the supremum, so $\sup S = M \in S$. $\qquad\square$

**Remark 1.** The number $\sup S$ may or may not belong to $S$. For instance, $[0, 1)$ does not contain its supremum, while $[0, 1]$ does. When $\sup S \in S$, it is (and is denoted) the maximum, $\max S$.

**Theorem 1.16** (Uniqueness of the supremum). *Let $S \subseteq \mathbb{R}$. If the supremum of $S$ exists, then it is unique.*

*Proof.* Suppose $B$ and $C$ both satisfy the defining properties of sup $S$:

(a) $B$ and $C$ are upper bounds of $S$.

(b) no number smaller than $B$ (respectively, $C$) is an upper bound of $S$.

If $B \neq C$, then without loss of generality $B < C$. By (b) for $C$, any number less than $C$ is *not* an upper bound of $S$; hence $B$ cannot be an upper bound, contradicting (a). Therefore $B = C$, and the supremum is uniquely determined. □

The greatest lower bound, or infimum, is the natural dual notion to the supremum. We now give its precise definition.

---

**Definition 1.11** (Infimum of a set). A real number $L$ is called the **greatest lower bound** (or **infimum**) of a nonempty set $S$, and is denoted by $\inf S$, if the following two conditions are satisfied:

(i) $L$ is a lower bound of $S$, that is, $L \leq x$ for every $x \in S$.

(ii) There is no number greater than $L$ that is also a lower bound of $S$.

If $\inf S$ belongs to the set $S$, it is often denoted by $\min S$.

---

Just as every nonempty set bounded above has a supremum, every nonempty set bounded below has an infimum. The following theorem establishes the existence and properties of the infimum.

---

**Theorem 1.17** (Existence of greatest lower bound). *Every nonempty set $S$ that is bounded below has a greatest lower bound. That is, there exists a real number $L = \inf S$. In this case, we have*

$$\inf S = -\sup(-S).$$

---

*Proof.* The set $-S$ consists of the elements of $S$ with opposite sign. Since $S$ is bounded below, it follows that $-S$ is bounded above. Therefore, the supremum $\sup(-S)$ exists; let it be $y^*$. Then, for every $a \in S$, we have $-a \leq y^*$, which implies $a \geq -y^*$. Hence, $-y^*$ is a lower bound for $S$.

We now prove that $-y^*$ is the greatest such lower bound. Suppose it is not. Then there exists some $\varepsilon > 0$ such that for all $a \in S$, we have

$$a \geq -y^* + \varepsilon \quad \text{or equivalently} \quad -a \leq y^* - \varepsilon.$$

But this contradicts the fact that $y^*$ is the least upper bound of $-S$. Therefore, no such $\varepsilon > 0$ exists, and thus $-y^*$ is the greatest lower bound of $S$. □

The natural numbers possess a remarkable property: they are unbounded above within the real numbers. This is known as the Archimedean property.

---

**Lemma 1.1** (Archimedean property of the natural numbers). *The set of natural numbers $\mathbb{N}$ is not bounded above.*

*Proof.* Assume, for contradiction, that $\mathbb{N}$ is bounded above. Since it is nonempty, it must have a least upper bound, say $b = \sup \mathbb{N}$.

Then $b - 1$ is not an upper bound of $\mathbb{N}$, so there exists at least one natural number $n_0 \in \mathbb{N}$ such that

$$n_0 > b - 1 \quad \Rightarrow \quad n_0 + 1 > b.$$

But since $n_0 \in \mathbb{N}$, it follows that $n_0 + 1 \in \mathbb{N}$ as well, which contradicts the assumption that $b$ is an upper bound of $\mathbb{N}$.

Therefore, no such least upper bound exists, and $\mathbb{N}$ is not bounded above. $\qquad\square$

**Theorem 1.18** (Archimedean property (general form)). *For every real number $x$, there exists a natural number $n \in \mathbb{N}$ such that*

$$n > x.$$

*Consequently, if $x > 0$ and $y$ is any real number, then there exists $n \in \mathbb{N}$ such that*

$$nx > y.$$

*Proof.* If this were not true, then some real number $x$ would be an upper bound of $\mathbb{N}$, which contradicts the Archimedean property proven in Lemma 1.1.

For the second part, let $x > 0$ and consider the number $y/x$. By the first part, there exists $n \in \mathbb{N}$ such that $n > y/x$, hence $nx > y$. $\qquad\square$

An important consequence of the Archimedean property is the following result. It shows how certain inequalities can force equality in the real numbers.

**Theorem 1.19** (Equality from vanishing upper bound). *If the real numbers $a, x, y$ satisfy*

$$a \leq x \leq a + \frac{y}{n}$$

*for every $n \in \mathbb{N}$, then $x = a$.*

*Proof.* Assume $x > a$. Then $x - a > 0$, and by the Archimedean property, there exists $n \in \mathbb{N}$ such that

$$n(x - a) > y \quad \Rightarrow \quad x > a + \frac{y}{n},$$

which contradicts the assumption. Therefore, $x = a$. $\qquad\square$

Combining the completeness property with induction, we can prove the existence and uniqueness of real roots.

The next proposition addresses $n$th roots of nonnegative real numbers.

**Proposition 1.6** (Existence and uniqueness of nth roots). *For every real number $a \geq 0$ and every $n \in \mathbb{N}$, there exists a unique positive real number $x$ such that*

$$x^n = a.$$

*This number is denoted by $\sqrt[n]{a}$ or $a^{1/n}$. In particular, there exists a unique positive real number $x$ such that $x^2 = 2$, and this number is irrational.*

*Proof.* We begin by proving the inequality

$$(1 + \varepsilon)^n < 1 + 3^n \varepsilon \quad \text{for } \varepsilon \in (0, 1). \tag{1.1}$$

We prove this by induction on $n$.

**Base case** $(n = 1)$: $(1 + \varepsilon)^1 = 1 + \varepsilon < 1 + 3\varepsilon$.

**Inductive step**: Assume the inequality holds for some $n$, i. e.,

$$(1 + \varepsilon)^n < 1 + 3^n \varepsilon.$$

Then

$$
\begin{aligned}
(1 + \varepsilon)^{n+1} &= (1 + \varepsilon)^n (1 + \varepsilon) \\
&< (1 + 3^n \varepsilon)(1 + \varepsilon) \\
&= 1 + (3^n + 1)\varepsilon + 3^n \varepsilon^2 \\
&< 1 + 3^{n+1} \varepsilon,
\end{aligned}
$$

where we used that $\varepsilon \in (0, 1)$ and hence $\varepsilon^2 < \varepsilon$.

We now prove the existence of the positive solution to $x^n = a$. Uniqueness follows since if $x < y$, then $x^n < y^n$ for $x, y > 0$.

If $a = 0$, then clearly $x = 0$ satisfies the equation.

Now assume $a > 0$. Define the set

$$E = \{t \in \mathbb{R} : t > 0, \ t^n < a\}.$$

This set is nonempty since $t_0 = \frac{a}{1+a} \in E$, which can be verified by induction on $n$. Also, $E$ is bounded above by $1 + a$, since:

- If $t \leq 1$, then clearly $t < 1 + a$.
- If $t > 1$, then $t^n > t > 1 \Rightarrow t^n < a < 1 + a \Rightarrow t < 1 + a$.

By the completeness axiom (Axiom 1.10), the set $E$ has a supremum in $\mathbb{R}$, denote it by $x = \sup E$. We will show that $x^n = a$.

**Case 1.** Suppose $x^n < a$. Choose $\varepsilon \in (0,1)$ such that

$$\varepsilon < \frac{a - x^n}{(3x)^n}.$$

Then by inequality (1.1)

$$x^n(1 + \varepsilon)^n < x^n(1 + 3^n\varepsilon) = x^n + (3x)^n\varepsilon < a.$$

Thus, $x(1 + \varepsilon) \in E$, which contradicts that $x = \sup E$. So $x^n \geq a$.

**Case 2.** Suppose $x^n > a$. Choose $\varepsilon \in (0,1)$ such that

$$\varepsilon < \frac{x^n - a}{3^n a}.$$

Then

$$a(1 + 3^n\varepsilon) < x^n \quad \Rightarrow \quad \frac{x^n}{1 + 3^n\varepsilon} > a.$$

By inequality (1.1), it follows that

$$a < \left(\frac{x}{1 + \varepsilon}\right)^n.$$

Since $\frac{x}{1+\varepsilon} < x$, there exists $t \in E$ such that

$$\left(\frac{x}{1 + \varepsilon}\right)^n < t^n < a,$$

which contradicts the definition of $x = \sup E$. Therefore, we conclude $x^n = a$.

Now, to prove that $\sqrt{2} \notin \mathbb{Q}$, suppose otherwise, that $x = \sqrt{2} = \frac{m}{n}$ with $m, n \in \mathbb{N}$, in lowest terms (i. e., $\gcd(m, n) = 1$). Then

$$x^2 = \frac{m^2}{n^2} = 2 \quad \Rightarrow \quad m^2 = 2n^2.$$

This implies $m^2$ is even, so $m$ is even. Let $m = 2k$. Then

$$(2k)^2 = 4k^2 = 2n^2 \quad \Rightarrow \quad n^2 = 2k^2,$$

so $n$ is even as well. But this contradicts the assumption that $m$ and $n$ are coprime. Hence, $\sqrt{2} \notin \mathbb{Q}$. $\square$

As a direct consequence of the previous results, we observe the existence of irrational numbers.

This highlights the richness of the real number system beyond the rationals.

**Corollary 1.2** (Existence of irrational numbers). *As a consequence of the previous results, we have*

$$\mathbb{Q} \subsetneq \mathbb{R}.$$

*That is, the set of rational numbers is a proper subset of the real numbers. Therefore, the set*

$$\mathbb{R} \smallsetminus \mathbb{Q}$$

*is nonempty and consists of real numbers that are not rational, called **irrational numbers** (see Figure 1.2).*

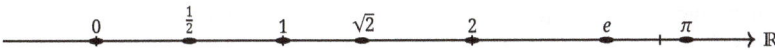

Figure 1.2: Representation of the real number line with important irrational and rational values marked.

For integer subsets, the situation regarding upper and lower bounds is even more straightforward.

The following proposition shows that the supremum (infimum) of such a set is always achieved.

**Proposition 1.7** (Supremum of a bounded integer set is maximum). *Let $A \subseteq \mathbb{Z}$ be a bounded above subset of the integers. Then the least upper bound (supremum) of A belongs to the set, i. e.,*

$$\sup A \in A,$$

*and thus it is the **maximum** of the set. Similarly, if A is bounded below, then the greatest lower bound (infimum) belongs to A, and it is the **minimum**.*

*Proof.* Since $A \subseteq \mathbb{Z}$ is bounded above, by the completeness axiom there exists a least upper bound $y = \sup A \in \mathbb{R}$.

We will show that $y \in A$. Since $y$ is the supremum, for any $\varepsilon > 0$ there exists an element $n \in A$ such that

$$y - \varepsilon < n \leq y.$$

Take $\varepsilon = 1$. Then

$$y - 1 < n \leq y.$$

Because $A \subseteq \mathbb{Z}$, all elements of $A$ are integers. So $n \in \mathbb{Z}$ and $n \leq y < n + 1$, which implies $y = n$. Hence, $y \in A$, and $y = \sup A = \max A$.

If $y \notin A$, then there would exist some $m \in A$ such that $y - 1 < n < m < y$, with $n, m \in \mathbb{Z}$, which implies two integers between $y - 1$ and $y$, a contradiction since the distance between distinct integers is at least one.

Therefore, the supremum belongs to $A$ and is its maximum. The argument for the infimum (minimum) is symmetric. $\qquad\Box$

There are useful relationships between the suprema and infima of a set and those of its negative.

We state and prove these identities below.

> **Proposition 1.8** (Supremum of the negative set). *Let $A \subseteq \mathbb{R}$. Define $-A = \{x \in \mathbb{R} : -x \in A\}$. Prove that*
>
> $$\sup(-A) = -\inf A \quad and \quad \inf(-A) = -\sup A.$$

*Proof.* We will prove the first identity; the second one follows analogously.

Assume that $A$ is bounded below and let $a = \inf A$. Then $x \geq a$ for every $x \in A$, and for every $\varepsilon > 0$ there exists $x^* \in A$ such that $x^* < a + \varepsilon$.

Multiplying these inequalities by $-1$, we obtain $x \leq -a$ for every $x \in -A$, and for every $\varepsilon > 0$ there exists $x^* \in -A$ such that $x^* > -a - \varepsilon$. Hence, $-a = \sup(-A)$.

If $A$ is not bounded below, then $-A$ is not bounded above and thus $\sup(-A) = -\inf A = +\infty$. $\qquad\Box$

Finally, we examine how the supremum and infimum behave under the union of sets.

The next result gives a precise formula in terms of the suprema and infima of the constituent sets.

> **Proposition 1.9** (Supremum and infimum of a union). *Let $A$ and $B$ be subsets of $\mathbb{R}$. Prove that*
>
> $$\sup(A \cup B) = \max\{\sup A, \sup B\}, \quad \inf(A \cup B) = \min\{\inf A, \inf B\}.$$

*Proof.* We will prove the first equality; the second one follows similarly.

Assume that both $A$ and $B$ are bounded above. Let $a = \sup A$ and $b = \sup B$, and assume without loss of generality that $a \leq b$. We claim that $\sup(A \cup B) = b$.

For every $x \in A \cup B$, we have $x \leq b$. Moreover, for every $\varepsilon > 0$, there exists $x^* \in B$ such that $x^* > b - \varepsilon$, and clearly $x^* \in A \cup B$. Therefore, the supremum of $A \cup B$ is $b$.

If either $A$ or $B$ is unbounded above, then $A \cup B$ is also unbounded above, and hence $\sup(A \cup B) = +\infty$, which agrees with the fact that $\max\{+\infty, c\} = +\infty$ for all $c \in \mathbb{R}$. $\qquad\Box$

### 1.6.1 Exercises

### True or false

Determine whether each statement is **true** or **false**. Justify your answer.

1. Every nonempty subset of $\mathbb{R}$ has a supremum.

   **Answer:** _____

2.  If a set has a maximum, then the maximum is also its supremum.
    **Answer:** _____
3.  The supremum of a set must always belong to the set.
    **Answer:** _____
4.  The completeness axiom ensures that bounded below sets have an infimum.
    **Answer:** _____
5.  If a set $A \subseteq \mathbb{Z}$ is bounded above, then $\sup A \in A$.
    **Answer:** _____

## Multiple choice: supremum and completeness

Choose the correct option (A, B, C, or D) for each question.

1.  The supremum of a set $S \subseteq \mathbb{R}$ is defined as
    A. any upper bound of $S$
    B. the largest element in $S$
    C. the smallest real number that is greater than or equal to all elements of $S$
    D. a number strictly greater than all elements of $S$
    **Answer:** _____
2.  Which of the following sets does **not** contain its supremum?
    A. $[0,1]$   B. $\{1/n : n \in \mathbb{N}\}$
    C. $(0,1)$   D. $\{0,1\}$
    **Answer:** _____
3.  The completeness axiom of the real numbers states that
    A. every nonempty set has a maximum
    B. every bounded set has a midpoint
    C. every nonempty bounded above set has a least upper bound in $\mathbb{R}$
    D. every sequence converges
    **Answer:** _____
4.  If $S \subseteq \mathbb{R}$ is bounded below, then
    A. $\inf S = -\sup S$
    B. $\inf S = \min S$ in all cases
    C. $\inf S = -\sup(-S)$
    D. $\inf S$ always belongs to $S$
    **Answer:** _____
5.  Let $A = \{x \in \mathbb{Z} : x \leq \sqrt{10}\}$. Then
    A. $\sup A = \sqrt{10}$
    B. $\sup A = 3 \in A$
    C. $\sup A = 4$
    D. $A$ has no supremum in $\mathbb{Z}$
    **Answer:** _____

## Matching

Match each concept in **Column A** with the correct description in **Column B**.

| Column A | Column B |
|---|---|
| A. Supremum of $S$ | 1. No element of $\mathbb{N}$ exceeds it. |
| B. Infimum of $S$ | 2. The number $-\inf S$. |
| C. Archimedean property | 3. Greatest lower bound of a set. |
| D. $\sup(-S)$ | 4. Nonempty bounded-above set has one. |
| E. Completeness axiom | 5. Least upper bound of a set. |

**Answers:** _____ – _____ – _____ – _____ – _____

### Fill in the blanks

1. A set is *bounded above* if every element is $\leq$ some real number called an _____.
2. For $A \subseteq \mathbb{R}$ one always has $\inf A = -$_____$(-A)$.
3. The Archimedean property implies: $\forall x > 0, \forall y \in \mathbb{R}, \exists n \in \mathbb{N}$ with $nx > y$, because $\mathbb{N}$ is not _____ above.
4. If $\sup S \in S$, we write $\sup S =$ _____ $S$.
5. For integer sets that are bounded above, the supremum is automatically an element of the set, hence it is the _____.

## 1.7 Absolute value and integer part of a number

In this section we introduce two foundational notions that recur throughout: absolute value and the integer part.

Absolute value measures distance from zero and underpins key estimation tools, most notably the triangle inequality.

The integer part bridges the continuous and the discrete: it is the greatest integer $\leq x$ (i. e., $[x] \leq x < [x] + 1$) and will be used for sharp bounds and constructive proofs.

### 1.7.1 Absolute value

We begin by introducing the concept of absolute value, which measures the distance of a real number from zero.

This notion is fundamental in both algebra and analysis.

**Definition 1.12** (Absolute value). Let $a \in \mathbb{R}$. The **absolute value** of $a$, denoted by $|a|$, is defined as

$$|a| = \begin{cases} a, & \text{if } a \geq 0, \\ -a, & \text{if } a < 0. \end{cases}$$

The absolute value of any real number is always nonnegative (see Figure 1.3 and Table 1.1).

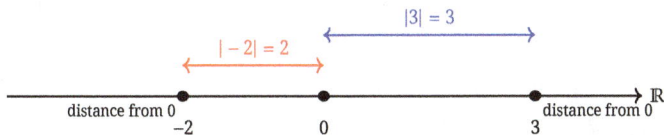

**Figure 1.3:** Geometric representation of the absolute value as distance from the origin on the real line.

**Table 1.1:** Fundamental properties of the absolute value in $\mathbb{R}$.

| Property | Statement |
|---|---|
| Nonnegativity | $\|x\| \geq 0$ for all $x \in \mathbb{R}$ |
| Zero property | $\|x\| = 0 \iff x = 0$ |
| Symmetry | $\|-x\| = \|x\|$ |
| Product rule | $\|xy\| = \|x\|\|y\|$ |
| Quotient rule | $\left\|\frac{x}{y}\right\| = \frac{\|x\|}{\|y\|}, y \neq 0$ |
| Distance formula | $\|x - y\| = \|y - x\|$ |
| Triangle inequality | $\|x + y\| \leq \|x\| + \|y\|$ |
| Reverse triangle inequality | $\left\|\,\|x\| - \|y\|\,\right\| \leq \|x - y\|$ |

One of the most significant properties of the absolute value is the triangle inequality. It describes how the absolute value behaves under addition.

**Proposition 1.10** (Triangle inequality). *For all real numbers $a, b \in \mathbb{R}$, the following inequality holds:*

$$\left|\,|a| - |b|\,\right| \leq |a + b| \leq |a| + |b|.$$

*Proof.* We will prove the two inequalities separately.

**Step 1:** Prove that

$$|a + b| \leq |a| + |b|.$$

This follows directly from the definition of absolute value and the following elementary inequality:

$$-(|a| + |b|) \leq a + b \leq |a| + |b|.$$

Indeed, since

$$-a \leq |a| \quad \text{and} \quad -b \leq |b| \quad \Rightarrow \quad -(a + b) \leq |a| + |b|,$$

we get

$$|a + b| \leq |a| + |b|.$$

**Step 2:** Prove that

$$\bigl||a| - |b|\bigr| \le |a + b|.$$

Using Step 1, we write

$$|a| = \bigl|(a + b) + (-b)\bigr| \le |a + b| + |b| \quad \Rightarrow \quad |a| - |b| \le |a + b|.$$

Similarly,

$$|b| = \bigl|(a + b) + (-a)\bigr| \le |a + b| + |a| \quad \Rightarrow \quad |b| - |a| \le |a + b|.$$

So in both cases

$$\bigl||a| - |b|\bigr| \le |a + b|.$$

Therefore, the full triangle inequality holds

$$\bigl||a| - |b|\bigr| \le |a + b| \le |a| + |b|. \qquad \square$$

### 1.7.2 Integer part

Let $x \in \mathbb{R}$. We define the set

$$A = \{n \in \mathbb{Z} : n \le x\},$$

that is, the set of all integers less than or equal to $x$.

Clearly, $A \subseteq \mathbb{Z}$ is bounded above in $\mathbb{R}$, so it has a supremum. By Proposition 1.7, its least upper bound belongs to $A$ and is its maximum.

This number is denoted by $[x]$, and it satisfies

$$[x] \le x < [x] + 1.$$

If this were not the case (i. e., if $x \ge [x] + 1$), then there would exist an integer strictly greater than $[x]$ and still less than or equal to $x$, contradicting the definition of $[x]$ as the greatest integer less than or equal to $x$.

> **Definition 1.13** (Integer part of a real number). For every real number $a \in \mathbb{R}$, the quantity $[a]$ denotes the **floor** or **integer part** of $a$, defined as the greatest integer less than or equal to $a$. It satisfies (see Figure 1.4)
>
> $$[a] \le a < [a] + 1.$$

A striking feature of the real line is that both the rational and the irrational numbers are dense in $\mathbb{R}$.

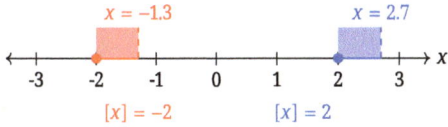

Figure 1.4: Graphical representation of the floor function (integer part).

The next proposition states this rigorously and supplies a proof. In particular, the argument employs the integer part (floor) and the Archimedean property.

**Proposition 1.11** (Density of $\mathbb{Q}$ and $\mathbb{R} \setminus \mathbb{Q}$ in $\mathbb{R}$). *Let $a, b \in \mathbb{R}$ with $a < b$. Then there exist:*
- *A rational number $c \in (a, b)$.*
- *An irrational number $d \in (a, b)$.*

*Proof.* To find a rational number between $a$ and $b$, we use the Archimedean property. Choose a natural number $s \in \mathbb{N}$ such that

$$\frac{1}{b-a} < s \quad \Rightarrow \quad \frac{1}{s} < b - a.$$

Now let $z = [sa] + 1 \in \mathbb{Z}$, where $[sa]$ is the floor (integer part) of $sa$. Then

$$z - 1 \le sa < z \quad \Rightarrow \quad a < \frac{z}{s} < a + \frac{1}{s} < b.$$

Hence, the rational number $c = \frac{z}{s} \in (a, b)$.

To show that an irrational number also lies in $(a, b)$, consider the interval

$$\left( \frac{a}{\sqrt{2}}, \frac{b}{\sqrt{2}} \right).$$

This is a nonempty interval since $a < b$. From the previous argument, there exists a rational number $u \in (\frac{a}{\sqrt{2}}, \frac{b}{\sqrt{2}})$.

Let

$$d = u \cdot \sqrt{2}.$$

Then

$$a < d < b.$$

Since $u \in \mathbb{Q}$ and $\sqrt{2} \notin \mathbb{Q}$, their product $d \notin \mathbb{Q}$. Thus, $d \in (a, b) \cap (\mathbb{R} \setminus \mathbb{Q})$, as required. □

## 1.7.3 Exercises

### True or false
Determine whether each statement is **true** or **false**. Justify your answer.

1. For any real number $x$, $|x| \geq x$.
   **Answer:** _____
2. If $|x| = |y|$, then $x = y$.
   **Answer:** _____
3. The integer part of a number $x$, denoted $[x]$, is always less than or equal to $x$.
   **Answer:** _____
4. For any integer $n$, $[n] = n$.
   **Answer:** _____
5. If $x$ is a real number, then $[x]$ is the largest integer less than $x$.
   **Answer:** _____

### Multiple choice: absolute value and integer part
Choose the correct option (A, B, C, or D) for each question.

1. The value of $|-5| + |3|$ is
   A. $-2$   B. $2$
   C. $8$    D. $-8$
   **Answer:** _____
2. Which of the following statements is always true for any real numbers $a$ and $b$?
   A. $|a + b| = |a| + |b|$
   B. $|a - b| = |a| - |b|$
   C. $|ab| = |a||b|$
   D. $|\frac{a}{b}| = \frac{|a|}{b}$ (for $b \neq 0$)
   **Answer:** _____
3. The integer part of $-3.7$ is
   A. $-3$   B. $-4$
   C. $3$    D. $4$
   **Answer:** _____
4. For any real number $x$, which of the following is true?
   A. $x - 1 < [x] \leq x$
   B. $[x] < x < [x] + 1$
   C. $[x] = x$
   D. $[x] \geq x$
   **Answer:** _____
5. Let $A = \{x \in \mathbb{R} : |x - 2| < 1\}$. The maximum integer in $A$ is
   A. $0$   B. $1$
   C. $2$   D. $3$
   **Answer:** _____

## Matching

Match each item in **Column A** with the best description in **Column B**.

| Column A | Column B |
| --- | --- |
| A. Triangle inequality | 1. Greatest integer $\leq x$. |
| B. Reverse triangle inequality | 2. $|x - y|$ equals their distance on the line. |
| C. Distance formula | 3. $\mid |x| - |y| \mid \leq |x - y|$. |
| D. Floor function $[x]$ | 4. $|x + y| \leq |x| + |y|$. |

**Answers:** \_\_\_\_\_ – \_\_\_\_\_ – \_\_\_\_\_ – \_\_\_\_\_

## Fill in the blanks

1. For all real numbers $a$, $b$, the inequality $|a + b| \leq$ _____ holds.
2. The reverse triangle inequality states $\mid |a| - |b| \mid \leq$ _____.
3. The floor satisfies $[x] \leq x <$ _____.
4. If $a \in (0, 1)$, then $|a - 1| =$ _____ $- a$.
5. For $n \in \mathbb{Z}$ and $x \in \mathbb{R}$, one always has $|x - n| \geq x -$ _____.

## 1.8 Topological structure of $\mathbb{R}$

In this section, we present several fundamental definitions and state (without proof) some results related to the topological structure of $\mathbb{R}$.

To formalize proximity in the real line, we introduce the concept of the $\varepsilon$-neighborhood of a point.

This notion serves as a building block for defining open sets and continuity.

**Definition 1.14** ($\varepsilon$-Neighborhood). Let $a \in \mathbb{R}$ and $\varepsilon > 0$. The $\varepsilon$-neighborhood of $a$, denoted by $\rho(a, \varepsilon)$, is defined as the open interval:

$$\rho(a, \varepsilon) = \{x \in \mathbb{R} : |x - a| < \varepsilon\}.$$

That is, $\rho(a, \varepsilon)$ consists of all real numbers whose distance from $a$ is strictly less than $\varepsilon$.

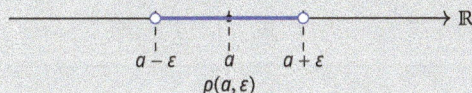

We now define interior points, which play a central role in topology.

A point is interior to a set if it is surrounded by a neighborhood contained entirely within the set.

**Definition 1.15** (Interior point). Let $A \subseteq \mathbb{R}$. A point $a \in A$ is called an **interior point** of $A$ if there exists an $\varepsilon$-neighborhood of $a$ that is entirely contained in $A$. That is,

$$p(a, \varepsilon) \subseteq A.$$

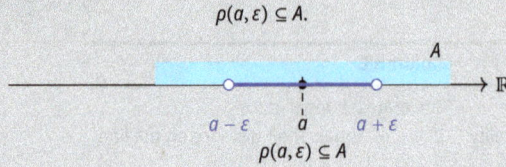

Not all points of a set need to have neighborhoods entirely within the set. Some points may stand alone—these are called isolated points.

**Definition 1.16** (Isolated point). A point $a \in A$ is called an **isolated point** if there exists $\varepsilon > 0$ such that

$$p(a, \varepsilon) \cap A = \{a\}.$$

Accumulation points, or limit points, capture where a set "clusters" in the real line. Their neighborhoods always contain infinitely many distinct points from the set.

**Definition 1.17** (Accumulation point). A point $a \in \mathbb{R}$ is called an **accumulation point** (or limit point) of a set $A \subseteq \mathbb{R}$ if for every $\varepsilon > 0$, we have

$$p(a, \varepsilon) \cap (A \setminus \{a\}) \neq \emptyset.$$

The most basic topological sets in $\mathbb{R}$ are the open and closed sets. These notions are defined in terms of interior points and complements. See Figure 1.5 for the simple case of intervals.

| Interval type | Graphical representation |
|---|---|
| *Unbounded intervals* | |
| $(a, +\infty) = \{x \in \mathbb{R} \mid x > a\}$ | |
| $[a, +\infty) = \{x \in \mathbb{R} \mid x \geq a\}$ | |
| $(-\infty, a) = \{x \in \mathbb{R} \mid x < a\}$ | |
| $(-\infty, a] = \{x \in \mathbb{R} \mid x \leq a\}$ | |
| *Bounded intervals* | |
| $[a, b] = \{x \in \mathbb{R} \mid a \leq x \leq b\}$ | |
| $(a, b) = \{x \in \mathbb{R} \mid a < x < b\}$ | |
| $[a, b) = \{x \in \mathbb{R} \mid a \leq x < b\}$ | |
| $(a, b] = \{x \in \mathbb{R} \mid a < x \leq b\}$ | |

**Figure 1.5:** Common interval notations and their graphical representations.

**Definition 1.18** (Open and closed sets). A subset $A \subseteq \mathbb{R}$ is called **open** if all of its points are interior points. It is called **closed** if its complement in $\mathbb{R}$ is open.

Open and closed sets exhibit several important properties under unions and intersections.

We summarize some of these basic results below.

**Proposition 1.12** (Properties of open and closed sets). *The sets $\mathbb{R}$ and $\emptyset$ are both open and closed. The finite intersection of open sets is open, and any union of open sets is open. The finite union of closed sets is closed, and any intersection of closed sets is also closed.*

A key property of accumulation points is that their neighborhoods are rich in points of the original set.

We make this precise in the following proposition.

**Proposition 1.13** (Neighborhoods of accumulation points). *If $a \in \mathbb{R}$ is an accumulation point of a set $A \subseteq \mathbb{R}$, then every neighborhood of $a$ contains infinitely many points of $A$.*

A classic characterization of closed sets is in terms of their accumulation points.

The next proposition makes this equivalence explicit.

**Proposition 1.14** (Closed sets and accumulation points). *A subset $A \subseteq \mathbb{R}$ is closed if and only if it contains all of its accumulation points.*

One of the most fundamental results in real analysis is the Bolzano–Weierstrass theorem.

It asserts that every infinite bounded set of real numbers has at least one accumulation point.

**Theorem 1.20** (Bolzano–Weierstrass theorem). *Every infinite and bounded subset of $\mathbb{R}$ has at least one accumulation point.*

## 1.8.1 Exercises

Determine whether each statement is **true** or **false**. Justify your answer.

1. Every point of a closed set is an accumulation point.
   **Answer:** _____
2. The empty set is both open and closed.
   **Answer:** _____
3. An isolated point is always an accumulation point.
   **Answer:** _____
4. The finite intersection of open sets is open.
   **Answer:** _____

5.  If a set contains all of its accumulation points, then it is closed.
    **Answer:** _____

## Multiple choice questions

Choose the correct option (A, B, C, or D) for each question.

1.  Which of the following correctly defines an accumulation point?
    A. $\exists \varepsilon > 0$ such that $\rho(a, \varepsilon) \subseteq A$
    B. $a$ is the limit of a sequence in $\mathbb{R}$
    C. For all $\varepsilon > 0$, $\rho(a, \varepsilon) \cap (A \setminus \{a\}) \neq \emptyset$
    D. $\exists \varepsilon > 0$ such that $\rho(a, \varepsilon) \cap A = \{a\}$
    **Answer:** _____

2.  Which of the following sets is both open and closed in $\mathbb{R}$?
    A. $(0, 1)$  B. $\mathbb{R}$
    C. $[0, 1)$  D. $\mathbb{Q}$
    **Answer:** _____

3.  Which statement best describes an isolated point?
    A. It is a limit point of a sequence in $A$
    B. There exists $\varepsilon > 0$ such that no other element of $A$ is in $\rho(a, \varepsilon)$
    C. It belongs to the complement of $A$
    D. It is always an interior point
    **Answer:** _____

4.  According to the Bolzano–Weierstrass theorem,
    A. every infinite and bounded set has an isolated point
    B. every infinite and bounded set has an accumulation point
    C. every closed set has finitely many accumulation points
    D. every open set is infinite
    **Answer:** _____

5.  Which of the following sets is **open** in $\mathbb{R}$?
    A. $[0, 1)$  B. $(0, 1)$
    C. $\{1\}$    D. $\mathbb{Q}$
    **Answer:** _____

## Matching

Match each term in **Column A** with the best description in **Column B**.

| Column A | Column B |
|---|---|
| A. Interior point | 1. Every neighborhood contains infinitely many A-points. |
| B. Accumulation point | 2. A set equal to its complement's complement. |
| C. Isolated point | 3. Some neighborhood lies entirely inside A. |
| D. Closed set | 4. Some neighborhood meets A only at the point itself. |

**Answers:** _____ – _____ – _____ – _____

### Fill in the blanks

1. The set $\rho(a, \varepsilon) = \{x \in \mathbb{R} : |x - a| < \varepsilon\}$ is an example of an _____ set in $\mathbb{R}$.

2. A point $a$ is *isolated* in $A$ iff $\exists \varepsilon > 0$ such that $\rho(a, \varepsilon) \cap A =$ _____.

3. A set is open precisely when all its points are _____ points.

4. A subset of $\mathbb{R}$ is closed iff it contains all its _____ points.

5. The Bolzano–Weierstrass theorem guarantees that every infinite bounded set has an _____ point.

## 1.9  Examples

**Example 1.1** (Supremum of a rational set). Let $A = \{\frac{n}{n+1} : n \in \mathbb{N}\}$. Find $\sup A$.

*Solution.*  We observe that

$$\frac{n}{n+1} = 1 - \frac{1}{n+1} < 1 \quad \text{for all } n \in \mathbb{N}.$$

So, 1 is an upper bound of $A$. We now show it is the least upper bound.

Let $\varepsilon > 0$. Choose $n \in \mathbb{N}$ such that

$$\frac{1}{n+1} < \varepsilon \quad \Rightarrow \quad 1 - \varepsilon < \frac{n}{n+1}.$$

Hence, for every $\varepsilon > 0$, there exists an element in $A$ greater than $1 - \varepsilon$, so

$$\sup A = 1. \qquad \square$$

**Example 1.2** (Infimum of reciprocals).  Let $B = \{\frac{1}{n} : n \in \mathbb{N}\}$. Determine $\inf B$.

*Solution.*  The set $B$ is clearly bounded below by 0 since $\frac{1}{n} > 0$ for all $n \in \mathbb{N}$.

We claim that $\inf B = 0$.

Let $\varepsilon > 0$. By the Archimedean property, there exists $n \in \mathbb{N}$ such that

$$\frac{1}{n} < \varepsilon.$$

Therefore, there are elements of $B$ arbitrarily close to 0, and no number greater than 0 can be a lower bound. So

$$\inf B = 0. \qquad \square$$

**Example 1.3** (Does the set have a maximum?).  Let $C = (0, 1) \subset \mathbb{R}$. Does $C$ have a maximum? What is $\sup C$?

*Solution.* Since the set $C = (0,1)$ is an open interval, it does not contain the number 1 although it is arbitrarily close to it.

Hence, $C$ has:

- No maximum (since for every $x \in C$, there exists $x' \in C$ such that $x < x' < 1$).
- Supremum: $\sup C = 1$ because 1 is an upper bound and any smaller number is not.

Thus,

$$\boxed{\sup C = 1}, \quad \text{but } \max C \text{ does not exist.} \qquad \square$$

**Example 1.4** (Infimum via negation). Let $A = \{-\frac{1}{n} : n \in \mathbb{N}\}$. Use the identity $\inf -A = -\sup(A)$ to compute $\sup A$.

*Solution.* We define the set

$$-A = \left\{ \frac{1}{n} : n \in \mathbb{N} \right\}.$$

We have proved already that $\inf -A = 0$, therefore using Proposition 1.8 we conclude that $\sup A = 0$. $\qquad \square$

**Example 1.5** (Supremum of union). Let $A = \{\frac{1}{n} : n \in \mathbb{N}\}$, and $B = \{2\}$. Compute $\sup(A \cup B)$.

*Solution.* We know

$$\sup A = 1 \quad \left( \text{since } \frac{1}{n} \to 0 \text{ and } \frac{1}{1} = 1 \right), \quad \sup B = 2.$$

Using Proposition 1.9 we get the result. $\qquad \square$

**Example 1.6** (Proof of AMGM for two variables). Prove that for all $a, b \geq 0$, the following inequality holds:

$$\frac{a+b}{2} \geq \sqrt{ab}.$$

*Solution.* We start from the identity

$$\left( \sqrt{a} - \sqrt{b} \right)^2 \geq 0.$$

Expanding:

$$a - 2\sqrt{ab} + b \geq 0 \quad \Rightarrow \quad a + b \geq 2\sqrt{ab}.$$

Dividing both sides by 2:

$$\frac{a+b}{2} \geq \sqrt{ab}.$$

Equality holds if and only if $a = b$. ☐

**Example 1.7** (Proof for fixed $n$). Prove that for $x > -1$, $x \neq 0$, and $n = 3$, the Bernoulli inequality holds

$$(1+x)^3 \geq 1 + 3x.$$

*Solution.* Expand the left-hand side:

$$(1+x)^3 = 1 + 3x + 3x^2 + x^3.$$

Since $x^2 \geq 0$ and $x^3 \geq -|x|^3$, we have

$$(1+x)^3 = 1 + 3x + \underbrace{3x^2 + x^3}_{\geq 0 \text{ if } x > -1}.$$

Hence,

$$(1+x)^3 \geq 1 + 3x \quad \text{for all } x > -1.$$

Equality holds only when $x = 0$. ☐

**Example 1.8** (Reverse triangle inequality). Prove that for all $x, y \in \mathbb{R}$, the following inequality holds:

$$\big||x| - |y|\big| \leq |x - y|.$$

*Solution.* We start from the triangle inequality

$$|x| = |(x - y) + y| \leq |x - y| + |y| \quad \Rightarrow \quad |x| - |y| \leq |x - y|.$$

Also

$$|y| = |(y - x) + x| \leq |y - x| + |x| = |x - y| + |x| \quad \Rightarrow \quad |y| - |x| \leq |x - y|.$$

Combining both

$$|x| - |y| \leq |x - y|, \quad |y| - |x| \leq |x - y| \quad \Rightarrow \quad \big||x| - |y|\big| \leq |x - y|. \qquad ☐$$

**Example 1.9** (Application to real line distance). Let $a, b, c \in \mathbb{R}$. Show that

$$|a - c| \leq |a - b| + |b - c|.$$

*Solution.* We rewrite the left-hand side:

$$|a - c| = |(a - b) + (b - c)|.$$

Then, by the triangle inequality,

$$|a - c| \le |a - b| + |b - c|.$$

$$\boxed{\text{Distance from } a \text{ to } c \text{ is at most the sum of intermediate distances.}}$$ □

**Example 1.10** (Induction: exponential inequality). Prove that for all $n \ge 5$, we have

$$2^n > n^2.$$

*Solution.*
**Base case:** $n = 5$:

$$2^5 = 32, \quad 5^2 = 25, \quad 32 > 25 \quad \checkmark$$

**Inductive hypothesis:** Assume $2^k > k^2$ for some $k \ge 5$.

**Show:** $2^{k+1} > (k + 1)^2$
We know

$$2^{k+1} = 2 \cdot 2^k > 2 \cdot k^2.$$

So it's enough to show

$$2k^2 > (k + 1)^2 \quad \Rightarrow \quad 2k^2 > k^2 + 2k + 1 \quad \Rightarrow \quad k^2 - 2k - 1 > 0.$$

The inequality $k^2 - 2k - 1 > 0$ holds for $k \ge 5$, so the inductive step is valid.

$$\boxed{2^n > n^2 \quad \text{for all } n \ge 5.}$$ □

**Example 1.11** (Induction: divisibility). Prove that $4^n - 1$ is divisible by 3 for all $n \in \mathbb{N}$.

*Solution.* Let $P(n) : 4^n - 1$ divisible by 3.

**Base case:** $n = 1 \Rightarrow 4^1 - 1 = 3$, divisible by 3. $\checkmark$

**Inductive step:** Assume $4^k - 1$ divisible by 3.
We show $4^{k+1} - 1$ is divisible by 3:

$$4^{k+1} - 1 = 4 \cdot 4^k - 1 = (4 \cdot 4^k - 4) + 3 = 4(4^k - 1) + 3.$$

By the inductive hypothesis, $4^k - 1$ divisible by $3 \Rightarrow$ the entire expression is divisible by 3.

$$4^n - 1 \equiv 0 \quad \text{mod 3 for all } n \in \mathbb{N}. \qquad \square$$

**Example 1.12** (Induction: geometric sum). Prove by induction that

$$\sum_{k=0}^{n} 2^k = 2^{n+1} - 1 \quad \text{for all } n \in \mathbb{N}.$$

*Solution.*
**Base case:** $n = 0 \Rightarrow \sum_{k=0}^{0} 2^k = 2^0 = 1$, and $2^{0+1} - 1 = 2 - 1 = 1$ ✓

**Inductive step:** Assume

$$\sum_{k=0}^{n} 2^k = 2^{n+1} - 1.$$

Then

$$\sum_{k=0}^{n+1} 2^k = \left( \sum_{k=0}^{n} 2^k \right) + 2^{n+1} = (2^{n+1} - 1) + 2^{n+1} = 2 \cdot 2^{n+1} - 1 = 2^{n+2} - 1.$$

So the result holds for $n + 1$, and

$$\sum_{k=0}^{n} 2^k = 2^{n+1} - 1. \qquad \square$$

**Example 1.13** (Inequality: $n! > 2^n$). Prove that for all integers $n \geq 4$, we have

$$n! > 2^n.$$

*Solution.*
**Base case:** $n = 4 \Rightarrow 4! = 24, 2^4 = 16 \Rightarrow 24 > 16$ ✓

**Inductive hypothesis:** Assume $k! > 2^k$ for some $k \geq 4$.

**Show:** $(k + 1)! > 2^{k+1}$
We use

$$(k + 1)! = (k + 1) \cdot k! > (k + 1) \cdot 2^k.$$

It suffices to show

$$(k + 1) \cdot 2^k > 2^{k+1} \quad \Rightarrow \quad k + 1 > 2.$$

This holds for all $k \geq 4$. Hence

$$(k+1)! > (k+1) \cdot 2^k > 2^{k+1}.$$

So,

$$\boxed{n! > 2^n \quad \text{for all } n \geq 4.}$$ □

**Example 1.14** (Divisibility: powers of 10). Prove that for all $n \in \mathbb{N}$,

$$10^n - 1 \quad \text{is divisible by 9.}$$

*Solution.* Let $P(n) : 10^n - 1 \equiv 0 \mod 9$.

**Base case:** $n = 1 \Rightarrow 10^1 - 1 = 9 \equiv 0 \mod 9$ ✓

**Inductive step:** Assume $10^k - 1 \equiv 0 \mod 9$

We prove $10^{k+1} - 1 \equiv 0 \mod 9$

$$10^{k+1} - 1 = 10 \cdot 10^k - 1 = 10(10^k - 1) + 9.$$

By I. H., $10^k - 1 \equiv 0 \Rightarrow 10(10^k - 1) \equiv 0 \Rightarrow \text{total} \equiv 9 \equiv 0 \mod 9$

$$\boxed{10^n - 1 \text{ divisible by 9 for all } n.}$$ □

**Example 1.15** (Recursive sequence with induction). Let $a_1 = 2$, and $a_{n+1} = 3a_n + 2$ for $n \geq 1$. Prove that

$$a_n = 3^n - 1 \quad \text{for all } n \in \mathbb{N}.$$

*Solution.*
**Base case:** $n = 1 \Rightarrow a_1 = 2, 3^1 - 1 = 2$ ✓

**Inductive hypothesis:** Assume $a_k = 3^k - 1$

**Inductive step:**

$$a_{k+1} = 3a_k + 2 = 3(3^k - 1) + 2 = 3^{k+1} - 3 + 2 = 3^{k+1} - 1.$$

Thus, the formula holds for all $n \in \mathbb{N}$.

$$\boxed{a_n = 3^n - 1.}$$ □

**Example 1.16** (Fibonacci inequality (strong induction)). Let $F_0 = 0$, $F_1 = 1$, $F_n = F_{n-1} + F_{n-2}$. Prove that $F_n < 2^n$ for all $n \geq 1$.

*Solution.*

**Base case:** $F_1 = 1 < 2^1 = 2$  ✓

**Inductive hypothesis:** Assume $F_k < 2^k$ for all $k \leq n$.

**We show:** $F_{n+1} < 2^{n+1}$.

Since

$$F_{n+1} = F_n + F_{n-1} < 2^n + 2^{n-1} = 2^{n-1}(2+1) = 3 \cdot 2^{n-1}.$$

Compare with $2^{n+1} = 4 \cdot 2^{n-1} \Rightarrow 3 \cdot 2^{n-1} < 4 \cdot 2^{n-1}$

$$\boxed{F_n < 2^n.}$$

□

## 1.10 Python with AI

**AI Request**

Use symbolic computation in Python to verify the classical formulas for the sums:

- $\sum_{k=0}^{n} k$
- $\sum_{k=0}^{n} k^2$
- $\sum_{k=0}^{n} k^3$

```
1  from sympy import symbols, Sum, Eq, simplify
2
3  # Declare symbolic variable k, n = symbols('k n')
4
5  # Define left-hand sides as symbolic sums
6
7  sum_k = Sum(k, (k, 0, n))
8
9  sum_k2 = Sum(k**2, (k, 0, n))
10
11 sum_k3=Sum(k**3, (k, 0, n))
12
13 # Define right-hand sides (formulas)
14
15 rhs_k = n*(n+1)/2
16
17 rhs_k2 = n*(n+1)*(2*n+1)/6
18
19 rhs_k3 =(n*(n+1)/2)**2
20
21 # Evaluate symbolic sums
22
23 lhs_k = sum_k.doit() lhs_k2 = sum_k2.doit() lhs_k3 = sum_k3.doit()
```

```
24
25 # Simplify both sides to compare
26
27 assert_k = Eq(simplify(lhs_k), simplify(rhs_k))
28
29 assert_k2=Eq(simplify(lhs_k2), simplify(rhs_k2))
30
31 assert_k3 =Eq(simplify(lhs_k3), simplify(rhs_k3))
32
33 # Output results
34
35 print("Sum of k: ", lhs_k, " == ", rhs_k, " > ", assert_k)
36 print("Sum of k^2: ", lhs_k2, " == ", rhs_k2, " > ", assert_k2)
37 print("Sum of k^3: ", lhs_k3, " == ", rhs_k3, " > ", assert_k3)
```

---

**AI Request**

Can you explain how to use the above code?

---

### How to Use This Code

This code uses symbolic summation via SymPy's `Sum(...).doit()` to compute the exact closed-form for each series. It then simplifies both sides to verify that the known formula matches the symbolic result, asserting the correctness of

$$\sum_{k=0}^{n} k = \frac{n(n+1)}{2}, \quad \sum_{k=0}^{n} k^2 = \frac{n(n+1)(2n+1)}{6}, \quad \sum_{k=0}^{n} k^3 = \left(\frac{n(n+1)}{2}\right)^2.$$

---

**AI Request**

Write a Python code to expand $(a + b)^n$ symbolically and display it with nice mathematics symbols in a notebook environment.

---

```
1 from sympy import symbols, expand, latex from IPython.display
2 import display, HTML
3
4 def expand_binomial_and_display_html(a_sym, b_sym, n_val):
5     """
6     Symbolically expands the expression (a + b)^n and displays the
       result in HTML
7     using MathJax rendering. Compatible with Google Colab or
       Jupyter Notebook.
8
9     Args:
10        a_sym (sympy.Symbol): The symbolic 'a' term.
11        b_sym (sympy.Symbol): The symbolic 'b' term.
12        n_val (int): The exponent 'n', must be a non-negative
       integer.
```

```
13          """
14          # Validate input
15          if not isinstance(n_val, int) or n_val < 0:
16              raise ValueError("n_val must be a non-negative integer.")
17
18          # Perform symbolic expansion
19          expression = (a_sym + b_sym)**n_val
20          expanded_expression = expand(expression)
21
22          # Convert to LaTeX format
23          latex_expression = latex(expanded_expression)
24          original_expression = latex(expression)
25
26          # Prepare HTML with MathJax rendering
27          html_output = f"""
28          <!DOCTYPE html>
29          <html>
30          <head>
31              <title>Binomial Expansion</title>
32              <script type="text/javascript" async
33                  src="https://cdnjs.cloudflare.com/ajax/libs/mathjax
        /2.7.7/MathJax.js?config=TeX-MML-AM_CHTML">
34              </script>
35              <style>
36                  body {{ font-family: Arial, sans-serif; margin: 10px;
        }}
37                  h2 {{ color: #333; }}
38                  p {{ margin-bottom: 5px; }}
39              </style>
40          </head>
41          <body>
42              <h2>Symbolic Expansion of
43 %
44 $${original_expression}$$</h2>
45              <p>The original expression is:
46 %
47 $${original_expression}$$</p>
48              <p>The expanded form is:</p>
49              <p>$$ {latex_expression}
50 %
51 $$</p>
52          </body>
53          </html>
54          """
55          # Display in notebook or Colab
56          display(HTML(html_output))
57
58
59 # --- Main Execution ---
```

```
60
61  # Define symbolic variables (customizable) x, y = symbols('x y')
62
63  # Example 1: Expand (x + y)^3 n1 = 3 print(f"--- Expansion for n =
64  {n1} ---") expand_binomial_and_display_html(x, y, n1)
65
66  # Example 2: Expand (x + y)^5 n2 = 5 print(f"\n--- Expansion for n
67  = {n2} ---") expand_binomial_and_display_html(x, y, n2)
```

---

**How to Use This Code**

This code is intended to be executed in a Python environment that supports HTML and MathJax rendering:
–   Jupyter Notebook
–   Google Colab

## Usage instructions

**Step 1: Define the function**

Paste the entire function definition and import statements into a notebook cell:

```
1  from sympy import symbols, expand, latex from IPython.display
2  import display, HTML
3
4  def expand_binomial_and_display_html(a_sym, b_sym, n_val):
5      ...
```

**Step 2: Define the variables**

Define your symbolic variables using sympy.symbols, for example:

```
1  x, y = symbols('x y')
```

**Step 3: Call the function**

To expand the binomial expression $(x + y)^3$, run

```
1  expand_binomial_and_display_html(x, y, 3)
```

To expand $(x + y)^5$,

```
1  expand_binomial_and_display_html(x, y, 5)
```

**Step 4: Customize inputs**

You may use any symbolic variables or expressions, such as

```
1  a, b = symbols('a b') expand_binomial_and_display_html(a, b, 4)
```

**Notes**
–   The exponent $n$ must be a nonnegative integer.
–   This function uses MathJax and will not render properly outside of a notebook or Colab.

---

## 1.11 Review exercises

**Exercise 1.** Which of the following sets is inductive?
(a) $A = \{1, 2, 3, 4, \ldots\}$
(b) $B = \{2, 4, 6, 8, \ldots\}$
(c) $C = \{x \in \mathbb{R} : x > 0\}$
(d) $D = \{x \in \mathbb{N} : x^2 < 100\}$

**Exercise 2.** Let $A = \{\frac{n}{n+1} : n \in \mathbb{N}\}$. Determine $\sup A$ and whether it belongs to $A$.

**Exercise 3.** Let $A = \{-\frac{1}{n} : n \in \mathbb{N}\}$. Compute $\inf A$ using the identity $\inf A = -\sup(-A)$.

**Exercise 4.** Find the integer part $[x]$ of each of the following numbers:
(a) $x = 3.999$
(b) $x = -5.001$
(c) $x = \sqrt{2}$
(d) $x = \pi$

**Exercise 5.** Which of the following sets are open in $\mathbb{R}$? Justify your answer.
(a) $(0, 1)$
(b) $[0, 1)$
(c) $\mathbb{Q}$ (the set of rational numbers)
(d) $\mathbb{R} \setminus \mathbb{Q}$ (the set of irrational numbers)

**Exercise 6.** Prove by mathematical induction that for all integers $n \geq 1$,

$$7 \text{ divides } 3^{2n} - 2^n.$$

In other words, prove that $3^{2n} - 2^n$ is divisible by 7.

**Exercise 7.** Let $S = \{\frac{1}{n} : n \in \mathbb{N}\}$. Find all accumulation points of $S$ in $\mathbb{R}$.

**Exercise 8.** Expand $(a+b)^5$ symbolically using the binomial theorem. Verify your result using a small Python script.

**Exercise 9.** Use matplotlib and numpy to create a plot of the interval $[1, 3)$ on the real line. Mark the endpoints appropriately (filled circle for included, hollow circle for excluded). Add axis labels and a title.

**Exercise 10** (Supremum and infimum). Find the supremum and the infimum of the following set:

$$A = \left\{ \frac{(-1)^n \cdot n}{n+1} : n \in \mathbb{N} \right\}.$$

Justify your answer.

**Exercise 11.** Let the set $A$ be defined as

$$A = \left\{ \frac{n}{n+1} : n = 1, 2, 3, \ldots, 100 \right\}.$$

(a) Using Python, generate a list containing the first 100 elements of $A$.
(b) Find $\max(A)$ and $\min(A)$. What can you deduce about $\sup A$ and $\inf A$?
(c) Extend the computation to $n = 10^5$ and numerically verify that $\sup A$ approaches a specific value.

**Exercise 12.** Consider the inequality

$$2^n > n^2, \quad \text{for } n \geq 5.$$

(a) Prove the inequality using mathematical induction.
(b) Write a Python script to verify the inequality for $n = 5, 6, \ldots, 20$.

**Exercise 13** (Density of rational numbers). Prove that between any two distinct real numbers $a < b$, there exist at least one rational number and one irrational number.

**Exercise 14** (Boundedness of sets). Consider the set

$$B = \{\sqrt{n+1} - \sqrt{n} : n \in \mathbb{N}\}.$$

Is the set $B$ bounded above and/or below? If so, determine its supremum and infimum.

**Exercise 15** (Python exercise). Write a Python function that receives a positive integer $n$ and returns a list containing the first $n$ terms of the following sequence:

$$a_n = \frac{(-1)^n \cdot n}{n+1}.$$

Then, use the function to compute the first 10 terms of the sequence.

## Fill in the blanks
Complete the following statements by filling in the blanks:
1. A set $A \subseteq \mathbb{R}$ is _____ above if there exists a real number $M$ such that for every $x \in A$, $x \leq M$.
2. The least upper bound of a set $A$ is called its _____.
3. Every nonempty set of real numbers that is bounded above has a _____ upper bound, according to the _____ axiom.
4. If $A$ is bounded below, then the greatest lower bound is called the _____ of $A$.
5. For any real number $a$, the _____ value, denoted by $|a|$, is always nonnegative.

6. The number $[x]$, called the _____ of $x$, satisfies the inequalities $[x] \leq x < [x] + 1$.
7. A set $A \subseteq \mathbb{R}$ is _____ if it contains all of its accumulation points.

## Matching

Match the concepts (1–5) with the correct descriptions (a–e):

1. Supremum
2. Archimedean property
3. Triangle inequality
4. Rational numbers ($\mathbb{Q}$)
5. Accumulation point

a. Numbers expressed as the quotient of two integers, where the denominator is not zero.

b. If $a, b \in \mathbb{R}$, then $|a + b| \leq |a| + |b|$.

c. For every real number, there exists a natural number greater than it.

d. A point $x$ such that every neighborhood around it contains infinitely many points of a set.

e. The least number that is greater than or equal to every element of a given set.

**Answers:** 1 – _____ 2 – _____ 3 – _____ 4 – _____ 5 – _____

# 2 Complex numbers

## Contents

Complex numbers extend the real number system by including solutions to equations like $x^2 + 1 = 0$. Defined as $z = a + ib$, where $a, b \in \mathbb{R}$ and $i^2 = -1$, they unify algebraic and geometric ideas.

This chapter introduces basic definitions, algebraic operations, geometric interpretation in the complex plane, polar and exponential forms, and key identities. Theory is complemented by worked examples, exercises, and Python code for symbolic and graphical computations.

## 2.1 Introduction

The concept of **complex numbers** represents one of the most elegant and powerful extensions of the number system. Initially introduced to solve equations such as $x^2 + 1 = 0$, which have no solutions in the real numbers, complex numbers have since become fundamental across a vast range of mathematical disciplines, from algebra and analysis to geometry, physics, and engineering.

Historically, the idea of imaginary quantities emerged in the 16th century when Italian mathematician *Gerolamo Cardano* encountered square roots of negative numbers while solving cubic equations. However, it was not until the work of *Rafael Bombelli* that rules for manipulating these "imaginary" numbers were formally established.

In the 18th century, *Leonhard Euler* introduced the symbol $i$ to represent $\sqrt{-1}$, and his famous identity

$$e^{i\pi} + 1 = 0$$

demonstrated the deep interconnections between algebra, trigonometry, and exponential functions.

The geometric interpretation of complex numbers was later developed by *Caspar Wessel, Jean-Robert Argand*, and *Carl Friedrich Gauss*, who independently proposed representing them as points or vectors in a two-dimensional plane, now known as the complex plane.

Gauss also provided the first rigorous proof of the **fundamental theorem of algebra**, which asserts that every nonconstant polynomial has at least one complex root, confirming the completeness of the complex number system.

https://doi.org/10.1515/9783112228289-002

Today, complex numbers are indispensable in fields such as signal processing, control theory, quantum mechanics, fluid dynamics, and electrical engineering. Their ability to unify algebraic and geometric perspectives makes them a cornerstone of modern mathematics.

This chapter introduces complex numbers from both theoretical and practical viewpoints, covering definitions, operations, polar forms, roots, and their visual representation. It is designed to build a strong foundation for further studies in advanced mathematics and applied sciences.

## 2.2 Basic definitions

By Theorem 1.11, it follows that there is no real number whose square is negative. We extend the number system in such a way that square roots exist for all real numbers.

---

**Definition 2.1** (Imaginary unit). We define a new number $i$ with the property
$$i^2 = -1.$$

---

With the imaginary unit established, we can now define complex numbers.

These numbers generalize the real numbers and allow for new algebraic operations.

---

**Definition 2.2** (Complex numbers). A **complex number** is an expression of the form
$$x = a + ib, \quad \text{where } a, b \in \mathbb{R}.$$

The number $a$ is called the **real part** of $x$ and is denoted by $\operatorname{Re}(x)$, while $b$ is called the **imaginary part** of $x$ and is denoted by $\operatorname{Im}(x)$.

---

As with real numbers, it is essential to specify when two complex numbers are considered equal.

This is determined by comparing their real and imaginary parts.

---

**Definition 2.3** (Equality of complex numbers). Two complex numbers $x = a + ib$ and $y = c + id$ are **equal** if and only if
$$a = c \quad \text{and} \quad b = d.$$

---

### 2.2.1 Exercises

Determine whether each statement is **true** or **false**. Justify your answer.
1. The equation $x^2 = -1$ has a solution in the real numbers.
   **Answer:** _____
2. The number $i$ is defined such that $i^2 = -1$.
   **Answer:** _____

3. A complex number must have both nonzero real and imaginary parts.
   **Answer:** _____
4. The number 4 is a complex number.
   **Answer:** _____
5. Two complex numbers $x = a + ib$ and $y = c + id$ are equal if and only if $a = c$ and $b = d$.
   **Answer:** _____

## Multiple choice questions

Choose the correct option (A, B, C, or D) for each question.

1. Which of the following defines the imaginary unit $i$?
   A. $i = \sqrt{-2}$
   B. $i = -1$
   C. $i^2 = -1$
   D. $i = \sqrt{1}$
   **Answer:** _____
2. What is the real part of the complex number $z = -3 + 2i$?
   A. $-3$   B. 2
   C. 3      D. $-2$
   **Answer:** _____
3. Which of the following numbers is not a complex number?
   A. $7 + 0i$   B. $-i$
   C. 3          D. None of the above
   **Answer:** _____
4. The imaginary part of the complex number $z = 5 - 7i$ is
   A. 5   B. $-7$
   C. 7   D. $-5$
   **Answer:** _____
5. Which of the following conditions guarantees that two complex numbers $x = a + ib$ and $y = c + id$ are equal?
   A. $a = c$ only        B. $b = d$ only
   C. $a = c$ and $b = d$   D. $x = y$ is not possible for complex numbers
   **Answer:** _____

## 2.3 Basic properties of complex numbers

Having defined complex numbers, we now turn to their fundamental algebraic properties. We will discuss operations such as addition, multiplication, and conjugation.

From the fundamental identity $i^2 = -1$, several important properties of complex numbers follow. Let $z_1 = a_1 + ib_1$ and $z_2 = a_2 + ib_2$. Then:

## Multiplication

The product of $z_1$ and $z_2$ is

$$z_1 z_2 = (a_1 + ib_1)(a_2 + ib_2) = a_1 a_2 + i a_1 b_2 + i b_1 a_2 + i^2 b_1 b_2 = a_1 a_2 - b_1 b_2 + i(a_1 b_2 + b_1 a_2).$$

## Addition

The sum of $z_1$ and $z_2$ is

$$z_1 + z_2 = (a_1 + a_2) + i(b_1 + b_2).$$

## Division

Division of two complex numbers can be performed as follows:

$$\frac{z_1}{z_2} = \frac{a_1 + ib_1}{a_2 + ib_2} = \frac{(a_1 + ib_1)(a_2 - ib_2)}{(a_2 + ib_2)(a_2 - ib_2)} = \frac{a_1 a_2 + b_1 b_2 + i(b_1 a_2 - a_1 b_2)}{a_2^2 + b_2^2}$$

$$= \frac{a_1 a_2 + b_1 b_2}{a_2^2 + b_2^2} + i\frac{b_1 a_2 - a_1 b_2}{a_2^2 + b_2^2}.$$

## Conjugate

If $z = a + ib$, then the complex number $\bar{z} = a - ib$ is called the conjugate of $z$.

## Zero product property

If for two complex numbers $z_1$ and $z_2$ it holds that $z_1 z_2 = 0$, then necessarily one of them must be zero. Indeed,

$$z_1 z_2 = a_1 a_2 - b_1 b_2 + i(a_1 b_2 + b_1 a_2) = 0 + i0.$$

Therefore, we obtain the following two equations:

$$a_1 a_2 - b_1 b_2 = 0 \quad \text{and} \quad a_1 b_2 + b_1 a_2 = 0.$$

Squaring both equations and adding them yields

$$(a_1^2 + b_1^2)(a_2^2 + b_2^2) = 0.$$

Hence, at least one of the factors must be zero. Suppose $a_1^2 + b_1^2 = 0$. This implies $a_1 = b_1 = 0$.

## Sum and product of conjugates

The sum of $z$ and $\bar{z}$ is $2a$ when $z = a + ib$, while the product of conjugates is $a^2 + b^2$. Therefore, both the sum and the product of a complex number and its conjugate are always real numbers.

The difference $z - \bar{z} = 2ib$ has a real part equal to zero. The quotient

$$\frac{z}{\bar{z}} = \frac{a^2 - b^2 + i2ab}{a^2 + b^2} = \frac{a^2 - b^2}{a^2 + b^2} + i\frac{2ab}{a^2 + b^2}$$

is therefore a complex number.

Complex conjugation preserves many arithmetic properties. The next theorem summarizes how conjugation interacts with addition, multiplication, and division.

**Theorem 2.1** (Conjugate of sum, difference, product, and quotient). *The conjugate of the sum, difference, product, or quotient of two complex numbers equals the corresponding sum, difference, product, or quotient of their conjugates. In other words, the following identities hold:*

$$\overline{x \pm y} = \bar{x} \pm \bar{y}, \quad \overline{xy} = \bar{x}\bar{y}, \quad \overline{\left(\frac{x}{y}\right)} = \frac{\bar{x}}{\bar{y}}.$$

*Proof.* Let $x = a_1 + ib_1$ and $y = a_2 + ib_2$ be two complex numbers. Their sum is $x + y = a_1 + a_2 + i(b_1 + b_2)$, and its conjugate is

$$\overline{x + y} = a_1 + a_2 - i(b_1 + b_2).$$

On the other hand, the sum of the conjugates is

$$\bar{x} + \bar{y} = (a_1 - ib_1) + (a_2 - ib_2) = a_1 + a_2 - i(b_1 + b_2),$$

which matches $\overline{x + y}$.

For the product, we compute

$$\overline{xy} = \overline{(a_1 + ib_1)(a_2 + ib_2)} = \overline{a_1 a_2 - b_1 b_2 + i(a_1 b_2 + b_1 a_2)} = a_1 a_2 - b_1 b_2 - i(a_1 b_2 + b_1 a_2) = \bar{x}\bar{y}.$$

Similarly, the identity for the quotient follows from applying the conjugate to both numerator and denominator. This completes the proof. $\square$

Polynomials with real coefficients have an important symmetry: their non-real roots occur in conjugate pairs.

The next lemma makes this precise.

**Lemma 2.1** (Conjugate root of a polynomial with real coefficients). *Let $P(z)$ be a polynomial given by*

$$P(z) = a_0 z^n + a_1 z^{n-1} + \cdots + a_n,$$

*where $z$ is a complex number, i. e., $z = a + ib$, and the coefficients $a_0, a_1, \ldots, a_n$ are real numbers. Then, if $z \in \mathbb{C}$ is a root of the polynomial, its conjugate $\bar{z}$ is also a root.*

| Operation | Definition/property |
|---|---|
| Complex number | $z = a + ib$, where $a, b \in \mathbb{R}$ |
| | $\text{Re}(z) = a$, $\text{Im}(z) = b$ |
| Equality | $a + ib = c + id$ if and only if $a = c$ and $b = d$ |
| Addition | $(a + ib) + (c + id) = (a + c) + i(b + d)$ |
| Multiplication | $(a + ib)(c + id) = (ac - bd) + i(ad + bc)$ |
| Conjugate | $\overline{a + ib} = a - ib$ |
| | $z + \bar{z} = 2a \in \mathbb{R}$ |
| | $z \cdot \bar{z} = a^2 + b^2 \in \mathbb{R}_{\geq 0}$ |
| Division | $\frac{a+ib}{c+id} = \frac{(a+ib)(c-id)}{c^2+d^2} = \frac{ac+bd}{c^2+d^2} + i\,\frac{bc-ad}{c^2+d^2}$ |
| Zero product property | If $z_1 z_2 = 0$, then $z_1 = 0$ or $z_2 = 0$ |
| Conjugation rules | $\overline{z_1 \pm z_2} = \overline{z_1} \pm \overline{z_2}$ |
| | $\overline{z_1 z_2} = \overline{z_1} \cdot \overline{z_2}$ |
| | $\overline{\left(\frac{z_1}{z_2}\right)} = \frac{\overline{z_1}}{\overline{z_2}}$ |

*Proof.* Since all coefficients $a_k$ are real, it follows that $\overline{a_k} = a_k$. Therefore, using the property of complex conjugates from Theorem 2.1, we have

$$\overline{P(z)} = \overline{a_0 z^n + \cdots + a_n} = a_0 \bar{z}^n + \cdots + a_n = P(\bar{z}).$$

Now suppose $z_1$ is a root of the polynomial with real coefficients, i. e., $P(z_1) = 0$. Then

$$\overline{P(z_1)} = P(\overline{z_1}) = 0,$$

which implies that $\overline{z_1}$ is also a root of the polynomial. This completes the proof. □

A cornerstone of algebra is the fundamental theorem, guaranteeing roots for every nonconstant complex polynomial. We state its main properties here.

**Theorem 2.2** (Fundamental theorem of algebra). *Let $P(z)$ be a polynomial given by*
$$P(z) = a_0 z^n + a_1 z^{n-1} + \cdots + a_n,$$
*where $z$ is a complex number (i. e., $z = a + ib$), and the coefficients $a_0, a_1, \ldots, a_n$ are complex numbers. Then this polynomial has exactly n roots (not necessarily distinct).*

*If $z_1, z_2, \ldots, z_k$ are the distinct roots of the polynomial with multiplicities $r_1, r_2, \ldots, r_k$, respectively, then it necessarily holds that*
$$r_1 + r_2 + \cdots + r_k = n.$$

*Moreover, for each root $z_i$, we have*
$$P(z_i) = P'(z_i) = \cdots = P^{(r_i-1)}(z_i) = 0 \quad \text{and} \quad P^{(r_i)}(z_i) \neq 0.$$

*Additionally, if there exists a complex number $z^* \in \mathbb{C}$ such that*
$$P(z^*) = P'(z^*) = \cdots = P^{(r^*)}(z^*) = 0 \quad \text{and} \quad P^{(r^*+1)}(z^*) \neq 0,$$
*then $z^*$ is a root of the polynomial of multiplicity $r^* + 1$.*

The behavior of polynomial roots in the complex plane leads to several noteworthy remarks. These include facts about multiplicities and solving equations with negative discriminant.

**Remark 2** (Multiplicity greater than degree implies zero polynomial). If a polynomial of degree $m$ has the sum of the multiplicities of its roots greater than $m$, then the polynomial must be the zero polynomial, that is, all its coefficients are equal to zero.

**Remark 3** (Solving a quadratic equation with negative discriminant). Consider the quadratic equation

$$ax^2 + bx + c = 0.$$

In the case where the discriminant $\Delta = b^2 - 4ac$ is negative, we can use complex numbers to compute the roots. We write $\Delta = |\Delta| i^2$, and the roots become

$$\lambda_{1,2} = \frac{-b \pm \sqrt{|\Delta| i^2}}{2a} = -\frac{b}{2a} \pm \frac{\sqrt{|\Delta|}}{2a} i.$$

We observe that the roots are complex conjugates of each other.

## 2.3.1 Solved exercises

### True or false questions
1. The equation $x^2 = -1$ has a solution in the real numbers.
   **Answer:** False
   *Explanation:* The square of any real number is nonnegative. Therefore, there is no real number whose square is negative.
2. The number $i$ is defined such that $i^2 = -1$.
   **Answer:** True
   *Explanation:* This is the definition of the imaginary unit $i$.
3. A complex number must have both nonzero real and imaginary parts.
   **Answer:** False
   *Explanation:* A complex number can have either a zero real part (e. g., $z = 0 + 3i$) or a zero imaginary part (e. g., $z = 5 + 0i$).
4. The number 4 is a complex number.
   **Answer:** True
   *Explanation:* Any real number can be written as $a + 0i$, so it is also a complex number.
5. Two complex numbers $x = a + ib$ and $y = c + id$ are equal if and only if $a = c$ and $b = d$.
   **Answer:** True
   *Explanation:* This is the definition of equality for complex numbers.

### Multiple choice questions
1. Which of the following defines the imaginary unit $i$?

A. $i = \sqrt{-2}$   B. $i = -1$
C. $i^2 = -1$   D. $i = \sqrt{1}$
**Answer:** C
*Explanation:* By definition, $i^2 = -1$.

2. What is the real part of the complex number $z = -3 + 2i$?
   A. $-3$   B. 2
   C. 3   D. $-2$
   **Answer:** A
   *Explanation:* The real part of $z = a + ib$ is $a$.

3. Which of the following numbers is not a complex number?
   A. $7 + 0i$   B. $-i$
   C. 3   D. None of the above
   **Answer:** D
   *Explanation:* All real numbers and purely imaginary numbers are special cases of complex numbers.

4. The imaginary part of the complex number $z = 5 - 7i$ is
   A. 5   B. $-7$
   C. 7   D. $-5$
   **Answer:** B
   *Explanation:* The imaginary part of $z = a + ib$ is $b$.

5. Which of the following conditions guarantees that two complex numbers $x = a + ib$ and $y = c + id$ are equal?
   A. $a = c$ only   B. $b = d$ only
   C. $a = c$ and $b = d$   D. Equality is not possible for complex numbers
   **Answer:** C
   *Explanation:* Two complex numbers are equal if and only if their real parts and imaginary parts are equal.

**Example 2.1.** Let $z_1 = 2 + 3i$ and $z_2 = -1 + 4i$. Compute:
(a) $z_1 + z_2$
(b) $z_1 \cdot z_2$

*Solution.* (a) Addition:

$$z_1 + z_2 = (2 + 3i) + (-1 + 4i) = (2 - 1) + i(3 + 4) = 1 + 7i.$$

(b) Multiplication:

$$z_1 \cdot z_2 = (2 + 3i)(-1 + 4i) = 2(-1) + 2(4i) + 3i(-1) + 3i(4i) = -2 + 8i - 3i + 12i^2$$
$$= -2 + 5i + 12(-1) = -2 + 5i - 12 = -14 + 5i. \qquad \square$$

**Example 2.2.** Compute the quotient:

$$\frac{3 + 4i}{1 - 2i}.$$

*Solution.* Multiply numerator and denominator by the conjugate of the denominator:

$$\frac{3+4i}{1-2i}\cdot\frac{1+2i}{1+2i}=\frac{(3+4i)(1+2i)}{(1)^2+(2)^2}=\frac{3+6i+4i+8i^2}{1+4}=\frac{3+10i-8}{5}=\frac{-5+10i}{5}=-1+2i.$$

□

**Example 2.3.** Let $P(x)=x^2+2x+5$. Find its roots using complex numbers.

*Solution.* Use the quadratic formula

$$x=\frac{-b\pm\sqrt{b^2-4ac}}{2a}.$$

Here, $a=1$, $b=2$, $c=5$, so

$$x=\frac{-2\pm\sqrt{4-20}}{2}=\frac{-2\pm\sqrt{-16}}{2}=\frac{-2\pm4i}{2}=-1\pm2i.$$

The roots are $-1+2i$ and $-1-2i$, which are complex conjugates. □

## 2.3.2 Exercises

Determine whether each statement is **true** or **false**. Justify your answer.
1. The product of a complex number and its conjugate is always a real number.
   **Answer:** _____
2. If $z=a+ib$, then $\bar{z}=a+ib$.
   **Answer:** _____
3. The conjugate of the product $z_1z_2$ equals the product of the conjugates $\bar{z}_1\bar{z}_2$.
   **Answer:** _____
4. If $z_1z_2=0$, then necessarily both $z_1=0$ and $z_2=0$.
   **Answer:** _____
5. The sum of a complex number and its conjugate is always purely imaginary.
   **Answer:** _____

## Multiple choice questions
Choose the correct option (A, B, C, or D) for each question.
1. Which of the following is the correct expression for the product of two complex numbers $z_1=a+ib$ and $z_2=c+id$?
   A. $ac+bd+i(ad+bc)$
   B. $ac-bd+i(ad+bc)$
   C. $ac+bd+i(ad-bc)$
   D. $ac-bd+i(ad-bc)$
   **Answer:** _____

2. What is the result of $\bar{z}$ if $z = a + ib$?
   A. $a + ib$   B. $-a - ib$
   C. $a - ib$   D. $-a + ib$
   **Answer:** _____

3. Which of the following identities is **not** generally true for complex numbers?
   A. $\overline{z_1 + z_2} = \bar{z}_1 + \bar{z}_2$
   B. $\overline{z_1 z_2} = \bar{z}_1 \bar{z}_2$
   C. $\overline{\frac{z_1}{z_2}} = \frac{\bar{z}_1}{\bar{z}_2}$
   D. $\bar{z}_1 + z_2 = z_1 + \bar{z}_2$
   **Answer:** _____

4. If $z = a + ib$, then $|z|^2$ is equal to
   A. $a^2 - b^2$   B. $a^2 + b^2$
   C. $a^2 + 2ab$   D. $(a + b)^2$
   **Answer:** _____

5. Given that $z_1 z_2 = 0$, which of the following must be true?
   A. $z_1 = 0$ or $z_2 = 0$
   B. $z_1 = z_2$
   C. Both $z_1$ and $z_2$ must be zero
   D. $|z_1| = |z_2|$
   **Answer:** _____

## Matching

Match each item in **Column A** with its best description in **Column B**.

| Column A | Column B |
| --- | --- |
| A. Conjugate of $z$ | 1. Largest integer $m$ with $P^{(m)}(z_0) = 0$ for a root $z_0$. |
| B. Zero-product property | 2. If $P$ has real coefficients and $z$ is a root, so is $\bar{z}$. |
| C. Multiplicity of a root | 3. (a) $\overline{z_1 + z_2} = \bar{z}_1 + \bar{z}_2$, (b) $\overline{z_1 z_2} = \bar{z}_1 \bar{z}_2$. |
| D. Conjugate-pair lemma | 4. At least one factor of $z_1 z_2 = 0$ equals 0. |
| E. Conjugation rules | 5. $\bar{z} = a - ib$ for $z = a + ib$. |

**Answers:** _____ – _____ – _____ – _____ – _____

## Fill in the blanks

1. For $z = a + ib$, one has $|z|^2 = $ _____.
2. If $z \neq 0$, its multiplicative inverse is given by $\frac{1}{z} = $ _____.
3. A quadratic with negative discriminant $b^2 - 4ac < 0$ has roots $-\frac{b}{2a} \pm $ _____ $i$.
4. The sum $z + \bar{z}$ is always _____; the difference $z - \bar{z}$ is always _____.
5. If $P(x) = x^2 + 2x + 5$, its complex roots are _____ and _____.

## 2.4 Geometric representation

Since every complex number $z = x + iy$ consists of two real numbers, that is a pair $(x, y)$, it can be represented geometrically (see Figure 2.1) as a point in the plane $\mathbb{R}^2$, or equivalently as a vector starting from the origin and ending at the point $(x, y)$.

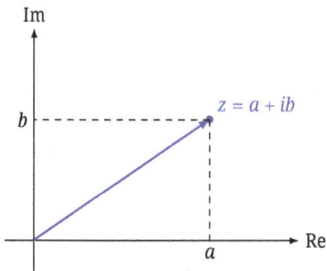

Figure 2.1: Geometric representation of a complex number $z = a + ib$.

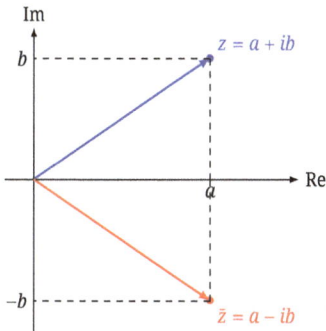

Figure 2.2: Conjugate of a complex number $z$ and its symmetry with respect to the real axis.

Complex numbers can be visualized as points or vectors in the plane. This geometric perspective leads naturally to the polar form.

**Definition 2.4** (Polar form, modulus, and argument of a complex number). Let $z = x + iy$ be a complex number. Since every complex number consists of two real components $(x, y)$, it can be geometrically represented as a point in the plane $\mathbb{R}^2$, or equivalently, as a vector from the origin to the point $(x, y)$. The conjugate of a complex number $z = x + iy$ is denoted by $\bar{z}$ and equals to $x - iy$ (see Figure 2.2).

Let $r$ denote the Euclidean distance of the point $(x, y)$ from the origin (i. e., the length of the corresponding vector), and let $\phi$ be the angle that this vector makes with the positive $x$-axis (see Figure 2.3). Then we can write

$$z = x + iy = r(\cos \phi + i \sin \phi),$$

where $x = r \cos \phi$ and $y = r \sin \phi$. This representation is called the polar form of the complex number $z$.

The quantity $r$ is called the modulus or absolute value of $z$ and is denoted by $|z|$. The angle $\phi$ is called the argument of $z$.

Therefore, the following relations hold:

$$r = \sqrt{x^2 + y^2}, \quad \tan \phi = \frac{y}{x}.$$

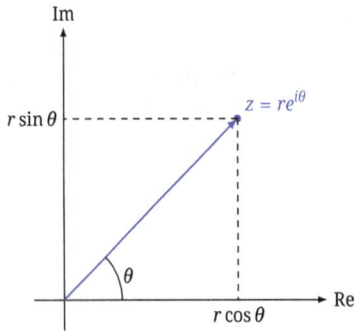

Figure 2.3: Polar representation of a complex number $z = re^{i\theta}$.

When adding two complex numbers $x = a_1 + ib_1$ and $y = a_2 + ib_2$, it is analogous to adding their corresponding vectors (see Figure 2.4).

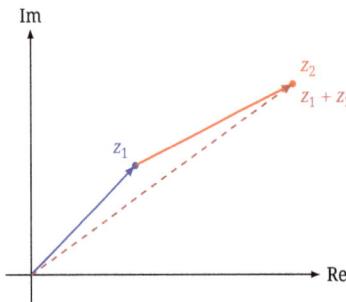

Figure 2.4: Vector addition of complex numbers $z_1$ and $z_2$.

Now consider the product of two complex numbers in polar form, see Figure 2.5:

$$z_1 z_2 = r_1 r_2 (\cos\theta_1 + i\sin\theta_1)(\cos\theta_2 + i\sin\theta_2)$$
$$= r_1 r_2 (\cos(\theta_1 + \theta_2) + i\sin(\theta_1 + \theta_2)).$$

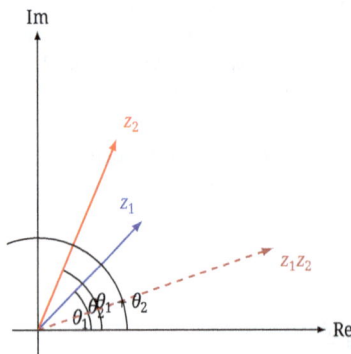

Figure 2.5: Complex multiplication as angle addition and magnitude multiplication.

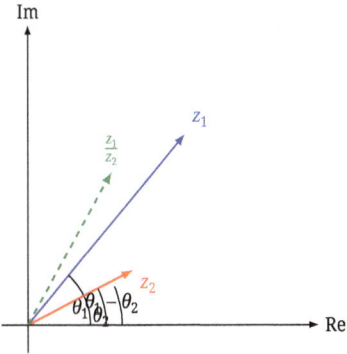

Im

$z_1$

$\frac{z_1}{z_2}$

$z_2$

$\theta_1$ $\theta$ $-\theta_2$

Re

**Figure 2.6:** Complex division as angle subtraction and modulus division.

The complex division is analogous (see Figure 2.6).

Here we used standard trigonometric identities. From this, we obtain the **De Moivre's formula** for powers of complex numbers:

$$z^n = r^n(\cos n\theta + i\sin n\theta) \quad \text{(De Moivre's formula)}.$$

This shows that any complex number $z = r(\cos\theta + i\sin\theta)$ has exactly $n$ distinct $n$th roots. These roots are given by

$$z_1 = \sqrt[n]{r}\left(\cos\frac{\theta}{n} + i\sin\frac{\theta}{n}\right)$$

$$\vdots$$

$$z_n = \sqrt[n]{r}\left(\cos\frac{\theta + (n-1)2\pi}{n} + i\sin\frac{\theta + (n-1)2\pi}{n}\right).$$

After the angle $\frac{\theta+(n-1)2\pi}{n}$, the values start repeating, forming a regular $n$-gon in the complex plane.

The modulus and argument of a complex number allow us to formulate important geometric properties.

Notably, the triangle inequality holds in the complex plane, just as it does for real numbers.

---

**Theorem 2.3** (Properties involving complex conjugates and the real part). *Let $z$ be a complex number. Then*

$$z\bar{z} = |z|^2 = r^2,$$

*where $r$ is the modulus of $z$.*

*Moreover, for any two complex numbers $z_1 = a_1 + ib_1$ and $z_2 = a_2 + ib_2$, we have*

$$z_1\bar{z}_2 + \bar{z}_1 z_2 = 2(a_1 a_2 + b_1 b_2) = 2\,\text{Re}(z_1\bar{z}_2),$$

*where $\text{Re}(z)$ denotes the real part of the complex number $z$.*

*Proof.* For $z = x + iy$, its complex conjugate is $\bar{z} = x - iy$. Then

$$z\bar{z} = (x + iy)(x - iy) = x^2 + y^2 = |z|^2 = r^2.$$

Next, let $z_1 = a_1 + ib_1$ and $z_2 = a_2 + ib_2$. Then

$$z_1\bar{z}_2 = (a_1 + ib_1)(a_2 - ib_2) = a_1a_2 + b_1b_2 + i(b_1a_2 - a_1b_2),$$
$$\Rightarrow \operatorname{Re}(z_1\bar{z}_2) = a_1a_2 + b_1b_2.$$

Similarly, $\bar{z}_1z_2 = (a_1 - ib_1)(a_2 + ib_2)$ yields the same real part. Therefore

$$z_1\bar{z}_2 + \bar{z}_1z_2 = 2(a_1a_2 + b_1b_2) = 2\operatorname{Re}(z_1\bar{z}_2). \qquad \square$$

**Theorem 2.4** (Triangle inequality for complex numbers). *Let $z_1$, $z_2$ be any complex numbers. Then the triangle inequality holds*

$$\left||z_1| - |z_2|\right| \le |z_1 + z_2| \le |z_1| + |z_2|.$$

*Proof.* To prove the right-hand inequality, we expand the square of the modulus:

$$|z_1 + z_2|^2 = (z_1 + z_2)(\bar{z}_1 + \bar{z}_2) = |z_1|^2 + |z_2|^2 + z_1\bar{z}_2 + \bar{z}_1z_2.$$

Therefore,

$$|z_1 + z_2|^2 - (|z_1| + |z_2|)^2 = 2(\operatorname{Re}(z_1\bar{z}_2) - |z_1||z_2|).$$

Since for any complex number $z = a + ib$ we have $\operatorname{Re}(z) = a \le |z| = \sqrt{a^2 + b^2}$, and also $|z_1\bar{z}_2| = |z_1||z_2|$, it follows that

$$\operatorname{Re}(z_1\bar{z}_2) - |z_1||z_2| \le 0,$$

and hence

$$|z_1 + z_2|^2 \le (|z_1| + |z_2|)^2 \quad \Rightarrow \quad |z_1 + z_2| \le |z_1| + |z_2|.$$

The left-hand inequality

$$\left||z_1| - |z_2|\right| \le |z_1 + z_2|$$

can be proven similarly, or derived from the right-hand inequality by substituting $z_2$ with $-z_2$.

For a geometric interpretation, see Figure 2.7. $\qquad \square$

By studying complex numbers and complex functions in greater depth, we can define the exponential function with a complex argument through the power series:

$$e^z = 1 + z + \frac{z^2}{2!} + \cdots + \frac{z^n}{n!} + \cdots,$$

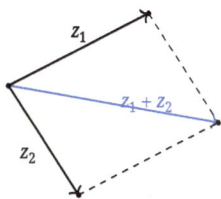

Figure 2.7: Graphical representation of the triangle inequality in the complex plane. The vectors $z_1$ and $z_2$ are added tip-to-tail to form the vector $z_1 + z_2$, demonstrating that $|z_1 + z_2| \leq |z_1| + |z_2|$.

where $z = a + ib$ is a complex number. Similarly, the sine and cosine functions for complex arguments are defined via the series:

$$\sin z = z - \frac{z^3}{3!} + \frac{z^5}{5!} - \cdots, \quad \cos z = 1 - \frac{z^2}{2!} + \frac{z^4}{4!} - \cdots$$

The properties of these functions remain valid even with complex arguments. Expanding $e^{iz}$ into a power series and regrouping terms, one can derive the identity

$$e^{iz} = \cos z + i \sin z \quad \text{(Euler's formula)}.$$

Using these properties, we obtain the following representations for the trigonometric functions:

$$\sin z = \frac{e^{iz} - e^{-iz}}{2i}, \quad \cos z = \frac{e^{iz} + e^{-iz}}{2} \quad \text{(Euler's identities)}. \tag{2.1}$$

Finally, in polar coordinates we have

$$z = r(\cos \theta + i \sin \theta) = re^{i\theta}.$$

More details can be found in references [2, 1].

## 2.4.1 Examples

**Example 2.4.** Convert the complex number $z = 1 + i\sqrt{3}$ into polar form.

*Solution.* To write $z = 1 + i\sqrt{3}$ in polar form

$$z = r(\cos \theta + i \sin \theta),$$

we compute

$$r = |z| = \sqrt{1^2 + (\sqrt{3})^2} = \sqrt{1 + 3} = \sqrt{4} = 2,$$

$$\tan \theta = \frac{\sqrt{3}}{1} = \sqrt{3} \quad \Rightarrow \quad \theta = \frac{\pi}{3}.$$

So the polar form is

$$z = 2\left(\cos\frac{\pi}{3} + i\sin\frac{\pi}{3}\right). \qquad \square$$

**Example 2.5.** Let $z_1 = 2(\cos\frac{\pi}{6} + i\sin\frac{\pi}{6})$ and $z_2 = 3(\cos\frac{\pi}{3} + i\sin\frac{\pi}{3})$. Compute $z_1 z_2$ using polar form.

*Solution.* Using the identity

$$z_1 z_2 = r_1 r_2 [\cos(\theta_1 + \theta_2) + i\sin(\theta_1 + \theta_2)],$$

we find

$$r = 2 \cdot 3 = 6,$$

$$\theta = \frac{\pi}{6} + \frac{\pi}{3} = \frac{\pi}{2}.$$

Therefore

$$z_1 z_2 = 6\left(\cos\frac{\pi}{2} + i\sin\frac{\pi}{2}\right) = 6i. \qquad \square$$

**Example 2.6.** Use De Moivre's formula to compute $(1 + i)^4$.

*Solution.* First, convert $z = 1 + i$ to polar form:

$$r = \sqrt{1^2 + 1^2} = \sqrt{2}, \quad \theta = \frac{\pi}{4} \Rightarrow z = \sqrt{2}\left(\cos\frac{\pi}{4} + i\sin\frac{\pi}{4}\right).$$

Now apply De Moivre's formula:

$$z^4 = (\sqrt{2})^4\left(\cos\left(4 \cdot \frac{\pi}{4}\right) + i\sin\left(4 \cdot \frac{\pi}{4}\right)\right) = 4(\cos\pi + i\sin\pi) = 4(-1 + 0i) = -4. \qquad \square$$

**Example 2.7.** Find all cube roots of $z = 8$.

*Solution.* We want to find all $w$ such that $w^3 = 8$. Since $8 = 8(\cos 0 + i\sin 0)$, we use the formula for $n$th roots

$$w_k = \sqrt[3]{8}\left(\cos\left(\frac{0 + 2k\pi}{3}\right) + i\sin\left(\frac{0 + 2k\pi}{3}\right)\right), \quad k = 0, 1, 2.$$

So

$$w_k = 2\left(\cos\left(\frac{2k\pi}{3}\right) + i\sin\left(\frac{2k\pi}{3}\right)\right).$$

For each $k$,

$$k = 0: \quad w_0 = 2(\cos 0 + i \sin 0) = 2,$$

$$k = 1: \quad w_1 = 2\left(\cos \frac{2\pi}{3} + i \sin \frac{2\pi}{3}\right) = 2\left(-\frac{1}{2} + i\frac{\sqrt{3}}{2}\right),$$

$$k = 2: \quad w_2 = 2\left(\cos \frac{4\pi}{3} + i \sin \frac{4\pi}{3}\right) = 2\left(-\frac{1}{2} - i\frac{\sqrt{3}}{2}\right).$$

Thus, the cube roots are

$$2, \quad -1 + i\sqrt{3}, \quad -1 - i\sqrt{3}. \qquad \square$$

**Example 2.8.** Express $z = e^{i\pi/4}$ in rectangular form.

*Solution.* Using Euler's formula

$$e^{i\theta} = \cos \theta + i \sin \theta \quad \Rightarrow \quad e^{i\pi/4} = \cos \frac{\pi}{4} + i \sin \frac{\pi}{4} = \frac{\sqrt{2}}{2} + i\frac{\sqrt{2}}{2}. \qquad \square$$

**Example 2.9.** Given $z_1 = 3 + 4i$ and $z_2 = 1 - 2i$, verify the triangle inequality

$$|z_1 + z_2| \leq |z_1| + |z_2|.$$

*Solution.* Compute:

$$z_1 + z_2 = (3 + 1) + i(4 - 2) = 4 + 2i,$$

$$|z_1 + z_2| = \sqrt{4^2 + 2^2} = \sqrt{16 + 4} = \sqrt{20} = 2\sqrt{5},$$

$$|z_1| = \sqrt{3^2 + 4^2} = \sqrt{9 + 16} = \sqrt{25} = 5,$$

$$|z_2| = \sqrt{1^2 + (-2)^2} = \sqrt{1 + 4} = \sqrt{5}.$$

Now check:

$$|z_1 + z_2| = 2\sqrt{5} \approx 4.47, \quad |z_1| + |z_2| = 5 + \sqrt{5} \approx 7.24 \Rightarrow 4.47 \leq 7.24.$$

The triangle inequality holds. $\qquad \square$

## 2.5 Python with AI

Modern computational tools make working with complex numbers both practical and interactive.

The following Python examples demonstrate basic calculations and visualizations in the complex plane.

---

**AI Request**

Write a Python code that:

- Accepts two complex numbers.
- Computes their sum, difference, product, and quotient.
- Converts them to polar and exponential forms.
- Plots the numbers and their operations in the complex plane.

---

```python
import matplotlib.pyplot as plt
import numpy as np
import cmath

# Input complex numbers
z1 = complex(2, 3)
z2 = complex(-1, 4)

# Arithmetic operations
z_sum = z1 + z2
z_diff = z1 - z2
z_prod =
z1 * z2
z_quot = z1 / z2

# Polar form
r1, theta1 = abs(z1), cmath.phase(z1)
r2, theta2 =
abs(z2), cmath.phase(z2)

# Exponential form
z1_exp = f"{r1:.2f} * exp(i * {theta1:.2f})"
z2_exp = f"{r2:.2f} * exp(i * {theta2:.2f})"

# Display results

print("z1 =", z1)
print("z2 =", z2)
print("Sum=", z_sum)
print("Difference =", z_diff)
print("Product =", z_prod)
print("Quotient =", z_quot)

print("z1 polar:", (r1,theta1), " =>", z1_exp)

print("z2 polar:",(r2, theta2), " => ", z2_exp)

# Plotting
def plot_complex_numbers(numbers, labels, colors):
    plt.figure(figsize=(6, 6))
```

```
41    ax = plt.gca()
42    ax.set_xlim(-10, 10)
43    ax.set_ylim(-10, 10)
44    ax.set_aspect('equal')
45    plt.axhline(0, color='gray', lw=1)
46    plt.axvline(0, color='gray', lw=1)
47
48    for z, label, color in zip(numbers, labels, colors):
49        plt.arrow(0, 0, z.real, z.imag, head_width=0.3, color=color
      , length_includes_head=True)
50        plt.text(z.real * 1.05, z.imag * 1.05, label, fontsize=12,
      color=color)
51
52    plt.xlabel('Real')
53    plt.ylabel('Imaginary')
54    plt.title('Complex Number Operations')
55    plt.grid(True)
56    plt.show()
57
58 # List of results to visualize
59
60 numbers = [z1, z2, z_sum, z_diff, z_prod, z_quot]
61
62 labels = ['z1','z2', 'z1 + z2', 'z1 - z2', 'z1 * z2', 'z1 / z2']
63
64 colors = ['blue', 'red', 'green', 'orange', 'purple', 'brown']
65
66 plot_complex_numbers(numbers, labels, colors)
```

**How to Use This Code**

This code demonstrates:
- How to define and operate with complex numbers in Python using built-in complex type.
- Conversion to polar form using abs and cmath.phase.
- Use of exponential form representation: $z = r \cdot \exp(i\theta)$.
- Use of matplotlib for plotting complex numbers as vectors in the complex plane.

The output will include textual results and a plot showing the original numbers and the results of operations such as sum, difference, product, and division.

## 2.5.1 Exercises

### True or false

Determine whether each statement is **true** or **false**. Justify your answer.

1. Every complex number has a unique representation in polar form with $\phi \in [0, 2\pi)$.

   **Answer:** _____

2. The modulus of a complex number $z = x + iy$ is given by $\sqrt{x^2 - y^2}$.

**Answer:** _____

3. If $z_1 = r_1 e^{i\theta_1}$ and $z_2 = r_2 e^{i\theta_2}$, then $z_1 z_2 = r_1 r_2 e^{i(\theta_1 + \theta_2)}$.

**Answer:** _____

4. The conjugate $\bar{z}$ of a complex number $z$ is reflected over the imaginary axis.

**Answer:** _____

5. The $n$th roots of a complex number form a regular polygon in the complex plane.

**Answer:** _____

## Multiple choice questions

Choose the correct option (A, B, C, or D) for each question.

1. Which of the following is the correct polar form of the complex number $z = x + iy$?

A. $z = \sqrt{x^2 + y^2} \cdot (\tan \phi + i)$

B. $z = r(\cos \phi + i \sin \phi)$

C. $z = r(\cos \phi - i \sin \phi)$

D. $z = r e^{i/x}$

**Answer:** _____

2. The modulus of $z = 3 - 4i$ is

A. 1     B. 7

C. 5     D. –5

**Answer:** _____

3. Which statement is true regarding complex multiplication?

A. The product rotates and stretches the vector.

B. The product only rotates the vector.

C. The product only scales the magnitude.

D. The product always results in a purely imaginary number.

**Answer:** _____

4. If $z = r e^{i\theta}$, then the conjugate $\bar{z}$ is

A. $r e^{i(-\theta)}$     B. $r e^{i(\pi - \theta)}$

C. $r e^{i\theta}$     D. $-r e^{-i\theta}$

**Answer:** _____

5. The number of distinct $n$th roots of a nonzero complex number $z$ is

A. 1     B. $n$

C. $2n$     D. infinite

**Answer:** _____

## Computational exercises

Solve the following problems:

1. Express the complex number $z = -1 + \sqrt{3}i$ in polar form.

2. Find the modulus and argument of $z = 1 - i$.

3. Compute the product $z_1 z_2$ if $z_1 = 2e^{i\frac{\pi}{4}}$ and $z_2 = 3e^{i\frac{\pi}{6}}$.

4. Find all cube roots of $z = 8e^{i\pi}$ and plot them on the complex plane.

5. Let $z = 2(\cos \frac{2\pi}{3} + i \sin \frac{2\pi}{3})$. Compute $z^3$ using De Moivre's theorem.

## Proof-based exercises

1. Prove that the conjugate of the product of two complex numbers is the product of their conjugates.
2. Show that for all complex numbers $z$, we have $|z|^2 = z\bar{z}$.
3. Prove that the triangle inequality holds: $|z_1 + z_2| \leq |z_1| + |z_2|$.
4. Let $z = re^{i\theta}$. Prove that $z^{-1} = \frac{1}{r}e^{-i\theta}$.
5. Prove that the $n$th roots of unity are the vertices of a regular $n$-gon inscribed in the unit circle.

**Exercise 16.** Write a Python function that takes two complex numbers $z_1$ and $z_2$ and returns:

– Their real and imaginary parts.
– Their conjugates.
– A Boolean indicating whether $z_1 \cdot \bar{z}_2$ is purely real.

**Exercise 17.** Plot the $n$-th roots of the complex number $z = 8(\cos(\pi) + i\sin(\pi))$ for $n = 3$ on the complex plane using `matplotlib`. Hint: Use Euler's formula and polar coordinates.

**Exercise 18.** Implement a Python program that accepts a complex number $z$ in rectangular form (e. g., $3 - 4i$) and:

1. Computes and displays its modulus and argument (angle).
2. Converts and prints it in exponential form $re^{i\theta}$.
3. Plots it in the complex plane.

**Exercise 19.** Let $z = 1 + i$. Write a Python code to compute and plot all 5th powers $z^k$ for $k = 1$ to 5 on the complex plane. Use colors or markers to distinguish each power.

## Fill in the blanks

Complete the following statements by filling in the blanks:

1. The imaginary unit $i$ is defined such that $i^2 = $ _____.
2. A complex number of the form $z = a + ib$ has a real part _____ and an imaginary part $b$.
3. Two complex numbers $z_1 = a + ib$ and $z_2 = c + id$ are equal if and only if $a = c$ and

   _____.
4. The complex conjugate of $z = a + ib$ is $\bar{z} = $ _____.
5. The modulus of a complex number $z = a + ib$ is given by $|z| = $ _____.
6. In polar form, a complex number $z$ is written as $z = $ _____, where $r$ is the modulus and $\theta$ is the argument.
7. The exponential form of a complex number $z$ is $z = re^{i\theta}$, which comes from _____'s formula.

## Matching

Match the concepts (1–5) with the correct descriptions (a–e):

**Column A**

1. Imaginary unit
2. Modulus
3. Conjugate
4. Polar form
5. Euler's formula

**Column B**

a. A representation of a complex number using angle and distance from the origin.
b. The number $e^{i\theta} = \cos\theta + i\sin\theta$.
c. Defined as $i^2 = -1$.
d. The distance of a complex number from the origin.
e. For $z = a + ib$, it is $a - ib$.

**Answers:** 1 – _____  2 – _____  3 – _____  4 – _____  5 – _____

# 3 Functions

## Contents

The concept of a function is one of the most fundamental and widely used ideas in mathematics. It provides a powerful framework for describing relationships between quantities, where each input corresponds to exactly one output. Functions appear in virtually every branch of mathematics and are essential in fields such as calculus, algebra, computer science, and physics.

In this chapter, we explore the core properties of functions, including injectivity (one-to-one mapping), surjectivity (onto mapping), and bijectivity (a combination of both). These concepts allow us to classify and understand how functions behave under different conditions and whether they possess inverses. Through definitions, examples, and visual illustrations, we will build a solid foundation for analyzing and applying functions in both theoretical and practical contexts.

## 3.1 Introduction

The concept of a **function** is one of the most fundamental and enduring ideas in mathematics. Its origins trace back to antiquity, where early mathematicians sought to describe relationships between quantities in geometry, astronomy, and natural phenomena. Although the formal definition of a function evolved over centuries, its intuitive use can be found in the works of ancient scholars such as *Archimedes* and *Apollonius*, who studied dependencies between geometric magnitudes.

The modern notion of a function began to take shape during the 17th century with the development of analytic geometry by *Rene Descartes* and *Pierre de Fermat*. These thinkers introduced the idea of representing relationships between variables algebraically, laying the groundwork for the coordinate system we now associate with functions.

However, it was *Gottfried Wilhelm Leibniz*, in the late 17th century, who first used the term "function" to describe a quantity related to a curve, such as its slope or area. Around the same time, *Isaac Newton* implicitly used functional relationships in his work on calculus, although he did not formalize the concept explicitly.

https://doi.org/10.1515/9783112228289-003

In the 18th century, *Leonhard Euler* gave the function its more recognizable form, using notation like $f(x)$ and classifying functions into algebraic, transcendental, and others. His work significantly influenced how functions were understood and taught across Europe.

By the 19th century, as mathematics became increasingly rigorous, *Peter Gustav Lejeune Dirichlet* provided the definition of a function that is widely accepted today: a rule that assigns to each element of a set $A$ exactly one element of a set $B$. This marked a crucial step toward the abstract formulation of functions, which has since become central to nearly every branch of mathematics.

Today, functions are not only essential in pure mathematics, including analysis, algebra, and topology, but also in applied fields such as physics, engineering, computer science, and economics. They serve as the foundation for modeling real-world phenomena, from population growth to signal processing, and underpin advanced concepts such as continuity, differentiability, and transformations.

This chapter explores the core properties of functions, including injectivity, surjectivity, bijectivity, and invertibility (see Table 3.1). Through precise definitions, illustrative examples, and visual representations, we aim to build a strong conceptual framework for understanding and applying functions in both theoretical and practical contexts.

**Table 3.1:** Detailed comparison of mapping, function, injective, surjective, and bijective functions, see Figures 3.1–3.9.

| Concept | Description | Symbolic condition | Example |
| --- | --- | --- | --- |
| Mapping | Any association from elements of $A$ to $B$, possibly with multiple or no outputs per input | No restriction | Assigning multiple outputs to an input: $a \mapsto \{b_1, b_2\}$ |
| Function | Each element in $A$ maps to *exactly one* element in $B$ | $\forall a \in A, \exists! b \in B : f(a) = b$ | $f(x) = x^2$ from $\mathbb{R} \to \mathbb{R}$ |
| Injective | Distinct inputs have distinct outputs | $f(a_1) = f(a_2) \Rightarrow a_1 = a_2$ | $f(x) = 2x$ on $\mathbb{R}$ |
| Surjective | Every element of $B$ is mapped by some element of $A$ | $\forall b \in B, \exists a \in A : f(a) = b$ | $f(x) = \tan x$ from $(-\frac{\pi}{2}, \frac{\pi}{2}) \to \mathbb{R}$ |
| Bijective | Both injective and surjective. Invertible | $\exists f^{-1} : B \to A$ such that $f^{-1}(f(a)) = a, f(f^{-1}(b)) = b$ | $f(x) = x + 3$ on $\mathbb{R} \to \mathbb{R}$ |

## 3.2 Basic definitions

Functions are fundamental objects in mathematics, providing a rigorous way to describe relations between sets. Before defining functions formally, let us clarify the essential terminology.

**Definition 3.1** (Function terminology). A **function** $f : A \to B$ assigns **exactly one** element of $B$ to each element of $A$.
– **Domain:** The set $A$, representing all possible inputs.
– **Codomain:** The set $B$, indicating the target outputs.
– **Range:** The actual outputs produced by $f$, that is, $\{f(a) : a \in A\}$.

A **mapping** from a set $A$ to a set $B$ is a rule that assigns each element of $A$ to an element of $B$. However, a **function** is a special kind of mapping in which *each* element of $A$ is assigned to *exactly one* element of $B$.

In other words, while every function is a mapping, not every mapping is a function.

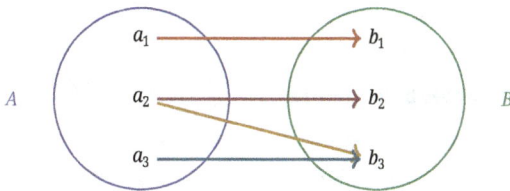

Figure 3.1: A mapping that is not a function (since $a_2$ maps to two values).

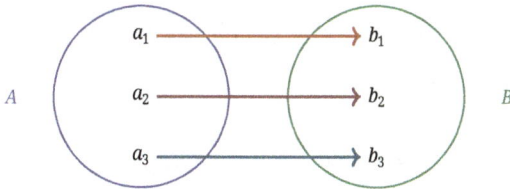

Figure 3.2: A valid function: each element of $A$ maps to exactly one element of $B$.

## 3.3 Onto vs. not onto functions

We next discuss the important property of surjectivity, which concerns whether all elements of the codomain are covered by a function.

**Definition 3.2** (Surjective function). A function $f : A \to B$ is called **onto** (or **surjective**) if every element of $B$ is the image of at least one element of $A$. In other words, for every $b \in B$, there exists $a \in A$ such that $f(a) = b$.

If there exists even one element in $B$ that is not "hit" by any element in $A$, then the function is **not onto**.

To better understand surjectivity, let us look at some concrete examples.

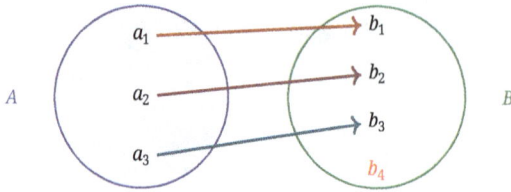

Figure 3.3: Not onto: $b_4 \in B$ is not mapped to by any element in $A$.

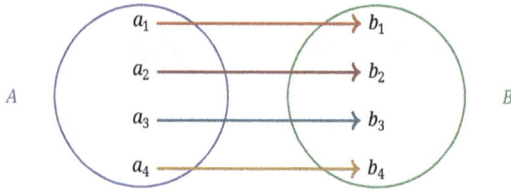

Figure 3.4: Onto function: every element in $B$ is mapped by some element in $A$.

**Example 1** (Onto function). Let $f : \mathbb{R} \to \mathbb{R}$ be defined by

$$f(x) = x^3.$$

This function is **onto** because for every $y \in \mathbb{R}$, there exists $x = \sqrt[3]{y} \in \mathbb{R}$ such that

$$f(x) = \left(\sqrt[3]{y}\right)^3 = y.$$

Therefore, every element of the codomain $\mathbb{R}$ is attained.

**Example 2** (Not onto function). Let $f : \mathbb{R} \to \mathbb{R}$ be defined by

$$f(x) = x^2.$$

This function is **not onto** because the range of $f$ is $[0, \infty)$, not all of $\mathbb{R}$. For example, there is no $x \in \mathbb{R}$ such that

$$f(x) = -1.$$

Hence, some elements of the codomain are not reached by the function.

## 3.4 Injective vs. not injective functions

Another key concept is injectivity, which ensures that distinct inputs always have distinct outputs.

**Definition 3.3** (Injective function). A function $f : A \to B$ is called **injective** (or **one-to-one**) if every distinct element of $A$ maps to a distinct element of $B$. That is, if $a_1 \neq a_2$, then $f(a_1) \neq f(a_2)$. Equivalently, if $f(a_1) = f(a_2)$, then $a_1 = a_2$.

If two or more distinct elements in $A$ map to the same element in $B$, then the function is **not injective**.

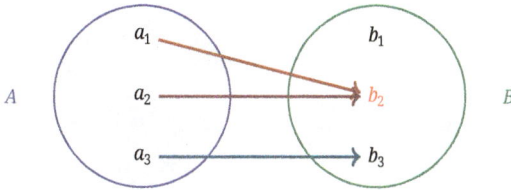

Figure 3.5: Not injective: $a_1 \neq a_2$ but $f(a_1) = f(a_2) = b_2$.

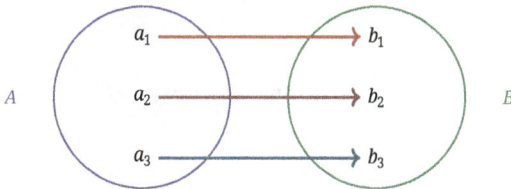

Figure 3.6: Injective: each element in $A$ maps to a unique element in $B$.

We now illustrate the concept of injectivity with the following examples.

**Example 1** (Injective function). Let $f : \mathbb{R} \to \mathbb{R}$ be defined by

$$f(x) = 2x + 1.$$

This function is **injective** because if

$$f(x_1) = f(x_2),$$

then

$$2x_1 + 1 = 2x_2 + 1 \quad \Rightarrow \quad 2x_1 = 2x_2 \quad \Rightarrow \quad x_1 = x_2.$$

Therefore, different inputs produce different outputs.

**Example 2** (Not injective function). Let $f : \mathbb{R} \to \mathbb{R}$ be defined by

$$f(x) = x^2.$$

This function is **not injective** because distinct inputs may produce the same output. For instance,

$$f(-2) = (-2)^2 = 4 = 2^2 = f(2),$$

but $-2 \neq 2$. Hence, the function fails the injectivity condition.

## 3.5 Composition of functions

The notion of composition allows us to combine functions into more complex transformations.

> **Definition 3.4** (Composition of functions). Let $f: A \to B$ and $g: B \to C$ be two functions. Their *composition* is
> $$g \circ f: A_1 \to C, \quad (g \circ f)(x) = g(f(x)),$$
> where
> $$A_1 = \{x \in A \mid f(x) \in \mathrm{dom}(g)\}$$
> is the largest subset of $A$ on which $g \circ f$ makes sense.

Therefore, in order to decide if the composition of two functions is meaningful, we should compute their ranges.

**Remark 4.** Composition of functions is *associative*: if $h: C \to D$, then

$$h \circ (g \circ f) = (h \circ g) \circ f.$$

In general $g \circ f \neq f \circ g$.

Let us now see some typical examples of function composition.

**Example 1** (Composition with a square-root). Let

$$f: \mathbb{R} \to \mathbb{R}, \quad f(x) = x^2 - 4, \quad g: [0, \infty) \to \mathbb{R}, \quad g(t) = \sqrt{t}.$$

Then

$$(g \circ f)(x) = g(f(x)) = \sqrt{x^2 - 4},$$

and for this to make sense we need

$$x^2 - 4 \geq 0 \quad \Longrightarrow \quad x \leq -2 \text{ or } x \geq 2.$$

Hence

$$\mathrm{dom}(g \circ f) = (-\infty, -2] \cup [2, \infty).$$

**Example 2** (Composition with a reciprocal). Let

$$f: (0, \infty) \to (0, \infty), \quad f(x) = \sqrt{x}, \quad g: \mathbb{R} \setminus \{0\} \to \mathbb{R}, \quad g(t) = \frac{1}{t}.$$

Then

$$(g \circ f)(x) = \frac{1}{\sqrt{x}},$$

and we require

$$x > 0 \quad \text{(from dom } f) \quad \text{and} \quad \sqrt{x} \neq 0 \quad \implies \quad x > 0.$$

Thus

$$\text{dom}(g \circ f) = (0, \infty).$$

**Example 3** (Composition with a logarithm). Let

$$f: \mathbb{R} \setminus \{1\} \to \mathbb{R} \setminus \{0\}, \quad f(x) = \frac{1}{x - 1}, \quad g: (0, \infty) \to \mathbb{R}, \quad g(t) = \ln t.$$

Then

$$(g \circ f)(x) = \ln\left(\frac{1}{x - 1}\right) = -\ln |x - 1|.$$

We need

$$x \neq 1 \quad \text{and} \quad \frac{1}{x - 1} > 0 \quad \implies \quad x - 1 > 0 \quad \implies \quad x > 1.$$

Therefore

$$\text{dom}(g \circ f) = (1, \infty).$$

**Example 4** (Composition not defined). Let

$$f: \{a, b\} \to \{1, 2\}, \quad f(a) = 1, \; f(b) = 2, \quad g: \{3, 4\} \to \{x, y\}, \quad g(3) = x, \; g(4) = y.$$

Since $f(\{a, b\}) = \{1, 2\}$ is *not* contained in $\text{dom}(g) = \{3, 4\}$, the composite

$$g \circ f: \{a, b\} \longrightarrow \{x, y\}$$

cannot be formed: for instance, $g(f(a)) = g(1)$ is undefined.

**Example 5** (Non-commutativity: **g ∘ f ≠ f ∘ g**). On $\mathbb{R}$, define

$$f(x) = x + 1, \quad g(x) = 2x.$$

Then, for every $x \in \mathbb{R}$,

$$(g \circ f)(x) = g(f(x)) = 2(x+1) = 2x + 2, \quad (f \circ g)(x) = f(g(x)) = (2x) + 1 = 2x + 1.$$

Hence

$$g \circ f \neq f \circ g \quad \text{(they differ by the constant 1).}$$

## 3.6 Bijective function and its inverse

Combining both injectivity and surjectivity leads to the important concept of bijectivity.

**Definition 3.5** (Bijective function). A function $f : A \to B$ is called **bijective** if it is both **injective** (one-to-one) and **surjective** (onto).

In that case, there exists a unique **inverse function** $f^{-1} : B \to A$ such that

$$f^{-1}(f(a)) = a \quad \text{for all } a \in A \quad \text{and} \quad f(f^{-1}(b)) = b \quad \text{for all } b \in B.$$

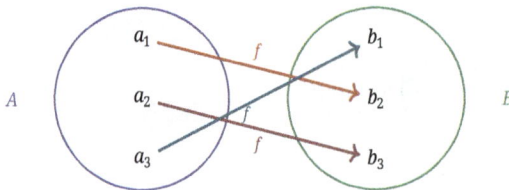

Figure 3.7: A bijective function. An inverse $f^{-1} : B \to A$ exists.

Let us demonstrate how to compute the inverse of a function with the following example.

**Example 1** (Bijective function). Let $f : \mathbb{R} \to \mathbb{R}$ be defined by

$$f(x) = x + 5.$$

This function is:

- **Injective:** If $f(x_1) = f(x_2)$, then $x_1 + 5 = x_2 + 5 \Rightarrow x_1 = x_2$.
- **Surjective:** For any $y \in \mathbb{R}$, let $x = y - 5$. Then $f(x) = x + 5 = y$, so every output is covered.

Thus, $f(x) = x + 5$ is bijective, and its inverse is $f^{-1}(x) = x - 5$.

**Example 2** (Not bijective function). Let $f : \mathbb{R} \to \mathbb{R}$ be defined by

$$f(x) = x^2.$$

This function is:

- **Not injective**: Since $f(-2) = f(2) = 4$, it fails the one-to-one condition.
- **Not surjective**: The codomain is $\mathbb{R}$, but the range is $[0, \infty)$, so negative values are not reached.

Hence, $f(x) = x^2$ is neither injective nor surjective, and therefore not bijective.

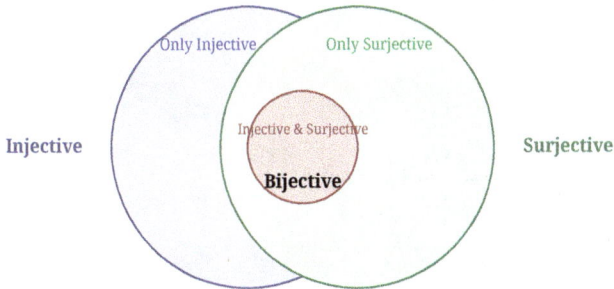

**Figure 3.8**: Relationship among injective, surjective, and bijective functions.

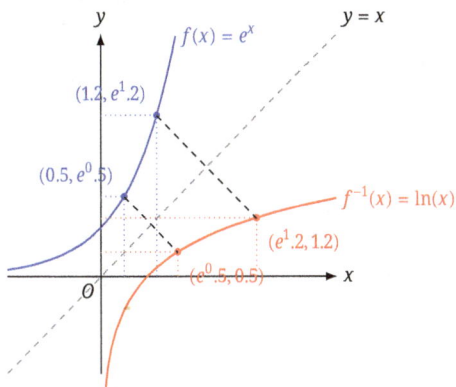

**Figure 3.9**: Symmetry of a function and its inverse with respect to the line $y = x$. The graph illustrates how $f(x) = e^x$ (blue) and $f^{-1}(x) = \ln(x)$ (red) are mirror images across the identity line $y = x$ (dashed gray). Specific points, like $(a, e^a)$ on $f(x)$ and its reflected counterpart $(e^a, a)$ on $f^{-1}(x)$, highlight this relationship.

## 3.7 Examples

**Example 3.1** (Inverse of a linear function). Let $f : \mathbb{R} \to \mathbb{R}$ be the function defined by

$$f(x) = 3x - 5.$$

Find the inverse function $f^{-1}$.

*Solution.* To find $f^{-1}$, solve the equation

$$y = 3x - 5$$

for $x$ in terms of $y$.

$$y = 3x - 5,$$
$$y + 5 = 3x,$$
$$x = \frac{y + 5}{3}.$$

Thus, the inverse function is

$$f^{-1}(x) = \frac{x + 5}{3}.$$

**Verification:**

$$f(f^{-1}(x)) = 3\left(\frac{x + 5}{3}\right) - 5 = x + 5 - 5 = x,$$
$$f^{-1}(f(x)) = \frac{3x - 5 + 5}{3} = \frac{3x}{3} = x. \qquad \square$$

**Example 3.2** (Inverse of a rational function). Let $f : \mathbb{R} \setminus \{2\} \to \mathbb{R} \setminus \{0\}$ be the function defined by

$$f(x) = \frac{1}{x - 2}.$$

**Find the inverse function $f^{-1}$.**

*Solution.* Start by solving the equation

$$y = \frac{1}{x - 2}$$

for $x$ in terms of $y$.

$$y = \frac{1}{x - 2},$$
$$x - 2 = \frac{1}{y},$$
$$x = \frac{1}{y} + 2.$$

Therefore, the inverse function is

$$f^{-1}(x) = \frac{1}{x} + 2, \quad x \neq 0.$$

**Verification:**

$$f(f^{-1}(x)) = f\left(\frac{1}{x} + 2\right) = \frac{1}{(\frac{1}{x} + 2) - 2} = \frac{1}{\frac{1}{x}} = x,$$
$$f^{-1}(f(x)) = f^{-1}\left(\frac{1}{x - 2}\right) = \frac{1}{\frac{1}{x-2}} + 2 = (x - 2) + 2 = x. \qquad \square$$

**Example 3.3** (Inverse of a cubic function). Let $f : \mathbb{R} \to \mathbb{R}$ be defined by

$$f(x) = x^3 + 1.$$

**Find the inverse function $f^{-1}$.**

*Solution.* Solve $y = x^3 + 1$ for $x$:

$$y - 1 = x^3 \quad \Longrightarrow \quad x = \sqrt[3]{y - 1}.$$

Thus,

$$f^{-1}(x) = \sqrt[3]{x - 1}.$$

**Verification:**

$$f(f^{-1}(x)) = \left(\sqrt[3]{x-1}\right)^3 + 1 = (x - 1) + 1 = x,$$
$$f^{-1}(f(x)) = \sqrt[3]{(x^3 + 1) - 1} = \sqrt[3]{x^3} = x.$$

□

**Example 3.4** (Inverse of an exponential function). Let $f : (-\infty, \infty) \to (0, \infty)$ be given by

$$f(x) = e^{2x}.$$

**Find the inverse function $f^{-1}$.**

*Solution.* Write $y = e^{2x}$ and solve for $x$:

$$\ln y = 2x \quad \Longrightarrow \quad x = \frac{1}{2} \ln y.$$

Therefore,

$$f^{-1}(x) = \frac{1}{2} \ln x, \quad x > 0.$$

**Verification:**

$$f(f^{-1}(x)) = e^{2\left(\frac{1}{2} \ln x\right)} = e^{\ln x} = x,$$
$$f^{-1}(f(x)) = \frac{1}{2} \ln(e^{2x}) = \frac{1}{2} (2x) = x.$$

□

**Example 3.5** (Inverse of a linear-fractional function). Let $f : \mathbb{R} \setminus \{4\} \to \mathbb{R} \setminus \{\frac{2}{3}\}$ be defined by

$$f(x) = \frac{2x + 3}{x - 4}.$$

**Find the inverse function $f^{-1}$.**

*Solution.* Set $y = \frac{2x+3}{x-4}$. Then

$$y(x-4) = 2x+3 \implies yx - 4y = 2x+3 \implies yx - 2x = 4y+3$$

$$\implies x(y-2) = 4y+3 \implies x = \frac{4y+3}{y-2}.$$

Hence

$$f^{-1}(x) = \frac{4x+3}{x-2}, \quad x \neq 2.$$

**Verification:**

$$f(f^{-1}(x)) = \frac{2\left(\frac{4x+3}{x-2}\right)+3}{\left(\frac{4x+3}{x-2}\right)-4} = \frac{\frac{8x+6+3(x-2)}{x-2}}{\frac{4x+3-4(x-2)}{x-2}} = \frac{8x+6+3x-6}{4x+3-4x+8} = \frac{11x}{11} = x,$$

$$f^{-1}(f(x)) = \frac{4\left(\frac{2x+3}{x-4}\right)+3}{\left(\frac{2x+3}{x-4}\right)-2} = \frac{\frac{8x+12+3(x-4)}{x-4}}{\frac{2x+3-2(x-4)}{x-4}} = \frac{8x+12+3x-12}{2x+3-2x+8} = \frac{11x}{11} = x. \qquad \square$$

## 3.8 Python and AI

It is interesting to calculate the domain and the range of the function via a Python code.
    Similarly, the computation of the inverse function and the corresponding plots using Python is interesting.

---

**AI Request**

I need a Python code to compute the inverse function given by the user and next verify the result. I also want the plot of the function and its inverse and the function $y = x$. Finally, I would like to see step-by-step the computation of the inverse function in HTML with MathJax.

---

### 3.8.1 Exercises

**True or false**

Determine whether each statement is **true** or **false**. Justify your answer.

1. A function $f : A \to B$ assigns exactly one element of $B$ to each element of $A$.
    **Answer:** _____

2. If a mapping from $A$ to $B$ has an element in $A$ that maps to two different elements in $B$, it is still a function.
    **Answer:** _____

3. The range of a function can be larger than its codomain.
    **Answer:** _____

4. A surjective function must map every element of $A$ to a unique element in $B$.
    **Answer:** _____

5. An injective function ensures that no two distinct elements in the domain have the same image.
   **Answer:** _____
6. Every bijective function has an inverse function.
   **Answer:** _____
7. The function $f(x) = x^2$ from $\mathbb{R} \to \mathbb{R}$ is injective.
   **Answer:** _____
8. The function $f(x) = e^x$ from $\mathbb{R} \to \mathbb{R}^+$ is surjective.
   **Answer:** _____

## Multiple choice questions

Choose the correct option (A, B, C, or D) for each question.
1. Which of the following best describes a function?
   A. A rule where some elements in the domain may not have outputs.
   B. A mapping where each input has at least one output.
   C. A mapping where each input has exactly one output.
   D. A mapping where each input has multiple outputs.
   **Answer:** _____
2. What defines a surjective function?
   A. Each element in the codomain is mapped by at least one element in the domain.
   B. Each element in the domain maps to a unique element in the codomain.
   C. Each element in the codomain is mapped by exactly one element in the domain.
   D. The domain and codomain are equal sets.
   **Answer:** _____
3. Which of the following functions is injective?
   A. $f(x) = x^2$ on $\mathbb{R} \to \mathbb{R}$
   B. $f(x) = |x|$ on $\mathbb{R} \to [0, \infty)$
   C. $f(x) = 3x + 2$ on $\mathbb{R} \to \mathbb{R}$
   D. $f(x) = \sin(x)$ on $\mathbb{R} \to [-1, 1]$
   **Answer:** _____
4. A function that is both injective and surjective is called
   A. continuous    B. invertible
   C. bijective     D. linear
   **Answer:** _____
5. Which of the following functions is bijective?
   A. $f(x) = x^2$ on $\mathbb{R} \to [0, \infty)$
   B. $f(x) = \ln(x)$ on $(0, \infty) \to \mathbb{R}$
   C. $f(x) = x^3$ on $\mathbb{R} \to \mathbb{R}$
   D. Both B and C
   **Answer:** _____

## Computational exercises

Solve the following problems:

1. Consider the function $f : \mathbb{R} \to \mathbb{R}$ defined by $f(x) = 2x + 5$. Prove whether this function is injective, surjective, and/or bijective.

2. Let $f : A \to B$ be a function where $A = \{1, 2, 3\}$ and $B = \{a, b, c\}$. Suppose $f(1) = a$, $f(2) = b, f(3) = c$. Is $f$ injective? Surjective? Bijective? Justify.

3. Given the function $f(x) = \frac{1}{x-2}$, determine:
   - Its domain.
   - Whether it is injective.
   - Whether it is surjective onto $\mathbb{R}$.

4. Draw a diagram showing a function from set $A = \{a_1, a_2, a_3\}$ to set $B = \{b_1, b_2, b_3\}$ that is injective but not surjective.

5. For the function $f(x) = x^3 - 3x$, determine if it is injective on $\mathbb{R}$. Justify using calculus or algebraic reasoning.

## 3.8.2 Python exercises

Answer the following questions using Python. You may use libraries like `matplotlib`, `numpy`, and `sympy` where appropriate. To construct Python codes for the following problems use AI accordingly. Test the code and make corrections.

1. Write a Python function to determine if a given function $f : A \to B$ is injective. Test it with

$$f(x) = x^3 \quad \text{on } A = [-2, 2], \quad B = [-8, 8].$$

2. Plot the graph of the function $f(x) = \sin(x)$ over the interval $[-2\pi, 2\pi]$. Based on the plot, explain why this function is:
   (a) Not injective.
   (b) Not surjective onto $\mathbb{R}$.

3. Create a Python script that checks whether the function $f(x) = e^x$ from $\mathbb{R} \to \mathbb{R}$ is injective and/or surjective. Justify your answer based on the output.

4. Use `matplotlib` to visualize the mapping between two finite sets:

$$A = \{1, 2, 3\}, \quad B = \{a, b, c\}, \quad f(1) = a, f(2) = b, f(3) = c.$$

Is this function bijective? Explain using your plot.

5. Write a Python function that accepts another function $f$, its domain $A$, and codomain $B$, and returns whether $f$ is bijective. Apply it to

$$f(x) = 2x + 1 \quad \text{with } A = B = \mathbb{R}.$$

Does your result match the mathematical analysis?

6. Given the function $f(x) = \frac{1}{x-2}$:
   (a) Plot the function.

(b) Determine its domain.

(c) Check if it is injective.

(d) Is it surjective onto ℝ? Why or why not?

## Fill in the blanks

Complete the following statements by filling in the blanks:

1. A function $f : A \rightarrow B$ assigns _____ element of $B$ to each element of $A$.

2. The _____ of a function is the set of all actual output values it produces.

3. A function is called _____ if every element in the codomain $B$ is mapped to by at least one element in $A$.

4. A function is called _____ if distinct elements in the domain have distinct images in the codomain.

5. A function that is both _____ and _____ is said to be bijective.

6. The _____ of a bijective function $f$ is a function $f^{-1}$ such that $f(f^{-1}(y)) = y$ and $f^{-1}(f(x)) = x$.

7. If a function is not _____, then it does not have an inverse.

## Matching

1. Function
2. Codomain
3. Injective
4. Surjective
5. Bijective

a. The set of all possible output values of a function.
b. A mapping where each input has exactly one output.
c. A function that is both injective and surjective.
d. Every element of the codomain is hit by at least one element from the domain.
e. Each distinct input maps to a distinct output.

**Answers:** 1 – _____   2 – _____   3 – _____   4 – _____   5 – _____

# 4 Limit of a function

## Contents

The concept of the **limit of a function** lies at the heart of mathematical analysis. It describes how a function behaves as its input approaches a particular value, even if that value is not in the domain. This idea enables us to define continuity, derivatives, and many other fundamental notions in calculus.

In this chapter, we present the formal $\varepsilon$-$\delta$ definition of a limit, explore one-sided and infinite limits, and examine key theorems such as the squeeze theorem and the algebra of limits. Through definitions, examples, and graphical illustrations, we aim to build a strong foundation for understanding and computing limits effectively.

## 4.1 Introduction

The concept of the **limit of a function** (see Table 4.1) is one of the cornerstones of mathematical analysis and forms the foundation upon which calculus was built. While its modern formalization came much later, the idea of approaching a value without necessarily reaching it has roots in ancient mathematics.

**Table 4.1:** Summary of limit definitions.

| Type of limit | Formal definition |
|---|---|
| Limit at a point | $\lim_{x \to \xi} f(x) = A \iff \forall \varepsilon > 0, \exists \delta > 0 \text{ s. t. } 0 < |x - \xi| < \delta \Rightarrow |f(x) - A| < \varepsilon$ |
| Infinite limit at a point | $\lim_{x \to \xi} f(x) = +\infty \iff \forall M > 0, \exists \delta > 0 \text{ s. t. } 0 < |x - \xi| < \delta \Rightarrow f(x) > M$ (similarly for $-\infty$) |
| Limit at infinity | $\lim_{x \to +\infty} f(x) = A \iff \forall \varepsilon > 0, \exists M > 0 \text{ s. t. } x > M \Rightarrow |f(x) - A| < \varepsilon$ (similarly for $x \to -\infty$) |
| Infinite limit at infinity | $\lim_{x \to +\infty} f(x) = +\infty \iff \forall M > 0, \exists N > 0 \text{ s. t. } x > N \Rightarrow f(x) > M$ (similarly for $-\infty$, or $x \to -\infty$) |
| Right-hand limit | $\lim_{x \to \xi+} f(x) = A \iff \forall \varepsilon > 0, \exists \delta > 0 \text{ s. t. } 0 < x - \xi < \delta \Rightarrow |f(x) - A| < \varepsilon$ |
| Left-hand limit | $\lim_{x \to \xi-} f(x) = A \iff \forall \varepsilon > 0, \exists \delta > 0 \text{ s. t. } 0 < \xi - x < \delta \Rightarrow |f(x) - A| < \varepsilon$ |

In antiquity, mathematicians such as *Archimedes* used intuitive notions of limits to compute areas and volumes through methods resembling integration. However, the precise definition of a limit emerged only in the 19th century, primarily through the work

https://doi.org/10.1515/9783112228289-004

of *Augustin-Louis Cauchy* and *Karl Weierstrass*, who sought to eliminate the vagueness of infinitesimals used by *Newton* and *Leibniz* in their development of calculus.

The modern $\varepsilon$-$\delta$ definition of a limit, introduced by Weierstrass, gave calculus a rigorous logical foundation. This definition allows us to precisely describe how a function behaves as its input approaches a certain point, even if that point lies outside the domain of the function.

This chapter explores the core definitions and properties of limits, including:
- The formal $\varepsilon$-$\delta$ definition of a limit at a point.
- One-sided limits and their role in understanding discontinuities.
- Infinite limits and limits at infinity.
- Algebraic properties of limits and the squeeze theorem.

These concepts are not only fundamental for defining continuity and differentiability but also serve as tools for analyzing the behavior of functions in both theoretical and applied settings.

Through precise definitions, illustrative diagrams, and example functions, this chapter aims to build a strong conceptual and computational foundation for working with limits, an indispensable skill in both pure and applied mathematics.

## 4.2 Definitions

We begin by defining the concept of a limit at a point, which is central to analysis. This definition formalizes the idea that a function can approach a specific value as $x$ approaches $\xi$.

---

**Definition 4.1** (Limit at a point (Figure 4.1)). Let $f : I \subseteq \mathbb{R} \to \mathbb{R}$ and let $\xi \in \mathbb{R}$ be a cluster point of $I$ (not necessarily $\xi \in I$). We say that $f$ tends to $A \in \mathbb{R}$ as $x \to \xi$ if for every $\varepsilon > 0$, there exists $\delta > 0$ such that

$$\left| f(x) - A \right| < \varepsilon \quad \text{whenever } |x - \xi| < \delta.$$

We write

$$\lim_{x \to \xi} f(x) = A.$$

---

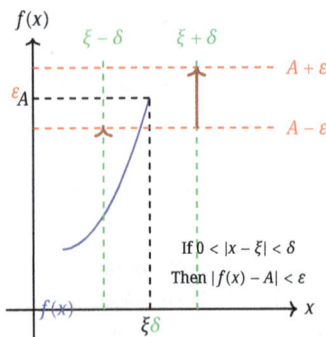

Figure 4.1: Limit at a point: For any $\varepsilon$-neighborhood around $A$, there exists a $\delta$-neighborhood around $\xi$ (excluding $\xi$) such that $f(x)$ falls within the $\varepsilon$-neighborhood.

The concept of infinite limits is fundamental in the study of real functions. It describes the behavior of a function as its values increase or decrease without bound near a given point.

**Definition 4.2** (Infinite limit at a point). We say that $f(x) \to +\infty$ (respectively $-\infty$) as $x \to \xi$ if for every $M > 0$, there exists $\delta > 0$ such that

$$f(x) > M \quad \left(\text{respectively } f(x) < -M\right) \quad \text{whenever } |x - \xi| < \delta.$$

We write

$$\lim_{x \to \xi} f(x) = \infty \quad \left(\text{respectively } \lim_{x \to \xi} f(x) = -\infty\right).$$

Limits at infinity describe the behavior of a function as the variable grows arbitrarily large or decreases without bound. This concept is essential for understanding the end behavior of real-valued functions.

**Definition 4.3** (Limit at infinity). We say that $f(x) \to A \in \mathbb{R}$ as $x \to +\infty$ (respectively $x \to -\infty$) if for every $\varepsilon > 0$, there exists $M > 0$ such that

$$\left|f(x) - A\right| < \varepsilon \quad \text{whenever } x > M \text{ (respectively } x < -M).$$

We write

$$\lim_{x \to \infty} f(x) = A \quad \left(\text{respectively } \lim_{x \to -\infty} f(x) = A\right).$$

Infinite limits at infinity describe functions whose values grow without bound as the input becomes arbitrarily large or small. This concept helps us understand extreme behaviors and divergence in mathematical analysis.

**Definition 4.4** (Infinite limit at infinity). We say that $f(x) \to +\infty$ (respectively $-\infty$) as $x \to +\infty$ if for every $M > 0$, there exists $N > 0$ such that

$$f(x) > M \quad \left(\text{respectively } f(x) < -M\right) \quad \text{whenever } x > N.$$

Similarly, we define

$$\lim_{x \to -\infty} f(x) = \infty.$$

One-sided limits are used to analyze the behavior of a function as the input approaches a point from only one direction. This concept is especially important at the endpoints of intervals and at points where a function may not be defined on both sides.

**Definition 4.5** (One-sided limits (see Figure 4.2)). Let $f : I \subseteq \mathbb{R} \to \mathbb{R}$ and $\xi \in \mathbb{R}$ be a boundary point of $I$. We say that $f(x) \to A \in \mathbb{R}$ as $x \to \xi^+$ (from the right) if for every $\varepsilon > 0$, there exists $\delta > 0$ such that

$$\left|f(x) - A\right| < \varepsilon \quad \text{whenever } \xi < x < \xi + \delta.$$

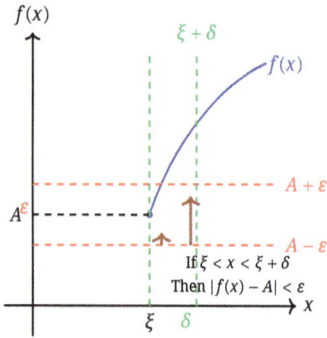

**Figure 4.2:** One-sided limit: For a right-hand limit, for any $\varepsilon$-neighborhood around $A$, there exists a $\delta$-interval to the right of $\xi$ such that $f(x)$ falls within the $\varepsilon$-neighborhood.

We write

$$\lim_{x \to \xi^+} f(x) = A.$$

Similarly, we define the left-hand limit $\lim_{x \to \xi^-} f(x) = A$ and the corresponding infinite one-sided limits.

The following theorem reveals the close relationship between two-sided and one-sided limits. It states that the existence of a two-sided limit is equivalent to the existence and agreement of both one-sided limits at the point.

**Theorem 4.1** (Equivalence of two-sided and one-sided limits). *The following equivalence holds:*

$$\lim_{x \to \xi} f(x) = A \quad \Longleftrightarrow \quad \left( \lim_{x \to \xi^+} f(x) = A \text{ and } \lim_{x \to \xi^-} f(x) = A \right).$$

*Proof.* If the two-sided limit exists and

$$\lim_{x \to \xi} f(x) = A,$$

then it is immediate that both one-sided limits also exist and are equal to $A$, that is,

$$\lim_{x \to \xi^+} f(x) = A \quad \text{and} \quad \lim_{x \to \xi^-} f(x) = A.$$

Conversely, suppose both one-sided limits exist and are equal to $A$:

$$\lim_{x \to \xi^+} f(x) = A \quad \text{and} \quad \lim_{x \to \xi^-} f(x) = A.$$

We want to prove that then

$$\lim_{x \to \xi} f(x) = A.$$

By assumption, for every $\varepsilon > 0$, there exist $\delta_1, \delta_2 > 0$ such that:

- If $\xi < x < \xi + \delta_1$ then $|f(x) - A| < \varepsilon$,
- If $\xi - \delta_2 < x < \xi$ then $|f(x) - A| < \varepsilon$.

Let $\delta = \min\{\delta_1, \delta_2\}$. Then, for all $x$ such that $|x - \xi| < \delta$, we have $|f(x) - A| < \varepsilon$. Therefore, the two-sided limit exists and equals $A$. $\qquad\square$

## 4.2.1 Examples

**Example 4.1** (Limit of the identity function). Show that

$$\lim_{x \to \xi} x = \xi.$$

*Solution.* Let $\varepsilon > 0$ and choose $\delta = \varepsilon$. Then, for $|x - \xi| < \delta$, we clearly have

$$|x - \xi| < \varepsilon,$$

which satisfies the definition of limit. Therefore, $\lim_{x \to \xi} x = \xi$. $\qquad\square$

**Example 4.2** (A function without a limit (Figure 4.4)). Show that the function $f(x) = \sin(\frac{1}{x})$ has no limit as $x \to 0$.

*Solution.* Assume, for contradiction, that $\lim_{x \to 0} \sin(\frac{1}{x}) = A \in \mathbb{R}$. Then, for every $\varepsilon > 0$, there must exist $\delta > 0$ such that

$$\left| \sin\left(\frac{1}{x}\right) - A \right| < \varepsilon, \quad \forall x \in (-\delta, \delta).$$

However, within any such interval lie points $x_n = \frac{1}{2\pi n}$ and $y_n = \frac{1}{2\pi n + \frac{\pi}{2}}$, for which

$$\sin\left(\frac{1}{x_n}\right) = 0, \quad \sin\left(\frac{1}{y_n}\right) = 1.$$

This contradicts the uniqueness of the limit. Hence, the limit does not exist. $\qquad\square$

**Example 4.3** (Infinite one-sided limits). Study the behavior of $f(x) = \frac{1}{1-x}$ as $x \to 1^-$ and $x \to 1^+$.

*Solution.* Let $x \to 1^-$. For any $M > 0$, choose $\delta = \frac{1}{M}$. Then, for $x \in (1 - \delta, 1)$, we have

$$f(x) = \frac{1}{1 - x} > M.$$

Hence, $\lim_{x \to 1^-} f(x) = +\infty$, and similarly $\lim_{x \to 1^+} f(x) = -\infty$. $\qquad\square$

**Example 4.4** (Limit at infinity). Find the limit of the function $f(x) = x$ as $x \to \pm\infty$.

*Solution.* We claim that

$$\lim_{x \to +\infty} x = +\infty \quad \text{and} \quad \lim_{x \to -\infty} x = -\infty.$$

Let $M > 0$ be an arbitrary positive real number. We need to find an $N > 0$ such that $x > N$ implies $x > M$.

Choose $N = M$. Then, for any $x > N$, we have $x > M$ by definition. Therefore, $f(x) = x > M$.

Since $M$ was arbitrary, this shows that for every $M > 0$, there exists an $N > 0$ (namely, $N = M$) such that $x > N$ implies $f(x) > M$.

By the definition of a limit approaching positive infinity, we conclude

$$\lim_{x \to +\infty} x = +\infty.$$

Now let $M < 0$ be an arbitrary negative real number. We need to find an $N < 0$ such that $x < N$ implies $x < M$.

Choose $N = M$. Then, for any $x < N$, we have $x < M$ by definition. Therefore, $f(x) = x < M$.

Since $M$ was arbitrary, this shows that for every $M < 0$, there exists an $N < 0$ (namely, $N = M$) such that $x < N$ implies $f(x) < M$.

By the definition of a limit approaching negative infinity, we conclude

$$\lim_{x \to -\infty} x = -\infty. \qquad \square$$

**Example 4.5** (Piecewise limit (Figure 4.3)). Let

$$f(x) = \begin{cases} x^2, & x \le \frac{1}{2}, \\ x, & x > \frac{1}{2}. \end{cases}$$

Show that

$$\lim_{x \to \frac{1}{2}^-} f(x) \ne \lim_{x \to \frac{1}{2}^+} f(x).$$

*Solution.* We compute

$$\lim_{x \to \frac{1}{2}^-} f(x) = \left(\frac{1}{2}\right)^2 = \frac{1}{4}, \quad \lim_{x \to \frac{1}{2}^+} f(x) = \frac{1}{2}.$$

Since the left and right limits differ, $\lim_{x \to \frac{1}{2}} f(x)$ does not exist. $\qquad \square$

$f(x)$

$\lim_{x\to 0.5^+} x$

$\lim_{x\to 0.5^-}$

$x$

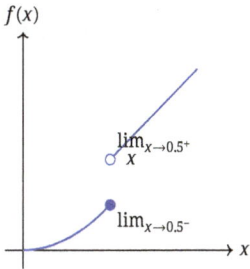

Figure 4.3: Discontinuity with unequal one-sided limits.

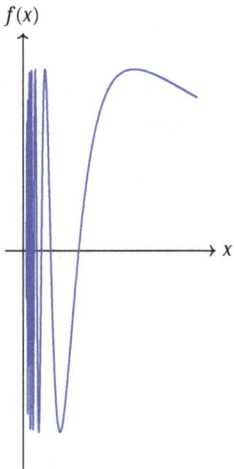

$f(x)$

$x$

Figure 4.4: No limit at $x = 0$: $f(x) = \sin(\frac{1}{x})$.

## 4.3 Algebraic properties of limits

The algebraic properties of limits allow us to combine limits of functions through addition, subtraction, multiplication, and division. These rules are fundamental tools for evaluating limits and analyzing the behavior of functions near a point.

**Theorem 4.2** (Algebra of limits). *If* $\lim_{x\to\xi} f(x) = A$ *and* $\lim_{x\to\xi} g(x) = B$, *then*

$$\lim_{x\to\xi}\big(f(x) \pm g(x)\big) = A \pm B, \quad \lim f(x)g(x) = AB, \quad \lim \frac{f(x)}{g(x)} = \frac{A}{B} \quad \text{if } B \neq 0.$$

*Proof.* We will first prove the property of the sum. As a given, we have that

$$\forall \varepsilon > 0, \, \exists \delta_1 > 0, \quad \text{such that if } |x - \xi| < \delta_1 \Rightarrow |f(x) - A| < \frac{\varepsilon}{2},$$

$$\forall \varepsilon > 0, \, \exists \delta_2 > 0, \quad \text{such that if } |x - \xi| < \delta_2 \Rightarrow |g(x) - B| < \frac{\varepsilon}{2}.$$

We choose $\delta = \min\{\delta_1, \delta_2\}$. Then both of the above statements hold with the same $\delta$. We can now add the inequalities term by term, and therefore we obtain

$$\forall \varepsilon > 0, \ \exists \delta = \min\{\delta_1, \delta_2\}, \quad \text{such that if } |x - \xi| < \delta \Rightarrow |f(x) + g(x) - (A + B)| < \varepsilon.$$

This, of course, is exactly what we wanted to show.

For the product, we proceed similarly:

$$\forall \varepsilon_1 > 0, \ \exists \delta_1 > 0, \quad \text{such that if } |x - \xi| < \delta_1 \Rightarrow |f(x) - A| < \varepsilon_1$$
$$\forall \varepsilon_2 > 0, \ \exists \delta_2 > 0, \quad \text{such that if } |x - \xi| < \delta_2 \Rightarrow |g(x) - B| < \varepsilon_2.$$

Assume that $B \neq 0$. Since $f(x) \to A$, for all $x$ such that $|x - \xi| \leq \delta_1$, we have $|f(x)| < M$ for some constant $M > 0$. We choose $\varepsilon_1 = \frac{\varepsilon/2}{|B|}$ and $\varepsilon_2 = \frac{\varepsilon/2}{M}$ for a given $\varepsilon > 0$, and set $\delta = \min\{\delta_1, \delta_2\}$. It follows that

$$|f(x)g(x) - AB| = |f(x)g(x) - f(x)B + f(x)B - AB|$$
$$\leq |f(x)| \cdot |g(x) - B| + |B| \cdot |f(x) - A|$$
$$\leq \varepsilon.$$

In the case where $A = B = 0$, the proof is simpler since

$$|f(x)g(x)| = |f(x)| \cdot |g(x)| \leq \varepsilon_1 \varepsilon_2.$$

By choosing $\varepsilon = \varepsilon_1 \varepsilon_2$, we get the desired result.

The case for the quotient is entirely similar, provided that $B \neq 0$, since one can view it as the product of the functions $f(x)$ and $\frac{1}{g(x)}$. $\square$

The squeeze theorem provides a powerful method for determining the limit of a function by comparing it to two other functions whose limits are known and equal. It is especially useful when direct computation of a limit is difficult.

**Theorem 4.3** (Squeeze theorem). *Let $\lim_{x \to x_0} f(x) = \lim_{x \to x_0} g(x) = k$, and assume that*

$$f(x) \leq h(x) \leq g(x)$$

*for all $x$ near $x_0$. Then*

$$\lim_{x \to x_0} h(x) = k.$$

*Proof.* From the definition of the limit, it follows that for every $\varepsilon > 0$, there exists some $\delta > 0$ such that

$$|f(x) - k| < \varepsilon, \quad |g(x) - k| < \varepsilon \quad \text{whenever } |x - x_0| < \delta.$$

Therefore, for every $\varepsilon > 0$, there exists $\delta > 0$ such that

$$-\varepsilon < f(x) - k \le h(x) - k \le g(x) - k < \varepsilon \quad \text{whenever } |x - x_0| < \delta.$$

This, of course, means that

$$\lim_{x \to x_0} h(x) = k. \qquad \square$$

The squeeze theorem can be applied in various contexts, including functions that approach zero or infinity. It offers a convenient way to deduce limits for functions bounded by others with known limiting behavior.

**Remark 5.** If $\lim_{x \to x_0} g(x) = 0$ and $|f(x)| \le g(x)$, then $\lim_{x \to x_0} f(x) = 0$. The squeeze theorem also applies for $x_0 = \pm\infty$ and infinite limits.

**Theorem 4.4** (Limit of a composition). *Let f and g be real-valued functions. Suppose that*

$$\lim_{x \to \xi} f(x) = A \quad \text{and} \quad \lim_{u \to A} g(u) = B.$$

*Furthermore, assume that there exists a deleted neighborhood of $\xi$ such that*

$$f(x) \ne A \quad \text{for all } x \text{ satisfying } 0 < |x - \xi| < \delta_0.$$

*Then*

$$\lim_{x \to \xi} g\big(f(x)\big) = B.$$

*Proof.* Let $\varepsilon > 0$ be given. Since $\lim_{u \to A} g(u) = B$, there exists $\eta > 0$ such that

$$0 < |u - A| < \eta \quad \text{implies} \quad |g(u) - B| < \varepsilon.$$

Since $\lim_{x \to \xi} f(x) = A$, for this $\eta > 0$, there exists $\delta_1 > 0$ such that

$$0 < |x - \xi| < \delta_1 \quad \text{implies} \quad |f(x) - A| < \eta.$$

By assumption, there exists $\delta_0 > 0$ such that $f(x) \ne A$ whenever $0 < |x - \xi| < \delta_0$. Now, let $\delta = \min(\delta_0, \delta_1) > 0$. Then, for all $x$ satisfying $0 < |x - \xi| < \delta$, we have:
- $f(x) \ne A$ (by choice of $\delta_0$).
- $|f(x) - A| < \eta$ (by choice of $\delta_1$).

Therefore, $0 < |f(x) - A| < \eta$, and so by the limit of $g$,

$$|g(f(x)) - B| < \varepsilon.$$

Hence, for every $\varepsilon > 0$, there exists $\delta > 0$ such that

$$0 < |x - \xi| < \delta \quad \text{implies} \quad |g(f(x)) - B| < \varepsilon.$$

This proves that

$$\lim_{x \to \xi} g(f(x)) = B.$$ □

**Remark 6** (The assumption $f(x) \neq A$ near $\xi$ is necessary). The additional assumption that $f(x) \neq A$ near $\xi$ (excluding $\xi$) is necessary. Without it, the conclusion may fail.
Define

$$f(x) = 0 \quad \text{for all } x \in \mathbb{R},$$

and

$$g(u) = \begin{cases} 0, & u \neq 0, \\ 1, & u = 0. \end{cases}$$

Then:

- $\lim_{x \to \xi} f(x) = 0 = A$ for any $\xi$,
- $\lim_{u \to 0} g(u) = 0 = B$, since $g(u) = 0$ for all $u \neq 0$,
- however, $g(f(x)) = g(0) = 1$ for all $x$, so

$$\lim_{x \to \xi} g(f(x)) = 1 \neq B.$$

This shows that the limit of the composition may not equal $B$ if $f(x) = A$ in every neighborhood of $\xi$, even if $\lim_{u \to A} g(u) = B$.

## 4.3.1 Examples

**Example 4.6** (Limit of a sum). Let

$$f(x) = 3x - 1, \quad g(x) = 2x^2 + 4,$$

and consider

$$\lim_{x \to 2} (f(x) + g(x)).$$

*Solution.* By the sum law,

$$\lim_{x \to 2} f(x) = 3 \cdot 2 - 1 = 5, \quad \lim_{x \to 2} g(x) = 2 \cdot 2^2 + 4 = 12.$$

Therefore

$$\lim_{x \to 2} (f(x) + g(x)) = 5 + 12 = 17.$$ □

**Example 4.7** (Limit of a difference). Let

$$f(x) = x^3, \quad g(x) = x + 5,$$

and consider

$$\lim_{x \to -1} (f(x) - g(x)).$$

*Solution.* By the difference law,

$$\lim_{x \to -1} f(x) = (-1)^3 = -1, \quad \lim_{x \to -1} g(x) = -1 + 5 = 4.$$

Hence

$$\lim_{x \to -1} (f(x) - g(x)) = -1 - 4 = -5. \qquad \square$$

**Example 4.8** (Limit of a product). Let

$$f(x) = x - 4, \quad g(x) = x^2 + 1,$$

and consider

$$\lim_{x \to 3} (f(x) g(x)).$$

*Solution.* By the product law,

$$\lim_{x \to 3} f(x) = 3 - 4 = -1, \quad \lim_{x \to 3} g(x) = 3^2 + 1 = 10.$$

Thus

$$\lim_{x \to 3} f(x) g(x) = (-1) \cdot 10 = -10. \qquad \square$$

**Example 4.9** (Limit of a quotient). Let

$$f(x) = 2x^2 - 2, \quad g(x) = x - 1,$$

and consider

$$\lim_{x \to 1} \frac{f(x)}{g(x)}.$$

*Solution.* We first check $g(1) = 0$, so we must factor

$$f(x) = 2(x^2 - 1) = 2(x - 1)(x + 1),$$

hence for $x \neq 1$

$$\frac{f(x)}{g(x)} = 2(x + 1).$$

Therefore

$$\lim_{x \to 1} \frac{f(x)}{g(x)} = \lim_{x \to 1} 2(x + 1) = 2 \cdot 2 = 4.$$  □

**Example 4.10** (Limit of a composition). Let

$$f(x) = x + 3, \quad g(u) = 2u - 1,$$

and consider

$$\lim_{x \to 1} g(f(x)).$$

*Solution.* First compute the inner limit:

$$\lim_{x \to 1} f(x) = 1 + 3 = 4.$$

Next compute the outer limit as $u \to 4$:

$$\lim_{u \to 4} g(u) = 2 \cdot 4 - 1 = 7.$$

By the algebraic property of composition,

$$\lim_{x \to 1} g(f(x)) = 7.$$  □

## 4.4 Python and AI

We can ask AI to solve theoretically an exercise and then write a suitable Python code. However, you should check carefully both of them!

---

**AI Request**

Study the limit of the function

$$f(x) = \frac{a_k x^k + \cdots + a_1 x + a_0}{c_r x^r + \cdots + c_1 x + c_0}$$

for all possible combinations of degrees $k, r \in \mathbb{N}$, assuming all coefficients are real. Give the Latex code with the solution step-by-step and explain how I can produce a PDF file. Write a Python code to determine the above limit.

---

---

**AI Request**

I want a Python code in Colab that calculates the left and right limits of a function. The results should be displayed with MathJax. Oh, and a graph too.

---

## 4.4.1 Exercises

**Exercise 20** (Polynomial rational limit). Study the limit of the function

$$f(x) = \frac{a_k x^k + \cdots + a_1 x + a_0}{c_r x^r + \cdots + c_1 x + c_0}$$

for all possible combinations of degrees $k, r \in \mathbb{N}$, assuming all coefficients are real.

**Exercise 21** (Limits to compute I). Compute the following:

$$\lim_{x \to 2}[x], \quad \lim_{x \to 0} \frac{\sqrt{x+3} - \sqrt{3}}{x}, \quad \lim_{x \to 2}\left(\frac{1}{x-2} - \frac{4}{x^2 - 4}\right), \quad \lim_{x \to +\infty} \frac{4x - 1}{\sqrt{x^2 + 2}}.$$

**Exercise 22** (Limits to compute II). Compute

$$\lim_{x \to 0^+} \frac{|x|}{x}, \quad \lim_{x \to 0^-} \frac{|x|}{x}.$$

### True or false
Determine whether each statement is **true** or **false**. Justify your answer.
1.  If $\lim_{x \to \xi} f(x) = A$, then $f(\xi) = A$.
    Answer: _____
2.  If $\lim_{x \to \xi^-} f(x) = \lim_{x \to \xi^+} f(x)$, then $\lim_{x \to \xi} f(x)$ exists.
    Answer: _____
3.  The limit $\lim_{x \to 0} \sin(\frac{1}{x})$ does not exist because the function oscillates rapidly near 0.
    Answer: _____
4.  If $\lim_{x \to \xi} f(x) = +\infty$, then the function approaches infinity as $x \to \xi$.
    Answer: _____
5.  If $\lim_{x \to \xi} f(x) = A$ and $\lim_{x \to \xi} g(x) = B$, then $\lim_{x \to \xi}(f(x)g(x)) = AB$.
    Answer: _____

### Multiple choice questions
Choose the correct option (A, B, C, or D) for each question.
1.  What is the value of $\lim_{x \to 0} \frac{\sin x}{x}$?
    A. 0          B. 1
    C. Does not exist    D. $\infty$
    Answer: _____
2.  Consider the function $f(x) = \frac{1}{x-1}$. Which of the following is true?
    A. $\lim_{x \to 1} f(x) = 1$          B. $\lim_{x \to 1^-} f(x) = +\infty$
    C. $\lim_{x \to 1^+} f(x) = -\infty$    D. $\lim_{x \to 1} f(x)$ does not exist
    Answer: _____
3.  Suppose $\lim_{x \to \xi} f(x) = A$ and $\lim_{x \to \xi} g(x) = B$ with $B \neq 0$. Then which of the following must be true?

A. $\lim_{x \to \xi}(f(x) + g(x)) = A + B$ B. $\lim_{x \to \xi}(f(x) - g(x)) = A - B$

C. $\lim_{x \to \xi} \frac{f(x)}{g(x)} = \frac{A}{B}$ D. All of the above

Answer: _____

4. What is the value of $\lim_{x \to \infty} \frac{x^2+1}{x^2-1}$?

A. 1     B. 0

C. $\infty$     D. Does not exist

Answer: _____

5. For the piecewise function

$$f(x) = \begin{cases} x^2, & x < 1, \\ x, & x > 1. \end{cases}$$

What is $\lim_{x \to 1} f(x)$?

A. 1                           B. 0

C. Does not exist   D. $\infty$

Answer: _____

## Computational exercises

Solve the following problems theoretically but also using Python and appropriate libraries such as numpy, matplotlib, and sympy.

1. Use sympy to compute

$$\lim_{x \to 2} \frac{x^2 - 4}{x - 2}.$$

Plot the function near $x = 2$ and explain what happens at that point.

2. Compute numerically and plot the function

$$f(x) = \frac{\sin x}{x}$$

for values of $x$ close to 0. Based on your graph and numerical results, estimate

$$\lim_{x \to 0} \frac{\sin x}{x}.$$

3. Consider the piecewise function

$$f(x) = \begin{cases} x + 1, & x < 1, \\ x^2, & x > 1. \end{cases}$$

Plot the function and compute

$$\lim_{x \to 1^-} f(x), \quad \lim_{x \to 1^+} f(x).$$

Does $\lim_{x \to 1} f(x)$ exist?

4. Use sympy to verify the squeeze theorem for

$$f(x) = x \cdot \sin\left(\frac{1}{x}\right)$$

around $x = 0$. Provide bounds and show that the limit is zero.
5. Investigate numerically and graphically

$$\lim_{x \to +\infty} \frac{3x^2 + 2x + 1}{x^2 + 1}.$$

What do you observe? Can you generalize this result?
6. Study the behavior of

$$f(x) = \frac{1}{x - 3}$$

as $x \to 3^-$ and $x \to 3^+$. Plot the function and describe the infinite one-sided limits.

## Python-based questions
Answer the following using a Python code.
1. Write a Python function to compute

$$\lim_{x \to \xi} f(x)$$

using sympy.limit. Test it with

$$f(x) = \frac{x^3 - 8}{x - 2}, \quad \xi = 2.$$

2. Create a script that plots the function

$$f(x) = \frac{\sin x}{x}$$

over the interval $[-\pi, \pi]$ and adds annotations showing:
(a) Where the function is undefined.
(b) The estimated value of the limit at $x = 0$.
3. Write a Python function that checks if a given function has a limit at a point by comparing left and right limits. Apply it to

$$f(x) = \begin{cases} x^2, & x < 1, \\ x + 1, & x > 1, \end{cases}$$

at $x = 1$.

4. Use `matplotlib` to visualize the epsilon-delta definition of a limit for

$$f(x) = x^2, \quad \xi = 2, \quad A = 4.$$

Show how for decreasing $\varepsilon$, the corresponding $\delta$ also decreases.

5. Write a Python script to evaluate and compare

$$\lim_{x \to \infty} \left(1 + \frac{1}{x}\right)^x$$

and

$$\lim_{x \to \infty} \left(1 + \frac{1}{x}\right)^{x^2}.$$

What do these limits represent?

6. Plot the function

$$f(x) = \sin\left(\frac{1}{x}\right)$$

near $x = 0$. Explain why the limit does not exist at $x = 0$ based on your plot and numerical observations.

## Fill in the blanks

Complete the following statements by filling in the blanks:

1. The limit of a function $f(x)$ as $x \to \xi$ is the value $A$ such that $f(x)$ gets arbitrarily close to $A$ when $x$ is sufficiently close to _____, but not equal to it.

2. If $\lim_{x \to \xi} f(x) = A$ and $\lim_{x \to \xi} g(x) = B$, then $\lim_{x \to \xi}(f(x)+g(x)) = $ _____.

3. A _____ limit considers only values of $x$ approaching from one side, either $x \to \xi^+$ or $x \to \xi^-$.

4. For the limit $\lim_{x \to \xi} f(x)$ to exist, the _____ limits from both sides must be equal.

5. We say $\lim_{x \to \xi} f(x) = +\infty$ if for every $M > 0$, there exists $\delta > 0$ such that $f(x) > M$ whenever $0 < |x - \xi| < $ _____.

6. The limit $\lim_{x \to \infty} f(x) = L$ means that $f(x)$ approaches $L$ as $x$ becomes arbitrarily _____.

7. If $f(x)$ oscillates without settling near any specific value as $x \to \xi$, then the limit at that point _____.

## Matching

1. Limit at a point
2. One-sided limit
3. Infinite limit
4. Limit at infinity
5. Discontinuity

a. The function grows without bound as $x$ approaches a certain value.
b. The function does not approach a single value due to jumps or oscillations.
c. The behavior of the function as $x$ becomes very large.
d. The value the function approaches as $x$ gets arbitrarily close to a finite point.
e. Only values from one side of a point are considered.

**Answers:** 1 – _____   2 – _____   3 – _____   4 – _____   5 – _____

# 5 Continuity

## Contents

This chapter delves into the fundamental concepts of real analysis, laying the groundwork for understanding the rigorous theory behind functions, limits, and continuity. We aim to present these ideas with clarity and precision, guiding the reader through the foundational definitions and essential theorems that form the bedrock of mathematical analysis. Our journey begins with the basic building blocks and progressively moves towards more intricate concepts, providing a comprehensive yet accessible introduction to this vital field of mathematics.

## 5.1 Introduction

Welcome to the rigorous world of real analysis, a cornerstone of modern mathematics that provides the formal framework for calculus and beyond. This chapter embarks on a journey through the fundamental concepts that underpin the behavior of real-valued functions, starting with the precise definitions of limits and continuity.

Historically, the need for such rigor became apparent in the 19th century. While calculus, developed by *Newton* and *Leibniz* in the 17th century, had achieved remarkable success in solving physical problems, its foundations were built upon intuitive, rather than strictly formal, notions. Concepts like "infinitesimally small quantities" and "approaching a limit" lacked the precise definitions necessary to resolve paradoxes and ensure the validity of new results.

Figures like *Augustin-Louis Cauchy* (1789–1857) began the arduous task of formalizing these concepts, introducing the $\varepsilon$-$\delta$ definitions that are central to our understanding of limits and continuity today. Later, *Karl Weierstrass* (1815–1897), often called the "father of modern analysis," refined these definitions and demonstrated their power in proving theorems that had previously rested on shaky ground. His insistence on arithmetical rigor, free from geometric intuition, transformed analysis into a truly independent and self-contained discipline.

In these pages, we will carefully construct this framework, starting with the $\varepsilon$-$\delta$ definition of continuity at a point, exploring its nuances, and distinguishing it from the related concept of a limit. We will then build upon this foundation to investigate the

https://doi.org/10.1515/9783112228289-005

properties of continuous functions, leading to profound results such as the intermediate value theorem and the extreme value theorem, which reveal the inherent structural characteristics of these functions on closed intervals. This journey into real analysis is not merely an exercise in abstraction; it is an exploration of the precise language and logical deduction required to truly understand the continuous world around us.

## 5.2 Definitions

In this section, we formally define what it means for a function to be continuous at a point, as well as the notions of right and left continuity. These foundational concepts are central to the study of real analysis and provide the rigorous framework for analyzing the behavior of functions near specific points. We also discuss subtle differences between continuity and the limit of a function and address special cases such as isolated points in the domain.

Continuity at a point is a central concept in real analysis, capturing the idea that small changes in the input lead to small changes in the output. It provides a formal way to describe when a function behaves smoothly at a given point.

**Definition 5.1** (Continuity at a point). A function $f : I \subseteq \mathbb{R} \to \mathbb{R}$ is called **continuous** at the point $\xi \in I$ if

$$\forall \varepsilon > 0, \ \exists \delta > 0 \text{ such that if } x \in I \text{ and } |x - \xi| < \delta, \text{ then } |f(x) - f(\xi)| < \varepsilon.$$

The value of $\delta$ depends on both $\varepsilon$ and the point $\xi$. The function $f$ is called **discontinuous** at $\xi \in I$ if it is not continuous at that point.

Right continuity describes the behavior of a function as the input approaches a point from values greater than that point. This concept is especially important when dealing with functions defined on intervals that are not open on both sides.

**Definition 5.2** (Right continuity). A function $f : I \subseteq \mathbb{R} \to \mathbb{R}$ is called **right continuous** at the point $\xi \in I$ if

$$\forall \varepsilon > 0, \ \exists \delta > 0 \text{ such that if } x \in I \text{ and } 0 < x - \xi < \delta, \text{ then } |f(x) - f(\xi)| < \varepsilon.$$

Left continuity characterizes how a function behaves as the input approaches a point from values less than that point. It is particularly useful for analyzing functions at the left endpoints of intervals or at discontinuities.

**Definition 5.3** (Left continuity). A function $f : I \subseteq \mathbb{R} \to \mathbb{R}$ is called **left continuous** at the point $\xi \in I$ if

$$\forall \varepsilon > 0, \ \exists \delta > 0 \text{ such that if } x \in I \text{ and } 0 < \xi - x < \delta, \text{ then } |f(x) - f(\xi)| < \varepsilon.$$

The relationship between the limit of a function and continuity at a point reveals subtle but important distinctions. This remark clarifies these differences and explains how continuity is defined even at isolated points of the domain.

**Remark 7.** Note that in the definition of continuity, we do *not* require that the point $\xi \in I$ is an accumulation point of the domain (unlike in the definition of the limit of a function). However, we do require that $\xi \in I$ (which is not required in the definition of a limit).

The definition of continuity of a function $f : I \subseteq \mathbb{R} \to \mathbb{R}$ at an accumulation point $\xi \in I$ is equivalent to the condition

$$\lim_{x \to \xi} f(x) = f(\xi) \quad \text{(see Definition 5.1).}$$

Similarly, this holds for right and left continuity at an accumulation point $\xi \in I$.

If the point $\xi \in I$ is an isolated point, then the definition of continuity still applies at $\xi$. Indeed, for every $\varepsilon > 0$, we can find a sufficiently small $\delta > 0$ such that the only $x \in I$ satisfying $|x - \xi| < \delta$ is $x = \xi$. Therefore, we have

$$\left| f(\xi) - f(\xi) \right| = 0 < \varepsilon,$$

and hence the function $f$ is considered continuous at the isolated point $\xi$.

In this case, although the function is continuous, its graph may not appear as a continuous curve—unlike what happens when the point is an accumulation point. □

The following exercises are designed to test your understanding of the fundamental definitions and properties of continuity and limits. They include true/false statements and multiple choice questions, encouraging careful reasoning and justification.

## 5.2.1 Exercises

### True or false
Determine whether each statement is **true** or **false**. Justify your answer.
1. A function can be continuous at an isolated point of its domain.
   **Answer:** _____
2. If a function is right continuous at $\xi$, then it is also left continuous at $\xi$.
   **Answer:** _____
3. If $f$ is continuous at $\xi$, then $\lim_{x \to \xi} f(x)$ exists and equals $f(\xi)$.
   **Answer:** _____
4. A function is continuous at $\xi$ only if $\xi$ is an accumulation point of its domain.
   **Answer:** _____
5. If $\lim_{x \to \xi} f(x) = f(\xi)$, then $f$ is continuous at $\xi$.
   **Answer:** _____

## Multiple choice questions

Choose the correct option (A, B, C, or D) for each question.

1. Which of the following is always true for a function $f$ continuous at $\xi$?
   A. $\xi$ is an accumulation point of the domain    B. $\lim_{x \to \xi} f(x)$ does not exist
   C. $f(\xi)$ must be zero                           D. $\lim_{x \to \xi} f(x) = f(\xi)$
   **Answer:** _____

2. If $\xi$ is an isolated point of the domain of $f$, then
   A. $f$ cannot be continuous at $\xi$    B. $\lim_{x \to \xi} f(x)$ exists
   C. $f$ is always continuous at $\xi$    D. $f(\xi) = 0$
   **Answer:** _____

3. A function is continuous at $\xi$ if and only if
   A. it is left continuous at $\xi$                      B. it is right continuous at $\xi$
   C. it is both left and right continuous at $\xi$    D. none of the above
   **Answer:** _____

4. The definition of continuity at a point differs from the definition of a limit because
   A. continuity requires $\xi$ to be in the domain
   B. limits require $\xi$ to be in the domain
   C. continuity does not require $\xi$ to be an accumulation point
   D. both A and C
   **Answer:** _____

5. Which of the following functions is continuous at $x = 0$?
   A. $f(x) = \frac{1}{x}$            B. $f(x) = \sin(\frac{1}{x})$
   C. $f(x) = 0$ for all $x$    D. $f(x) = \begin{cases} 1, & x \neq 0, \\ 0, & x = 0. \end{cases}$
   **Answer:** _____

## 5.3 Basic theorems on continuity

In this section, we explore essential properties of continuous functions, including their behavior under algebraic operations, composition, and transformations such as the absolute value and maximum/minimum.

Continuity at a point can be characterized in terms of one-sided continuity. The following proposition formalizes this equivalence for accumulation points of the domain.

**Proposition 5.1** (Equivalence of left and right continuity). *A function $f : I \subseteq \mathbb{R} \to \mathbb{R}$ is continuous at an accumulation point $\xi \in I$ **if and only if** it is both left continuous and right continuous at that point.*

*Proof.* The proof follows directly from Remark 7 and Theorem 4.1.  □

As a basic example, we consider the continuity of the identity function. This simple case illustrates how the definition of continuity operates in practice.

> **Proposition 5.2** (Continuity of the identity function). *Show that the function $f(x) = x$ is continuous at every point $\xi \in \mathbb{R}$.*

*Proof.*

$$\lim_{x \to \xi} f(x) = \xi = f(\xi).$$

Therefore, $f$ is continuous at every point of $\mathbb{R}$. □

In order to check the continuity of more complicated functions, one can divide them into simpler functions. Then, using the following theorems, we arrive easily at the desired characterization. The continuity of sums, products, and quotients is a natural extension of the algebraic properties of limits. The following proposition demonstrates that these operations preserve continuity, provided the denominator does not vanish.

> **Proposition 5.3** (Continuity of sums, products, and quotients). *Let $f$, $g$ be functions continuous at $\xi \in \mathbb{R}$. Prove that*
> $$f \pm g, \quad fg, \quad and \quad \frac{f}{g} \quad (if\ g(\xi) \neq 0)$$
> *are also continuous at $\xi$.*

*Proof.* Since $f$ and $g$ are continuous at $\xi$, we have

$$\lim_{x \to \xi} f(x) = f(\xi) \quad \text{and} \quad \lim_{x \to \xi} g(x) = g(\xi).$$

Then, using standard limit theorems,

$$\lim_{x \to \xi} (f(x) \pm g(x)) = f(\xi) \pm g(\xi),$$

$$\lim_{x \to \xi} (f(x)g(x)) = f(\xi)g(\xi),$$

$$\lim_{x \to \xi} \frac{f(x)}{g(x)} = \frac{f(\xi)}{g(\xi)}, \quad \text{provided } g(\xi) \neq 0.$$

Hence, $f + g$, $fg$, and $\frac{f}{g}$ are continuous at $\xi$. □

The composition of continuous functions inherits the property of continuity. This theorem provides a fundamental tool for building new continuous functions from existing ones.

> **Theorem 5.1** (Continuity of a composition). *Let $y = f(x)$ be a function continuous at $\xi$, and let $F(y)$ be continuous at $f(\xi)$. Then the composition $F \circ f$ is continuous at $\xi$.*

*Proof.* Let $\varepsilon > 0$. We want to find $\delta > 0$ such that

$$|x - \xi| < \delta \quad \Rightarrow \quad |F(f(x)) - F(f(\xi))| < \varepsilon.$$

Since $F$ is continuous at $f(\xi)$, there exists $\delta' > 0$ such that

$$|y - f(\xi)| < \delta' \quad \Rightarrow \quad |F(y) - F(f(\xi))| < \varepsilon.$$

Because $f$ is continuous at $\xi$, for this $\delta'$ there exists $\delta'' > 0$ such that

$$|x - \xi| < \delta'' \quad \Rightarrow \quad |f(x) - f(\xi)| < \delta'.$$

Thus, choosing $\delta = \delta''$, we get

$$|x - \xi| < \delta \quad \Rightarrow \quad |F(f(x)) - F(f(\xi))| < \varepsilon,$$

as required. $\qquad\square$

Taking the absolute value of a continuous function yields another continuous function. This proposition uses the reverse triangle inequality to establish the result.

**Proposition 5.4** (Continuity of $|f|$). *Let $f : I \subseteq \mathbb{R} \to \mathbb{R}$ be continuous at $x_0 \in I$. Then $|f|$ is also continuous at $x_0$.*

*Proof.* Since $f$ is continuous at $x_0$, for every $\varepsilon > 0$ there exists $\delta > 0$ such that for all $x \in I$, if $|x - x_0| < \delta$, then

$$|f(x) - f(x_0)| < \varepsilon.$$

We want to show that $|f|$ is continuous at $x_0$, i. e., for every $\varepsilon > 0$ there exists $\delta > 0$ such that

$$\big||f(x)| - |f(x_0)|\big| < \varepsilon \quad \text{whenever } |x - x_0| < \delta.$$

This follows directly from the reverse triangle inequality

$$\big||f(x)| - |f(x_0)|\big| \leq |f(x) - f(x_0)|.$$

Therefore, whenever $|f(x) - f(x_0)| < \varepsilon$, it also holds that

$$\big||f(x)| - |f(x_0)|\big| < \varepsilon.$$

Hence, $|f|$ is continuous at $x_0$. $\qquad\square$

Certain standard operations, such as taking the maximum, minimum, or positive and negative parts of continuous functions, also preserve continuity. The following proposition explains why these constructions yield continuous functions.

**Proposition 5.5** (Continuity of max, min, positive and negative parts). *Let $f$, $g$ be continuous functions on an interval $I \subseteq \mathbb{R}$. Then the functions $\max\{f, g\}$, $\min\{f, g\}$, $f^+ = \max\{f, 0\}$ and $f^- = -\min\{f, 0\}$ are also continuous on $I$.*

*Proof.* We use the identities:

$$\max\{f(x), g(x)\} = \frac{1}{2}(f(x) + g(x)) + \frac{1}{2}|f(x) - g(x)|,$$
$$\min\{f(x), g(x)\} = \frac{1}{2}(f(x) + g(x)) - \frac{1}{2}|f(x) - g(x)|.$$

The functions $f(x) + g(x)$ and $f(x) - g(x)$ are continuous as sums and differences of continuous functions. Moreover, $|f(x) - g(x)|$ is continuous since the absolute value of a continuous function is continuous (by Proposition 5.4).

Hence, both $\max\{f, g\}$ and $\min\{f, g\}$ are compositions and sums of continuous functions, and are therefore continuous on $I$.

We observe that

$$f^+(x) = \max\{f(x), 0\} = \frac{1}{2}(f(x) + |f(x)|),$$
$$f^-(x) = -\min\{f(x), 0\} = \frac{1}{2}(-f(x) + |f(x)|).$$

Again, since $f(x)$ is continuous and $|f(x)|$ is continuous by Proposition 5.4, both $f^+(x)$ and $f^-(x)$ are composed of continuous operations and thus are continuous. □

### 5.3.1 Examples

Next, we give some interesting examples concerning the notion of continuity.

**Example 5.1** (Left but not right continuity). Let
$$f(x) = \begin{cases} x^2, & \text{if } x \leq \frac{1}{2}, \\ x, & \text{if } x > \frac{1}{2}. \end{cases}$$
Show that $f$ is left continuous but not right continuous at $x = \frac{1}{2}$. Is it continuous there?

*Solution.* The left-hand limit at $x = \frac{1}{2}$ equals $f(\frac{1}{2}) = (\frac{1}{2})^2 = \frac{1}{4}$, while the right-hand limit is $1/2 \neq 1/4$. Thus, the function is not continuous at $x = \frac{1}{2}$. □

**Example 5.2** (Discontinuity of the Dirichlet function). Consider the function
$$D(x) = \begin{cases} 1, & \text{if } x \in \mathbb{Q}, \\ 0, & \text{otherwise.} \end{cases}$$
Show that $D$ is not continuous at any point of $\mathbb{R}$.

*Solution.* Every interval contains both rational and irrational numbers. Hence, there is no neighborhood of any point where the values of $D(x)$ are close to $D(\xi)$. So $D$ is discontinuous everywhere. □

**Example 5.3** (Continuous extension of $x \sin \frac{1}{x}$). Define $f(x) = x \sin \frac{1}{x}$ for $x \neq 0$. Show that $f$ has a limit at $x = 0$ and can be extended continuously at $x = 0$.

*Solution.* Since $|\sin(1/x)| \leq 1$, we have

$$\left| x \sin \frac{1}{x} \right| \leq |x| \quad \Rightarrow \quad \lim_{x \to 0} x \sin \frac{1}{x} = 0.$$

Define

$$\hat{f}(x) = \begin{cases} x \sin \frac{1}{x}, & x \neq 0, \\ 0, & x = 0. \end{cases}$$

Then $\hat{f}$ is continuous everywhere. □

### 5.3.2 Exercises

#### True or false
Determine whether each statement is **true** or **false**. Justify your answer.

1. If a function is left continuous but not right continuous at a point, then it is not continuous there.
   
   **Answer:** _____

2. If $f$ and $g$ are continuous at $\xi$, then $f/g$ is also continuous at $\xi$.
   
   **Answer:** _____

3. The composition of two continuous functions is always continuous.
   
   **Answer:** _____

4. If $f(x)$ is continuous on $\mathbb{R}$, then $|f(x)|$ is also continuous on $\mathbb{R}$.
   
   **Answer:** _____

5. The function $f(x) = \begin{cases} 1, & x \in \mathbb{Q}, \\ 0, & x \notin \mathbb{Q} \end{cases}$ is discontinuous everywhere.
   
   **Answer:** _____

#### Multiple choice questions
Choose the correct option (A, B, C, or D) for each question.

1. Which of the following is a sufficient condition for a function $f$ to be continuous at $\xi$?
   
   A. $f$ is defined only at $\xi$   B. $f$ is both left and right continuous at $\xi$
   
   C. $f(\xi) = 0$   D. $\lim_{x \to \xi} f(x)$ does not exist

**Answer:** _____

2. Let $f(x) = x^2$ and $g(x) = \frac{1}{x}$. Then the function $h(x) = f(g(x))$ is
   A. continuous everywhere    B. continuous only at $x = 1$
   C. discontinuous at $x = 0$    D. undefined for all $x$
   **Answer:** _____

3. Which of the following statements must be true if $f$ and $g$ are both continuous at $\xi$?
   A. $\max\{f, g\}$ is continuous at $\xi$    B. $\min\{f, g\}$ is continuous at $\xi$
   C. $f + g$ is continuous at $\xi$    D. All of the above
   **Answer:** _____

4. Consider the function $f(x) = x \sin(\frac{1}{x})$ for $x \neq 0$. Can this function be made continuous at $x = 0$?
   A. No, because $\sin(1/x)$ oscillates    B. Yes, by defining $f(0) = 1$
   C. Yes, by defining $f(0) = 0$    D. No, because the limit does not exist
   **Answer:** _____

5. Suppose $f(x)$ is continuous at $\xi$ and $F(y)$ is continuous at $f(\xi)$. Then which of the following is necessarily continuous at $\xi$?
   A. $f(x) + F(x)$    B. $F(f(x))$
   C. $\frac{F(x)}{f(x)}$          D. None of the above
   **Answer:** _____

## 5.4 Consequences and applications of continuity

This section explores key consequences and applications of continuity in real analysis. We demonstrate how the assumption of continuity yields powerful structural results: from determining the global behavior of functions defined on dense subsets, to proving fundamental theorems like the intermediate and extreme value theorems. These results provide the theoretical foundation for understanding zeros, boundedness, and fixed points of functions, all of which play a central role in both theoretical and applied mathematics.

---

**Proposition 5.6** (Zero function on $\mathbb{Q}$ implies zero on $\mathbb{R}$). *Let $f$ be a continuous function such that $f(x) = 0$ for all $x \in \mathbb{Q}$. Prove that $f(x) = 0$ for all $x \in \mathbb{R}$.*

---

*Solution.* For any $c \in \mathbb{R}$, choose a sequence of rational numbers $q_n \to c$. Since $f$ is continuous, we have

$$\lim_{n \to \infty} f(q_n) = f(c),$$

but $f(q_n) = 0$ for all $n$, so $f(c) = 0$. Therefore, $f$ is identically zero on $\mathbb{R}$.      □

**Proposition 5.7** (Additive and continuous implies linear). *Let* $f : \mathbb{R} \to \mathbb{R}$ *be a continuous function such that*

$$f(x + y) = f(x) + f(y), \quad \forall x, y \in \mathbb{R}.$$

*Show that* $f(x) = cx$ *for some constant* $c$.

*Solution.* Let $f(1) = c$. By induction, $f(n) = cn$ for $n \in \mathbb{N}$. Then

$$f(0) = f(0 + 0) = f(0) + f(0) \quad \Rightarrow \quad f(0) = 0.$$

Also, $f(-n) = -f(n) = -cn$ for $n \in \mathbb{N}$, so $f(n) = cn$ for all $n \in \mathbb{Z}$. For rationals

$$f\left(\frac{m}{n}\right) = \frac{1}{n} f(m) = c\frac{m}{n}.$$

Finally, since $f$ is continuous and rationals are dense in $\mathbb{R}$, it follows that $f(x) = cx$ for all $x \in \mathbb{R}$. $\quad\square$

**Proposition 5.8** (Exponential function from functional equation). *Let* $f : \mathbb{R} \to \mathbb{R}$ *be a continuous function such that*

$$f(x + y) = f(x)f(y) \quad \text{and} \quad f(x) > 0 \quad \text{for all } x, y \in \mathbb{R}.$$

*Show that there exists a constant* $c \in \mathbb{R}$ *such that*

$$f(x) = e^{cx}.$$

*Solution.* Define $g(x) = \ln f(x)$. Since $f(x) > 0$ for all $x$, $g(x)$ is well-defined and continuous.

Then

$$g(x + y) = \ln(f(x + y)) = \ln(f(x)f(y)) = \ln(f(x)) + \ln(f(y)) = g(x) + g(y).$$

Thus, $g$ is additive and continuous on $\mathbb{R}$. From standard results, this implies

$$g(x) = cx \quad \text{for some } c \in \mathbb{R}.$$

Hence,

$$f(x) = e^{g(x)} = e^{cx}. \quad\square$$

The following proposition underscores a crucial property of continuous functions: the preservation of their sign around points where they do not vanish, an essential aspect in analyzing local behavior and stability.

**Proposition 5.9** (Sign preservation of continuous functions). *Let* $f$ *be continuous at* $\xi$ *and suppose* $f(\xi) \neq 0$. *Show that there exists a neighborhood around* $\xi$ *in which* $f(x)$ *has the same sign as* $f(\xi)$.

*Solution.* Assume $f(\xi) > 0$. Since $f$ is continuous at $\xi$, for $\varepsilon = f(\xi) > 0$, there exists $\delta > 0$ such that for all $x$ with $|x - \xi| < \delta$, we have

$$|f(x) - f(\xi)| < \varepsilon = f(\xi) \quad \Rightarrow \quad f(x) > 0.$$

Similarly, if $f(\xi) < 0$, let $\varepsilon = -f(\xi) > 0$, then again for some $\delta > 0$, we get

$$|f(x) - f(\xi)| < \varepsilon \quad \Rightarrow \quad f(x) < 0.$$

Thus, $f$ preserves its sign in a neighborhood of $\xi$. □

The following theorem is one of the cornerstones of real analysis, guaranteeing that continuous functions defined on closed intervals are not only bounded but also achieve their extreme values.

> **Theorem 5.2** (Extreme value theorem). *A continuous function on a closed interval is bounded and, in addition, attains both its maximum and minimum value at least once.*

*Proof.* We divide the interval $[a, b]$ into $n$ equal subintervals of length $\delta$, such that for $x, y$ with $|x - y| < \delta$ we have $|f(x) - f(y)| < \varepsilon$ for a given $\varepsilon > 0$. This is possible because $f(x)$ is continuous on $[a, b]$.

Choose any $x \in [a, b]$ and a sequence of points $y_1, y_2, \ldots, y_n$ with $y_n = b$, each belonging to the subintervals we constructed. Then we have

$$-\varepsilon < f(x) - f(y_1) < \varepsilon \quad \vdots \quad -\varepsilon < f(y_{n-1}) - f(b) < \varepsilon.$$

Adding the above inequalities yields

$$-n\varepsilon + f(b) < f(x) < f(b) + n\varepsilon,$$

which shows that $f(x)$ is bounded.

Next, we will prove that there exists $x^* \in [a, b]$ such that

$$f(x^*) = \sup_{x \in [a,b]} f(x) = M.$$

The equality $\sup_{x \in [a,b]} f(x) = M$ means that $f(x)$ approaches $M$ arbitrarily closely, i.e., for every $\varepsilon > 0$, there exists $x \in [a, b]$ such that $M - \varepsilon < f(x)$.

Consider the function

$$F(x) = \frac{1}{M - f(x)}.$$

If there is no $x^* \in [a, b]$ such that $f(x^*) = M$, then $F$ is continuous on $[a, b]$ and thus bounded. However, this contradicts the fact that for every $\varepsilon > 0$ we can find $x \in [a, b]$

such that $M - \varepsilon < f(x)$, making the denominator of $F(x)$ arbitrarily small and thus $F(x)$ arbitrarily large—so $F$ is unbounded.

Therefore, we must conclude that there exists $x^* \in [a, b]$ such that

$$f(x^*) = \sup_{x \in [a,b]} f(x) = M.$$

The proof for the minimum follows similarly. □

The following theorem, commonly known as Bolzano's theorem, is central to real analysis, establishing the intuitive fact that continuous functions with opposite signs at the endpoints of an interval must cross the horizontal axis at least once.

---

**Theorem 5.3** (Intermediate value theorem (Bolzano's theorem)). *If $f$ is continuous on the interval $[a, b]$ and $f(a)f(b) < 0$, then there exists some $x \in [a, b]$ such that $f(x) = 0$.*

---

*Proof.* Since $f$ is continuous on $[a, b]$, we can divide the interval into $n$ equal subintervals of length $\delta$, such that for any two points $x, y$ in the same subinterval, we have $|f(x) - f(y)| < \varepsilon$.

If $f$ does not vanish at any of the endpoints of these subintervals, then in at least one subinterval, the sign of $f$ must change. Let $x, y$ be two points in such a subinterval where $f(x) > 0$ and $f(y) < 0$. Then

$$|f(x)| < |f(x) - f(y)| < \varepsilon.$$

Assume for contradiction that $\frac{1}{f(x)}$ is continuous and therefore bounded. However, we have shown that for any $\varepsilon > 0$, there exists $x \in [a, b]$ such that $\frac{1}{f(x)} > \frac{1}{\varepsilon}$, which contradicts the assumption of boundedness and thus continuity.

The only reason $\frac{1}{f(x)}$ would not be continuous is if $f(x) = 0$ for some $x \in [a, b]$. Therefore, $f(x) = 0$ for some $x$ in the interval, as claimed. □

The following theorem is a powerful generalization of Bolzano's theorem, affirming that continuous functions achieve every intermediate value between any two points, thus providing a fundamental tool for analyzing their behavior.

---

**Theorem 5.4** (Generalized intermediate value theorem). *Let $I \subseteq \mathbb{R}$ be an interval and let*

$$f : I \longrightarrow \mathbb{R}$$

*be continuous. If $a, b \in I$ and*

$$f(a) \leq L \leq f(b) \quad or \quad f(b) \leq L \leq f(a),$$

*then there exists $c$ between $a$ and $b$ (i. e., $c \in [\min\{a, b\}, \max\{a, b\}]$) such that*

$$f(c) = L.$$

---

*Proof.* Set up the auxiliary function

$$g(x) = f(x) - L.$$

Since $f$ is continuous on $I$, so is $g$. Let

$$\alpha = \min\{a, b\}, \quad \beta = \max\{a, b\},$$

so $[\alpha, \beta] \subseteq I$. By hypothesis either

$$f(a) \leq L \leq f(b) \quad \Longrightarrow \quad g(\alpha) = f(\alpha) - L \leq 0 \leq f(\beta) - L = g(\beta),$$

or the same inequalities with $a, b$ interchanged. In either case

$$g(\alpha) \cdot g(\beta) \leq 0.$$

By the (classical) intermediate value Theorem 5.3, there is some

$$c \in [\alpha, \beta]$$

with

$$g(c) = 0.$$

Hence

$$f(c) = g(c) + L = L,$$

and $c$ lies between $a$ and $b$, as required. $\quad\square$

The nested intervals property, stated below, is fundamental in real analysis, capturing the completeness of the real numbers by ensuring the existence and uniqueness of a limit point for a shrinking sequence of nested intervals.

**Theorem 5.5** (Nested intervals property). *Let*

$$[a_0, b_0] \supset [a_1, b_1] \supset [a_2, b_2] \supset \cdots$$

*be a sequence of closed intervals in $\mathbb{R}$ such that*

$$b_n - a_n \longrightarrow 0 \quad (n \to \infty).$$

*Then there exists a unique point $x^* \in \bigcap_{n=0}^{\infty} [a_n, b_n]$.*

*Proof.* Since each interval $[a_n, b_n]$ is nonempty and

$$a_0 \leq a_1 \leq a_2 \leq \cdots \quad \text{and} \quad b_0 \geq b_1 \geq b_2 \geq \cdots,$$

the sequences $(a_n)$ and $(b_n)$ are monotone bounded, hence converge

$$\lim_{n \to \infty} a_n = \alpha, \quad \lim_{n \to \infty} b_n = \beta.$$

Because $a_n \le b_n$ for all $n$, we have $\alpha \le \beta$. But

$$b_n - a_n \longrightarrow 0 \quad \Longrightarrow \quad \beta - \alpha = 0,$$

so $\alpha = \beta =: x^*$. Thus

$$x^* = \lim_{n \to \infty} a_n \quad \text{and} \quad x^* = \lim_{n \to \infty} b_n.$$

Since each $[a_n, b_n]$ is closed, taking limits shows

$$x^* \in [a_n, b_n] \quad \text{for every } n,$$

so $x^* \in \bigcap_{n=0}^{\infty} [a_n, b_n]$. Uniqueness follows because if $y$ also lay in all $[a_n, b_n]$, then $|y - x^*| \le b_n - a_n \to 0$, forcing $y = x^*$. □

The following theorem rigorously establishes the convergence properties of the bisection method, a classical and reliable numerical technique used to approximate solutions of equations, and provides an explicit error bound for its approximations.

---

**Theorem 5.6** (Bisection method). *Let $f \in C([a, b])$ and assume $f(a) \cdot f(b) \le 0$. Let $(x_n)$ be the sequence of approximations generated by the bisection method. Then:*
1. *Either there exists $N \in \mathbb{N}$ such that $x_N = x^*$,*
2. *Or $x_n \to x^*$ as $n \to \infty$.*

*Moreover, for all $n \in \mathbb{N}$, the following error estimate holds:*

$$|x^* - x_n| \le \frac{b - a}{2^n}.$$

---

*Proof.* Since $f(a)f(b) \le 0$ and $f$ is continuous on $[a, b]$, the intermediate value theorem (Bolzano's theorem) guarantees at least one root $x^* \in [a, b]$ with $f(x^*) = 0$.

The bisection procedure constructs a nested sequence of closed intervals

$$[a_0, b_0] \supset [a_1, b_1] \supset [a_2, b_2] \supset \cdots,$$

where $[a_0, b_0] = [a, b]$ and for each $n \ge 0$:
1. Compute the midpoint $x_{n+1} = \frac{a_n + b_n}{2}$.
2. If $f(a_n)f(x_{n+1}) \le 0$, set $[a_{n+1}, b_{n+1}] = [a_n, x_{n+1}]$; otherwise, set $[a_{n+1}, b_{n+1}] = [x_{n+1}, b_n]$.

In either case, the continuity of $f$ ensures $f$ changes sign over $[a_{n+1}, b_{n+1}]$, so $x^* \in [a_{n+1}, b_{n+1}]$.

Since the length of the interval halves at each step,

$$b_n - a_n = \frac{b_0 - a_0}{2^n},$$

and $x_n \in [a_{n-1}, b_{n-1}]$, it follows that

$$|x^* - x_n| \leq \max\{|x^* - a_{n-1}|, |x^* - b_{n-1}|\} \leq \frac{b_{n-1} - a_{n-1}}{2} = \frac{b_0 - a_0}{2^n}.$$

Hence the error bound $|x^* - x_n| \leq (b - a)/2^n$ holds.

Moreover, the nested intervals $[a_n, b_n]$ satisfy $b_n - a_n \to 0$. By Theorem 5.5, there is a unique point in their intersection. Since each interval contains a root, this unique point must be $x^*$. Thus either for some $N$, $x_N = x^*$, or else $x_n \to x^*$ as $n \to \infty$, completing the proof. □

The following theorem is a key result in analysis known as the fixed point theorem, which ensures the existence of at least one point where a continuous function intersects the line $y = x$, a property extensively applied across various mathematical disciplines.

> **Theorem 5.7** (Fixed point theorem). *Let $f : [a, b] \to I \subseteq [a, b]$ be a continuous function. Then there exists a point $x \in [a, b]$ such that $f(x) = x$.*

*Proof.* If $f(a) = a$ or $f(b) = b$, then the result holds trivially. Otherwise, suppose (without loss of generality) that $f(a) > a$ and $f(b) < b$. Define the continuous function $h(x) = f(x) - x$. We observe that $h(a) > 0$ and $h(b) < 0$, so $h(a)h(b) < 0$. By the intermediate value theorem, there exists some $c \in [a, b]$ such that $h(c) = 0$, i. e., $f(c) = c$. Hence, $c$ is a fixed point of $f$. □

We will need the following notion of continuity later on.

> **Definition 5.4** (Uniformly Continuous Function). *Let $f : A \to \mathbb{R}$ be a function defined on a subset $A \subseteq \mathbb{R}$. We say that $f$ is uniformly continuous on $A$ if and only if*
> $$\forall \varepsilon > 0, \ \exists \delta > 0 \text{ such that for all } x, y \in A, \text{ if } |x - y| < \delta, \text{ then } |f(x) - f(y)| < \varepsilon.$$

## 5.5 Python and AI

---

**AI Request**

Write a Python code that applies the bisection method to a given function.

---

```
1  import numpy as np
2  import matplotlib.pyplot as plt
3  from sympy
4  import symbols, lambdify
5
6  # --- 1. Input from user --- x=symbols('x') function_str =
```

```
7  input("Enter the function f(x): ") a= float(input("Enter the left
8  endpoint a: ")) b=float(input("Enter the right endpoint b: "))
9  tolerance=float(input("Enter the tolerance (e.g., 0.001): "))
10 max_iterations = int(input("Enter the maximum number of
11 iterations: "))
12
13 # --- 2. Convert string to function ---
14 f = lambdify(x,
15 function_str, 'numpy')
16
17 # --- 3. Check Bolzano condition ---
18 if f(a) * f(b) >= 0:
19     print("Bolzano's condition is not satisfied (f(a)*f(b) < 0
       required).")
20     exit()
21
22 # --- 4. Bisection method with history tracking ---
23
24 c_values = []
25 iteration = 0
26 a_n, b_n = a, b
27 print("\n--- Bisection
28 Method Steps ---")
29
30 while (b_n - a_n) / 2 > tolerance and iteration < max_iterations:
31     c_n = (a_n + b_n) / 2
32     c_values.append(c_n)
33     print(f"Step {iteration + 1}: c_{iteration + 1} = {c_n:.6f}, f(
       c_{iteration + 1}) = {f(c_n):.6f}")
34
35     if f(c_n) == 0:
36         break
37
38     if f(a_n) * f(c_n) < 0:
39         b_n = c_n
40     else:
41         a_n = c_n
42
43     iteration += 1
44
45 # --- 5. Final result ---
46
47 if len(c_values) > 0:
48     root_estimate = c_values[-1]
49     print(f"\nApproximate root found: {root_estimate:.6f}")
50 else:
51     print("No root found within the given number of iterations.")
52
53 # --- 6. Plotting convergence --- plt.figure(figsize=(10, 5))
```

```
54 plt.plot(range(1, len(c_values) + 1),c_values, 'bo-',
55 label='Approximations $c_n$')
56 plt.axhline(y=root_estimate,
57 color='r', linestyle='--', label='Final Estimate')
58 plt.xlabel('Iteration n') plt.ylabel('Approximated value $c_n$')
59 plt.title('Convergence of the Bisection Method') plt.grid(True)
60 plt.legend()
61 plt.tight_layout()
62 plt.show()
```

### 5.5.1 Exercises

#### True or false

Determine whether each statement is **true** or **false**. Justify your answer.

1. If $f : \mathbb{R} \to \mathbb{R}$ is continuous and $f(x) = 0$ for all $x \in \mathbb{Q}$, then $f(x) = 0$ for all $x \in \mathbb{R}$.
   **Answer:** _____

2. If $f(x + y) = f(x) + f(y)$ for all $x, y \in \mathbb{R}$ and $f$ is continuous, then $f(x) = cx$ for some constant $c \in \mathbb{R}$.
   **Answer:** _____

3. If $f$ is continuous at $\xi$ and $f(\xi) > 0$, then there exists a neighborhood around $\xi$ where $f(x) > 0$.
   **Answer:** _____

4. Every continuous function on a closed interval $[a, b]$ attains both a maximum and a minimum value.
   **Answer:** _____

5. If $f : [a, b] \to [a, b]$ is continuous, then it has at least one fixed point.
   **Answer:** _____

#### Multiple choice questions

Choose the correct option (A, B, C, or D) for each question.

1. Suppose $f : \mathbb{R} \to \mathbb{R}$ is continuous and satisfies $f(x + y) = f(x)f(y)$ for all $x, y \in \mathbb{R}$, with $f(x) > 0$. Which of the following must be true?
   A. $f(x) = e^{cx}$   B. $f(x) = cx$
   C. $f(x) = 0$   D. $f(x) = 1$
   **Answer:** _____

2. Let $f$ be continuous on $[a, b]$ and suppose $f(a) < 0 < f(b)$. Then, by the intermediate value theorem,
   A. there exists $c \in [a, b]$ such that $f(c) = 0$   B. $f$ is strictly increasing on $[a, b]$
   C. $f$ is unbounded on $[a, b]$   D. $f$ has no fixed point
   **Answer:** _____

3. If $f(x) = x \sin(\frac{1}{x})$ for $x \neq 0$, can $f$ be made continuous at $x = 0$?
   A. Yes, by defining $f(0) = 0$   B. Yes, by defining $f(0) = 1$
   C. No, because $\sin(1/x)$ oscillates   D. No, because the limit does not exist
   **Answer:** _____

4. Let $f : [0,1] \to [0,1]$ be continuous. Which of the following must be true?
   A. $f$ is differentiable    B. $f$ is injective
   C. $f$ has a fixed point    D. $f$ is constant
   **Answer:** _____

5. Suppose $f$ is continuous on $[a, b]$. Which of the following is NOT guaranteed?
   A. $f$ is bounded on $[a, b]$        B. $f$ attains its maximum on $[a, b]$
   C. $f$ is differentiable on $[a, b]$    D. $f$ attains its minimum on $[a, b]$
   **Answer:** _____

## 5.6 Examples

**Example 5.4** (Continuity of a piecewise function). Examine the continuity of the function

$$f(x) = \begin{cases} x, & x > 0, \\ x^2, & x \le 0. \end{cases}$$

*Solution.* For $x > 0$, the function is linear and continuous. For $x \le 0$, the function is a polynomial and hence continuous. We check continuity at $x = 0$:

$$\lim_{x \to 0^-} f(x) = 0, \quad \lim_{x \to 0^+} f(x) = 0, \quad f(0) = 0.$$

Thus, $f$ is continuous at 0 and therefore continuous on $\mathbb{R}$. □

**Example 5.5** (Discontinuity of a piecewise constant function). Examine the continuity of the function

$$f(x) = \begin{cases} 1, & x \ge 0, \\ -1, & x < 0. \end{cases}$$

*Solution.* For $x \ne 0$, the function is constant on each interval, hence continuous. At $x = 0$

$$\lim_{x \to 0^-} f(x) = -1, \quad \lim_{x \to 0^+} f(x) = 1, \quad f(0) = 1.$$

The left and right limits differ, so $f$ is discontinuous at $x = 0$. □

**Example 5.6** (Removable discontinuity). Examine the continuity of the function

$$f(x) = \begin{cases} \frac{x^2-4}{x+2}, & x \ne -2, \\ x^2, & x = -2. \end{cases}$$

*Solution.* For $x \ne -2$,

$$f(x) = \frac{(x - 2)(x + 2)}{x + 2} = x - 2.$$

So, for $x \neq -2, f(x) = x - 2$, which is continuous. At $x = -2$

$$\lim_{x \to -2} f(x) = -4, \quad f(-2) = (-2)^2 = 4.$$

Hence, there is a removable discontinuity at $x = -2$. □

**Example 5.7** (Discontinuity of the floor function). Examine the continuity of the floor function
$$f(x) = [x].$$

*Solution.* The floor function is discontinuous at all integer values of $x$, since

$$\lim_{x \to n^-} f(x) = n - 1, \quad \lim_{x \to n^+} f(x) = n, \quad f(n) = n.$$

The left and right limits differ, so $f$ is discontinuous at all integers. □

**Example 5.8** (Continuity with parameter). Find the value of $k \in \mathbb{R}$ such that the function is continuous at $x = 2$

$$f(x) = \begin{cases} \frac{\sqrt{7x+2}-\sqrt{6x+4}}{x-2}, & x \geq -\frac{2}{7}, x \neq 2, \\ k, & x = 2. \end{cases}$$

*Solution.* We compute the limit as $x \to 2$:

$$\lim_{x \to 2} \frac{\sqrt{7x + 2} - \sqrt{6x + 4}}{x - 2} = \lim_{x \to 2} \frac{\sqrt{16} - \sqrt{16}}{x - 2} = \frac{0}{0} \quad \text{(indeterminate).}$$

Multiply numerator and denominator by the conjugate:

$$\frac{\sqrt{7x + 2} - \sqrt{6x + 4}}{x - 2} \cdot \frac{\sqrt{7x + 2} + \sqrt{6x + 4}}{\sqrt{7x + 2} + \sqrt{6x + 4}} = \frac{7x + 2 - (6x + 4)}{(x - 2)(\sqrt{7x + 2} + \sqrt{6x + 4})}$$

$$= \frac{x - 2}{(x - 2)(\sqrt{7x + 2} + \sqrt{6x + 4})}.$$

Cancel $x - 2$:

$$\frac{1}{\sqrt{7x + 2} + \sqrt{6x + 4}} \quad \Rightarrow \quad \lim_{x \to 2} \frac{1}{\sqrt{16} + \sqrt{16}} = \frac{1}{4 + 4} = \frac{1}{8}.$$

To ensure continuity at $x = 2$, set $k = \frac{1}{8}$. □

**Example 5.9** (Fixed point of $\phi(x) = 2^{-x}$). Prove that the function $\phi(x) = 2^{-x}$ has a fixed point in the interval $[0, 1]$.

*Solution.* We have $\phi'(x) = -\ln 2 \cdot 2^{-x} < 0$ for all $x \in \mathbb{R}$, which implies that $\phi$ is strictly decreasing.

Also, we compute the endpoints

$$\phi(0) = 1, \quad \phi(1) = \frac{1}{2}.$$

Thus,

$$\phi([0,1]) = \left[\frac{1}{2}, 1\right] \subseteq [0,1].$$

This means that $\phi : [0,1] \to [0,1]$ is a continuous mapping from a closed interval into itself.

By Theorem 5.7, $\phi$ has at least one fixed point in $[0,1]$; i. e., there exists $x \in [0,1]$ such that $\phi(x) = x$. $\qquad\qquad$ $\square$

## 5.7 Review exercises

### Fill in the blanks
Complete each sentence with the most appropriate word or phrase.
1. A function $f$ is said to be _____ at a point $\xi$ if for every $\varepsilon > 0$, there exists a $\delta > 0$ such that $|x - \xi| < \delta$ implies $|f(x) - f(\xi)| < \varepsilon$.
   Answer: _____
2. If $f(x+y) = f(x)+f(y)$ for all $x, y \in \mathbb{R}$ and $f$ is continuous, then $f(x) =$ _____.
   Answer: _____
3. The _____ value theorem states that a continuous function on a closed interval must attain both a maximum and a minimum value.
   Answer: _____
4. If $f(a) \cdot f(b) < 0$ and $f$ is continuous on $[a, b]$, then by the _____ value theorem, there exists some $c \in [a, b]$ such that $f(c) = 0$.
   Answer: _____
5. A function $f : [a, b] \to [a, b]$ that is continuous has at least one _____.
   Answer: _____

### Matching
Match each term on the left with its correct description on the right.

1. Continuity at a point
2. Intermediate value theorem
3. Extreme value theorem
4. Fixed point theorem
5. Additive functional equation
6. Multiplicative functional equation

A. $f(x + y) = f(x)f(y) \Rightarrow f(x) = e^{cx}$
B. Ensures existence of a fixed point
C. $\lim_{x \to \xi} f(x) = f(\xi)$
D. Guarantees a zero in an interval
E. Function attains max and min on $[a, b]$
F. $f(x + y) = f(x) + f(y) \Rightarrow f(x) = cx$

Answers: 1. _____ 2. _____ 3. _____ 4. _____ 5. _____ 6. _____

## Theoretical and computational problems

Solve each problem step-by-step.

1. Prove that the function

$$f(x) = \begin{cases} x^2, & x < 1, \\ 2x - 1, & x > 1 \end{cases}$$

   is continuous at all points except $x = 1$. Is the discontinuity at $x = 1$ removable?

2. Let $f(x) = x \cdot \sin\left(\frac{1}{x}\right)$ for $x \neq 0$. Define $f(0)$ so that $f$ becomes continuous at $x = 0$.

3. Use the definition of continuity to prove that the function $f(x) = 3x + 2$ is continuous at every real number.

4. Show that if $f$ is continuous at $x = a$, then $|f|$ is also continuous at $x = a$.

5. Give an example of a function that is:
   - Continuous nowhere.
   - Continuous only at one point.
   - Continuous everywhere except at a single point.

## Python-based exercises

Use Python libraries like `sympy`, `numpy`, and `matplotlib` to solve these problems.

1. Use `sympy` to compute

$$\lim_{x \to 3} \frac{x^2 - 9}{x - 3}.$$

   Plot the function near $x = 3$ and explain what happens at that point.

2. Plot the function

$$f(x) = \frac{\sin x}{x}$$

   for values close to zero and estimate $\lim_{x \to 0} f(x)$ numerically.

3. Consider the piecewise function

$$f(x) = \begin{cases} x^2, & x < 2, \\ 2x, & x > 2. \end{cases}$$

   Compute the left and right limits at $x = 2$ and determine whether the function can be made continuous there.

4. Use `matplotlib` to visualize the epsilon-delta definition of continuity for $f(x) = x^2$ at $x = 1$.

5. Write a Python function to check if a given function is continuous at a point using symbolic computation. Test it on

$$f(x) = \frac{x^3 - 8}{x - 2} \quad \text{at } x = 2.$$

6. Plot the function

$$f(x) = \sin\left(\frac{1}{x}\right)$$

near $x = 0$ and explain why the limit does not exist as $x \to 0$.

# 6 Differentiation

## Contents

The concept of differentiation, central to calculus, emerged in the late 17th century through the independent work of **Isaac Newton** and **Gottfried Wilhelm Leibniz**. While Newton developed fluxions to describe motion, Leibniz introduced the notation $dy/dx$ that we still use today. Their work built on earlier ideas from Fermat, Descartes, and others on tangents and rates of change. Although initially lacking rigor, the derivative was later formalized using limits by Cauchy and Weierstrass in the 19th century. This chapter begins with the precise definition of the derivative and explores its fundamental properties, blending historical insight with mathematical precision.

## 6.1 Introduction

The development of calculus, and in particular the concept of differentiation, stands as one of the most profound achievements in the history of mathematics and science. The derivative—a measure of how a function changes as its input changes—emerged from the need to solve fundamental problems in geometry, physics, and astronomy: finding tangents to curves, computing velocities and accelerations, and determining maxima and minima.

The foundations of differentiation were laid independently in the late 17th century by two towering figures: **Sir Isaac Newton** (1643–1727) and **Gottfried Wilhelm Leibniz** (1646–1716). Newton, working in England, developed his method of "fluxions," where he viewed functions as evolving quantities (fluents) and their derivatives as rates of change (fluxions). His work was primarily motivated by problems in mechanics and celestial motion. Meanwhile, in continental Europe, Leibniz introduced a more formal and systematic notation—including the iconic symbols $\frac{dy}{dx}$ and $d$—that emphasized infinitesimal differences and greatly facilitated computation. Although a bitter priority dispute marred their legacies, modern scholarship acknowledges both as co-founders of calculus.

Earlier mathematicians had already approached the idea of tangents and instantaneous change. The ancient Greeks, particularly **Archimedes**, used methods resembling limits to compute areas and tangents. In the 17th century, **Pierre de Fermat** devised a technique for finding maxima and minima and tangents to curves by considering in-

https://doi.org/10.1515/9783112228289-006

finitesimal increments—a method strikingly close to differentiation. **Rene Descartes** also contributed geometric methods for finding tangents, while **John Wallis** and **Isaac Barrow** (Newton's teacher) further developed analytical approaches.

Despite its power, early calculus lacked rigorous foundations. The use of "infinitesimals"—quantities smaller than any positive number yet not zero—was criticized, most famously by Bishop **George Berkeley**, who mocked them as "ghosts of departed quantities." It was not until the 19th century that mathematicians such as **Augustin-Louis Cauchy**, **Bernard Bolzano**, and **Karl Weierstrass** formalized the concept of the limit, thereby placing differentiation on a solid logical foundation. Cauchy, in particular, defined the derivative using limits in his lectures at the Ecole Polytechnique, closely resembling the definition we use today:

$$f'(a) = \lim_{h \to 0} \frac{f(a+h) - f(a)}{h}.$$

This chapter introduces the theory of differentiation starting from this rigorous limit-based definition. We explore its relationship with continuity, examine key examples—both differentiable and non-differentiable—and derive fundamental properties of derivatives. Through historical lens and mathematical precision, we aim to convey not only the *how* but also the *why* behind this central tool of mathematical analysis.

The journey from geometric intuition to formal definition reflects the evolution of mathematical thought itself—a journey this chapter invites you to continue.

## 6.2 Definition and a consequence on continuity

**Definition 6.1** (Differentiability, Figure 6.1). A function $f$ is said to be **differentiable at a point** $a$ if the following limit exists:

$$\lim_{h \to 0} \frac{f(a+h) - f(a)}{h}.$$

This limit, denoted by $f'(a)$, is called the **derivative of** $f$ at $a$. If $f$ is differentiable at every point in a set $I \subseteq \mathbb{R}$, we say that $f$ is **differentiable on** $I$.

**Remark 8.** Having in mind Theorem 4.1, the function $f$ is differentiable at some point $a$ if the left and right limits exist and are equal

$$\lim_{h \to 0^{\pm}} \frac{f(a+h) - f(a)}{h}.$$

The theorem presented below establishes a fundamental link between differentiability and continuity, emphasizing that differentiability at a point inherently ensures continuity there, although continuity alone is not sufficient for differentiability.

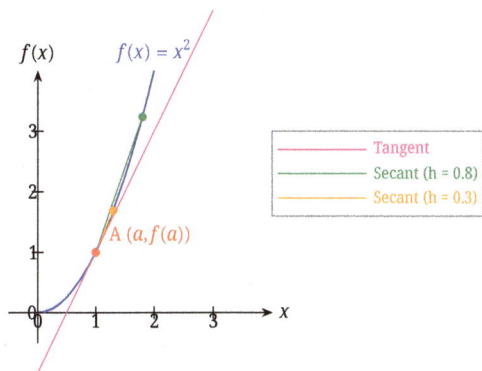

**Figure 6.1:** As $h \to 0$, the secant line approaches the tangent line at point A. This defines the derivative geometrically: $f'(a) = \lim_{h \to 0} \frac{f(a+h)-f(a)}{h}$.

**Theorem 6.1** (Differentiability implies continuity). *If $f$ is differentiable at $a$, then $f$ is continuous at $a$. The converse does not generally hold.*

*Proof.* We want to prove that

$$\lim_{x \to a}(f(x) - f(a)) = 0$$

or equivalently,

$$\lim_{h \to 0}(f(a + h) - f(a)) = 0.$$

Multiplying and dividing by $h$,

$$\lim_{h \to 0} \frac{f(a + h) - f(a)}{h} \cdot h = f'(a) \cdot \lim_{h \to 0} h = 0.$$

Thus, $f$ is continuous at $a$. For the converse, consider the function $f(x) = |x|$. This function is continuous everywhere on $\mathbb{R}$ but not differentiable at 0. $\square$

**Remark 9.** Examining the definition of the derivative, we observe that when a function $f$ is differentiable at a point $a$, there exists a function $u(h)$ such that

$$u(h) \to 0 \quad \text{as } h \to 0,$$

and

$$f(a + h) - f(a) = h \cdot (f'(a) + u(h)).$$

## 6.3 Python and AI

**AI Request**

I want a Python code that checks if a function is differentiable at a point. I want to check the left and right limits and see the graph of the function.

**AI Request**

Provide any historical information you know about the concept of the derivative of a function.

## 6.4 Examples

**Example 6.1** (Non-differentiability of the absolute value at zero). The function $f(x) = |x|$ is **not differentiable at** 0 because the limit

$$\lim_{h \to 0} \frac{|h|}{h}$$

depends on the sign of $h$. Specifically, as $h \to 0^+$, the limit equals 1, while as $h \to 0^-$, the limit equals $-1$. Thus, the left and right derivatives at 0 are not equal.

**Example 6.2** (Differentiability of $x^2 \sin \frac{1}{x}$ at zero). Consider the function

$$f(x) = \begin{cases} x^2 \sin \frac{1}{x}, & \text{if } x \neq 0, \\ 0, & \text{if } x = 0. \end{cases}$$

This function **is differentiable at** 0 since

$$f'(0) = \lim_{x \to 0} \frac{x^2 \sin \frac{1}{x} - 0}{x} = \lim_{x \to 0} x \sin \frac{1}{x} = 0.$$

**Example 6.3** (Functional equation: differentiability and characterization). Let $f$ be a function such that $f(x + h) = f(x)f(h)$ and $f(0) = f'(0) = 1$. We will prove that $f'(x) = f(x)$ for all $x \in \mathbb{R}$. Indeed,

$$f(x + h) - f(x) = f(x)f(h) - f(x) = f(x)\big(f(h) - 1\big) = f(x)\big(f(h) - f(0)\big).$$

Thus,

$$f'(x) = \lim_{h \to 0} \frac{f(x + h) - f(x)}{h} = f(x) \lim_{h \to 0} \frac{f(h) - f(0)}{h} = f(x)f'(0) = f(x).$$

**Example 6.4** (Parity and differentiability). If $f$ is **even** and differentiable, then $f'$ is **odd**. Similarly, if $f$ is **odd** and differentiable, then $f'$ is **even**. Indeed, suppose $f$ is even and differentiable. Then

$$f'(-x) = \lim_{h \to 0} \frac{f(-x + h) - f(-x)}{h}$$

$$= \lim_{h \to 0} \frac{f(x-h) - f(x)}{h}$$

$$= -\lim_{k \to 0} \frac{f(x+k) - f(x)}{k} \qquad (k = -h)$$

$$= -f'(x).$$

Similarly, if $f$ is odd and differentiable,

$$f'(-x) = \lim_{h \to 0} \frac{f(-x+h) - f(-x)}{h}$$

$$= -\lim_{h \to 0} \frac{f(x-h) - f(x)}{h}$$

$$= \lim_{k \to 0} \frac{f(x+k) - f(x)}{k} \qquad (k = -h)$$

$$= f'(x). \qquad \qquad \square$$

**Example 6.5** (Derivative of the cubic root function). Compute the derivative of $f(x) = \sqrt[3]{x}$ using the definition after first proving that it is a continuous function for all $x_0 \geq 0$.

*Solution.* To prove continuity at $x_0 > 0$, let $\varepsilon > 0$. We need to find $\delta > 0$ such that

$$\left| \sqrt[3]{x} - \sqrt[3]{x_0} \right| \leq \varepsilon \quad \text{whenever } |x - x_0| < \delta.$$

But

$$\left| \sqrt[3]{x} - \sqrt[3]{x_0} \right| = \frac{|x - x_0|}{x^{2/3} + x^{1/3}x_0^{1/3} + x_0^{2/3}} \leq \frac{|x - x_0|}{x_0^{2/3}}$$

if $x$ is close to $x_0$ and positive. So, choose $\delta$ much smaller than $\varepsilon x_0^{2/3}$ so that $x$ remains positive.

To show right continuity at $x_0 = 0$, for every $\varepsilon > 0$, choose $\delta \leq \varepsilon^3$. Then, for $0 \leq x < \delta$,

$$\sqrt[3]{x} < \varepsilon,$$

as required.

To prove differentiability, compute the difference:

$$f(x+h) - f(x) = \sqrt[3]{x+h} - \sqrt[3]{x}$$

$$= \frac{(\sqrt[3]{x+h} - \sqrt[3]{x})((x+h)^{2/3} + x^{1/3}(x+h)^{1/3} + x^{2/3})}{(x+h)^{2/3} + x^{1/3}(x+h)^{1/3} + x^{2/3}}$$

$$= \frac{(x+h) - x}{(x+h)^{2/3} + x^{1/3}(x+h)^{1/3} + x^{2/3}}.$$

Thus,

$$\frac{f(x+h) - f(x)}{h} = \frac{1}{(x+h)^{2/3} + x^{1/3}(x+h)^{1/3} + x^{2/3}}.$$

Taking the limit as $h \to 0$ gives

$$\lim_{h \to 0} \frac{f(x+h) - f(x)}{h} = \frac{1}{3}x^{-2/3}. \qquad \square$$

**Example 6.6** (Using the definition of derivative). Use the definition of derivative to find $f'(x)$ if $f(x) = x^2 + 3x$.

*Solution.* Recall that

$$f'(x) = \lim_{h \to 0} \frac{f(x+h) - f(x)}{h}.$$

We compute

$$f(x+h) = (x+h)^2 + 3(x+h)$$
$$= x^2 + 2xh + h^2 + 3x + 3h$$
$$f(x+h) - f(x) = (x^2 + 2xh + h^2 + 3x + 3h) - (x^2 + 3x)$$
$$= 2xh + h^2 + 3h$$
$$\Rightarrow \quad \frac{f(x+h) - f(x)}{h} = \frac{2xh + h^2 + 3h}{h} = 2x + h + 3.$$

Taking the limit as $h \to 0$,

$$f'(x) = \lim_{h \to 0} (2x + h + 3) = 2x + 3. \qquad \square$$

## 6.5 Exercises

### True or false
Determine whether each statement is **true** or **false**. Justify your answer.
1.  If a function is differentiable at a point, then it must be continuous there.
    **Answer:** _____
2.  The function $f(x) = |x|$ is differentiable at $x = 0$.
    **Answer:** _____
3.  If a function has left-hand and right-hand limits at a point that are not equal, then it is not differentiable there.
    **Answer:** _____
4.  The derivative of a function at a point gives the slope of the tangent line at that point.
    **Answer:** _____
5.  If a function is odd and differentiable, then its derivative is also odd.
    **Answer:** _____

## Multiple choice questions

Choose the correct option (A, B, C, or D) for each question.

1. Which function is differentiable at $x = 0$?
   A. $f(x) = |x|$   B. $f(x) = \sqrt[3]{x}$
   C. $f(x) = \frac{1}{x}$   D. $f(x) = \frac{|x|}{x}$
   **Answer:** _____

2. If $f(x) = x^2 \sin(\frac{1}{x})$ for $x \neq 0$ and $f(0) = 0$, then
   A. $f$ is not continuous at 0   B. $f$ is differentiable at 0
   C. $f'(0) = 1$                  D. $f$ is not differentiable at 0
   **Answer:** _____

3. If $f(x) = \sqrt[3]{x}$, then $f'(x) =$
   A. $\frac{1}{3}x^{2/3}$   B. $\frac{3}{x^{2/3}}$
   C. $\frac{1}{3}x^{-2/3}$   D. $3x^{-2/3}$
   **Answer:** _____

4. Which of the following is necessarily true for a differentiable function $f$?
   A. $f$ is continuous everywhere                          B. $f$ has a limit at all points
   C. $f$ has a tangent line at each point in its domain   D. $f'$ is always continuous
   **Answer:** _____

5. If $f(x + h) = f(x)f(h), f(0) = 1$, and $f'(0) = 1$, then
   A. $f'(x) = 1$   B. $f'(x) = f(x)$
   C. $f'(x) = x$   D. $f'(x) = f(h)$
   **Answer:** _____

## Matching exercise

Match each concept with its appropriate description.

| | |
|---|---|
| 1. Differentiability at a point | A. Function with a corner or cusp |
| 2. Non-differentiability of $|x|$ | B. Derivative exists at that point |
| 3. $f(x) = x^2 \sin \frac{1}{x}$ at 0 | C. Continuous and differentiable at 0 |
| 4. Derivative of odd function | D. Is even |
| 5. Derivative of even function | E. Is odd |

**Answers: 1.** _____   **2.** _____   **3.** _____   **4.** _____   **5.** _____

## Fill in the blanks

Complete each sentence with the most appropriate word or phrase.

1. A function is differentiable at a point if a certain _____ exists.
   **Answer:** _____

2. Differentiability implies _____ but not vice versa.
   **Answer:** _____

3. The function $f(x) = |x|$ is not differentiable at _____.
   **Answer:** _____

4.  If $f$ is even and differentiable, then $f'$ is _____.

    **Answer:** _____

5.  The derivative of $\sqrt[3]{x}$ is _____.

    **Answer:** _____

## Theoretical and computational problems

Solve each problem step-by-step.

1.  Use the definition of the derivative to show that if $f(x) = \sqrt{x}$, then $f'(x) = \frac{1}{2\sqrt{x}}$.
2.  Use the limit definition to compute the derivative of $f(x) = x^3 + 3x$.
3.  Show that if $f(x) = x^3 \sin\frac{1}{x}$, $f(0) = 0$, then $f$ is differentiable at $x = 0$.

## Python-based exercises

Use Python (`sympy`, `numpy`, `matplotlib`) to solve the following:

1.  Plot $f(x) = |x|$ and numerically estimate $f'(0)$. Explain the discrepancy with theory.
2.  Use `sympy` to differentiate $f(x) = \sqrt{x}$ and compare with manual computation.
3.  Use symbolic computation to verify that if $f(x + h) = f(x)f(h)$ with $f(0) = 1$, then $f'(x) = f(x)$.
4.  Plot $f(x) = x^2 \sin(1/x)$ near 0. Use numerical methods to explore whether it is differentiable at $x = 0$.
5.  Write a Python function that checks whether a given function is even, odd, or neither and test it on $f(x) = x^3$, $f(x) = \cos x$, and $f(x) = |x|$.

# 7 Differentiation: theorems and techniques

**Contents**

Differentiation is a central concept in calculus, providing a precise way to describe how functions change. The derivative measures the instantaneous rate of change and appears throughout mathematics, science, and engineering.

In this chapter, we review the fundamental theorems and techniques of differentiation, from the basic rules to more advanced results like the chain rule and the product rule. Each theorem is accompanied by a proof to deepen understanding.

Illustrative examples and exercises are included to help you apply the rules, develop intuition, and build computational skills. By mastering these techniques, you will be prepared to tackle more complex problems in calculus and beyond.

## 7.1 Introduction

Differentiation stands at the very heart of calculus, serving as a fundamental tool for understanding change, motion, and rates in mathematics and its applications. The concept of derivative, introduced to capture the notion of instantaneous rate of change, has evolved into a powerful framework that underpins much of modern science and engineering.

In this chapter, we embark on a systematic exploration of the core theorems and techniques of differentiation. We begin with the most basic results, i. e., computing derivatives of elementary functions and establishing the foundational rules such as linearity, the product and quotient rules, and the celebrated chain rule. The proofs, which are grounded in the definition of the derivative, illuminate the underlying logic and offer insight into why these results hold.

Beyond the fundamental rules, we demonstrate their application through a range of illustrative examples, both theoretical and computational. These examples highlight not only how to compute derivatives efficiently, but also how differentiation reveals subtle properties of functions—such as smoothness, local linearity, and the existence of tangents.

A special emphasis is placed on methods for handling more complex cases, including the differentiation of composite and implicit functions, as well as the use of computational tools like Python's sympy. Through carefully chosen exercises and problems,

https://doi.org/10.1515/9783112228289-007

you will be encouraged to practice the techniques, justify results, and explore the deep connections between differentiability and continuity.

By the end of this chapter, you will have developed a strong foundation in both the theory and practice of differentiation, equipping you with essential tools for further study in calculus, analysis, and beyond.

## 7.2 Basic results

In this section, we present several fundamental rules and results that enable the systematic computation of derivatives. These rules—including linearity, product, power, reciprocal, and chain rules—form the backbone of differential calculus, allowing derivatives of complex functions to be computed efficiently by breaking them down into simpler components.

**Theorem 7.1** (Derivatives of basic functions). *If $f(x) = c$ (constant), then $f'(a) = 0$ for all a. The derivative of $f(x) = x$ is $f'(a) = 1$ for all $a \in \mathbb{R}$.*

*Proof.* Let $f(x) = c \in \mathbb{R}$. Then

$$f'(a) = \lim_{h \to 0} \frac{c - c}{h} = 0.$$

If $f(x) = x$, then

$$f'(a) = \lim_{h \to 0} \frac{a + h - a}{h} = \lim_{h \to 0} \frac{h}{h} = 1. \qquad \square$$

Again, in order to compute the derivative of more complicated functions, one has simply to work on simpler pieces.

**Theorem 7.2** (Linearity and the product rule). *If $f$ and $g$ are differentiable at a, then so is $f \pm g$, with*

$$(f \pm g)'(a) = f'(a) \pm g'(a).$$

*Also, the product $fg$ is differentiable at a and*

$$(fg)'(a) = f'(a)g(a) + f(a)g'(a).$$

*Proof.* The proof for the sum is a direct application of the definition. For the product, consider

$$(fg)'(a) = \lim_{h \to 0} \frac{f(a + h)g(a + h) - f(a)g(a)}{h}$$

$$= \lim_{h \to 0} \left( \frac{f(a + h)[g(a + h) - g(a)]}{h} + \frac{[f(a + h) - f(a)]g(a)}{h} \right)$$

$$= f(a)g'(a) + f'(a)g(a). \qquad \square$$

**Theorem 7.3** (Power rule). *Let $f(x) = x^n$ with $n \in \mathbb{N}$. Then, for any $a \in \mathbb{R}$,*

$$f'(a) = na^{n-1}.$$

*Proof.* The proof is by induction on $n$. The result holds for $n = 1$. Suppose it holds for $n = k$; we will prove it for $n = k + 1$.

Assume that if $f(x) = x^k$, then $f'(x) = kx^{k-1}$. Set $g(x) = x^{k+1} = x^k x$. By the product rule (Theorem 7.2),

$$g'(x) = (x^k)'x + x^k \cdot 1 = kx^{k-1}x + x^k = (k+1)x^k. \qquad \square$$

**Theorem 7.4** (Derivative of the reciprocal function). *If $g$ is differentiable at $a$ and $g(a) \neq 0$, then $\frac{1}{g}$ is differentiable at $a$, and*

$$\left(\frac{1}{g}\right)'(a) = -\frac{g'(a)}{[g(a)]^2}.$$

*Proof.* Applying the definition of the derivative to $\frac{1}{g}$, we have

$$\lim_{h \to 0} \frac{1/g(a+h) - 1/g(a)}{h} = \lim_{h \to 0} \frac{g(a) - g(a+h)}{h\,g(a+h)\,g(a)}$$

$$= -\lim_{h \to 0} \frac{g(a+h) - g(a)}{h} \cdot \lim_{h \to 0} \frac{1}{g(a+h)g(a)}$$

$$= -\frac{g'(a)}{[g(a)]^2}. \qquad \square$$

**Theorem 7.5** (Chain rule). *If $g$ is differentiable at $a$ and $f$ is differentiable at $g(a)$, then $f \circ g$ is differentiable at $a$ and*

$$(f \circ g)'(a) = f'(g(a))\,g'(a).$$

*Proof.* We use Remark 9 and Theorem 6.1. Set $w(x) = f(g(x))$. Since $g$ is differentiable at $a$, it is also continuous at $a$, hence $g(a + h) - g(a) \to 0$ as $h \to 0$. To show that $w$ is differentiable at $a$, consider $w(a + h) - w(a)$. For $k = g(a + h) - g(a)$,

$$w(a+h) - w(a) = f(g(a+h)) - f(g(a))$$
$$= f(g(a) + k) - f(g(a))$$
$$= k(f'(g(a)) + u(k))$$
$$= h(g'(a) + v(h))(f'(g(a)) + u(k)).$$

Therefore,

$$\frac{w(a+h) - w(a)}{h} = (g'(a) + v(h))(f'(g(a)) + u(k)).$$

Taking the limit as $h \to 0$ termwise, we get the desired result since $k \to 0$ and $g$ is continuous (since it is differentiable). □

**Remark 10.** To compute the derivative of a function of the form $f(x)^{g(x)}$, we write it as $e^{g(x)\ln f(x)}$. □

## 7.3 Examples on differentiation

**Example 7.1.** The function

$$f(x) = \begin{cases} e^{-1/x^2}, & \text{if } x \neq 0, \\ 0, & \text{if } x = 0 \end{cases}$$

is infinitely differentiable ($C^\infty$), and in fact $f^{(n)}(0) = 0$ for every $n$. Here, $f^{(n)}$ denotes the $n$-th derivative of $f$. □

**Example 7.2** (Product rule application). Differentiate $f(x) = x^2 \cdot \cos x$ using the product rule.

*Solution.* Let $u(x) = x^2$ and $v(x) = \cos x$. Then

$$f'(x) = u'(x)v(x) + u(x)v'(x).$$

Compute derivatives

$$u'(x) = 2x, \quad v'(x) = -\sin x.$$

So

$$f'(x) = 2x \cdot \cos x + x^2 \cdot (-\sin x) = 2x \cos x - x^2 \sin x.$$ □

**Example 7.3** (Chain rule application). Find the derivative of $f(x) = \sin(x^2)$.

*Solution.* Let $u(x) = x^2$, so $f(x) = \sin(u(x))$. By the chain rule,

$$f'(x) = \cos(u(x)) \cdot u'(x) = \cos(x^2) \cdot 2x.$$

Hence

$$f'(x) = 2x \cos(x^2).$$ □

**Example 7.4** (Reciprocal rule application). Differentiate $f(x) = \frac{1}{x^3+1}$.

*Solution.* Let $g(x) = x^3 + 1$. Then

$$f(x) = \frac{1}{g(x)} \quad \Rightarrow \quad f'(x) = -\frac{g'(x)}{[g(x)]^2}.$$

Compute $g'(x) = 3x^2$, so

$$f'(x) = -\frac{3x^2}{(x^3 + 1)^2}. \qquad \square$$

**Example 7.5** (Derivative of $x^x$). Compute the derivative of $f(x) = x^x$.

*Solution.* We will use Remark 10. Let

$$f(x) = x^x = e^{x \ln x}, \quad x > 0.$$

Then, by the chain rule,

$$f'(x) = e^{x \ln x} \frac{d}{dx} (x \ln x) = x^x (\ln x + 1). \qquad \square$$

## 7.4 Python and AI

**Example 7.6** (Python-based derivative computation). Use sympy to compute the derivative of $f(x) = e^{x^2}$ symbolically.

```
1  import sympy as sp
2
3  x = sp.symbols('x')
4  f = sp.exp(x**2)
5
6  # Compute derivative
7  f_prime = sp.diff(f, x)
8
9  print("f'(x) =", f_prime)
```

## 7.5 Exercises

### True or false
Determine whether each statement is **true** or **false**. Justify your answer.
1.   If a function is continuous at a point, then it is also differentiable there.
     **Answer:** _____

2.  The sum of two differentiable functions is always differentiable.
    **Answer:** _____

3.  The function $f(x) = x^{1/3}$ is differentiable at $x = 0$.
    **Answer:** _____

4.  If $f$ is differentiable at $a$, then the graph of $f$ has a tangent at $(a, f(a))$.
    **Answer:** _____

5.  The derivative of a constant function is always zero.
    **Answer:** _____

## Multiple choice questions

Choose the correct option (A, B, C, or D) for each question.

1.  Which of the following functions is **not** differentiable at $x = 0$?
    A. $f(x) = x^2$       B. $f(x) = |x|$
    C. $f(x) = \sin x$    D. $f(x) = e^x$
    **Answer:** _____

2.  If $f(x) = \frac{1}{x}$, what is $f'(x)$?
    A. $-x^{-2}$    B. $x^{-2}$
    C. $-x^2$    D. $x^2$
    **Answer:** _____

3.  The chain rule is used when
    A. adding two functions                 B. multiplying two functions
    C. differentiating a composition    D. finding a limit
    **Answer:** _____

4.  The derivative of $f(x) = \sin(3x)$ is
    A. $3\cos(3x)$      B. $\cos(3x)$
    C. $-3\sin(3x)$    D. $3\sin(x)$
    **Answer:** _____

5.  If $f(x) = e^{x^2}$, then $f'(x)$ equals
    A. $2xe^{x^2}$    B. $e^{2x}$
    C. $e^{x^2}$       D. $xe^{x^2}$
    **Answer:** _____

## Matching exercise

Match each concept with its corresponding property.

1. Power rule                           A. $(f \circ g)'(x) = f'(g(x))g'(x)$
2. Product rule                         B. $f'(x) = nx^{n-1}$
3. Chain rule                           C. $f'(x) = 0$
4. Derivative of a constant    D. $f'(x) = f'(x)g(x) + f(x)g'(x)$
**Answers:** 1. _____    2. _____    3. _____    4. _____

## Fill in the blanks

Fill in each blank with the appropriate word or expression.

1. The derivative of $f(x) = x^n$ is _____.

   **Answer:** _____

2. Differentiability at a point implies _____ at that point.

   **Answer:** _____

3. The chain rule is used to differentiate _____ of functions.

   **Answer:** _____

4. The function $f(x) = |x|$ is not differentiable at _____.

   **Answer:** _____

5. If $f$ is differentiable and $f'(x) = 0$ for all $x$, then $f$ is a _____ function.

   **Answer:** _____

## Theoretical and computational problems

Provide complete solutions.

1. Use the definition of the derivative to find $f'(a)$ for $f(x) = x^2 + 2x$.
2. Show that $f(x) = |x|$ is not differentiable at $x = 0$.
3. Use the chain rule to differentiate $f(x) = (2x + 1)^5$.
4. Compute the derivative of $f(x) = \frac{\sin x}{x}$ for $x \neq 0$.
5. Prove that if $f(x)$ is constant, then $f'(x) = 0$ for all $x$.

## Python-based exercises

Use Python (`sympy`, `numpy`, `matplotlib`) to solve the following:

1. Use `sympy` to symbolically compute the derivative of $f(x) = \ln(x^2 + 1)$.
2. Plot $f(x) = |x|$ and show numerically that the derivative at $x = 0$ does not exist.
3. Use `sympy` to verify the product rule for $f(x) = x^2$ and $g(x) = \sin x$.
4. Write a Python function that returns the numerical derivative of any function at a given point.
5. Use `numpy` and `matplotlib` to plot the functions $f(x) = x^3$ and $f'(x) = 3x^2$ on the same graph.

# 8 Local extrema

## Contents

The study of local extrema—maximum and minimum values of functions—is a central theme in calculus. Understanding where and how these extrema occur provides key insights into the behavior of functions, both theoretically and in applications.

In this chapter, we introduce the concepts of local and global extrema and explain how derivatives are used to find them. We present fundamental results such as Fermat's theorem and the second derivative test, alongside classic theorems like Rolle's and the mean value theorem.

Through examples, proofs, and exercises you will learn effective methods for identifying, classifying, and interpreting critical points and extrema of functions.

## 8.1 Introduction

The analysis of local extrema—points where a function attains local maximum or minimum values—is fundamental in calculus and mathematical modeling. Identifying where a function reaches its highest or lowest values is essential in a wide variety of fields, from optimization and economics to physics and engineering.

This chapter explores the theoretical foundations and practical methods for detecting and classifying extrema of real-valued functions. We begin by defining what it means for a function to have a maximum, a minimum, or a local extremum, emphasizing the importance of both global and local perspectives. The distinction between critical points and extrema is clarified, and we discuss why not every critical point corresponds to a local maximum or minimum.

The derivative emerges as a powerful tool in this context: Fermat's theorem shows that interior extrema occur at points where the derivative is zero (or does not exist). Building on this, we examine how to systematically search for candidate points, i. e., critical points and endpoints, where local extrema may be found. The second derivative test provides a convenient criterion for classifying these points as maxima, minima, or points of inflection, while counterexamples illustrate its limitations.

Classical results such as Rolle's theorem and the mean value theorem reveal deep connections between local behavior and the global properties of functions. Through worked examples and visualizations, we demonstrate how these theorems can be applied to analyze and interpret the behavior of functions in concrete settings.

https://doi.org/10.1515/9783112228289-008

Table 8.1: Summary of points to check for local extrema.

| Check for local extrema at: | Condition |
|---|---|
| Critical points | $f'(x) = 0$ or $f'(x)$ does not exist |
| Endpoints | Endpoints of the domain (if the domain is closed and bounded) |

By the end of this chapter, you will have developed a solid understanding of how to use derivatives to locate and classify extrema (see Table 8.1), equipping you with essential tools for solving optimization problems and for further study in analysis and applied mathematics.

**Definition 8.1** (Maximum, minimum, local maximum, local minimum). Let $f$ be a function and $A \subseteq \mathbb{R}$. A point $x \in A$ is called a **maximum point** of $f$ in $A$ if $f(x) \geq f(y)$ for every $y \in A$. The number $f(x)$ is called the **maximum value** of $f$ on $A$. An interior point $x \in A$ is called a **local maximum** if there exists $\delta > 0$ sufficiently small so that $x$ is a maximum point on $(x - \delta, x + \delta)$. Similarly for minimum points. A **local extremum** is a point that is either a local maximum or a local minimum.

The following theorem, commonly known as Fermat's theorem, provides an essential criterion linking local extrema with differentiability, identifying critical points as positions where the derivative vanishes. This result forms the foundation for optimization problems in calculus.

**Theorem 8.1** (Fermat's theorem (interior extremum)). *Suppose $f$ is defined on $(a, b)$ and has a local extremum at $x$. If $f$ is differentiable at $x$, then $f'(x) = 0$.*

*Proof.* Suppose $f$ has a local maximum at $x$ (the minimum case is similar). Then for every $h > 0$ we have $f(x) \geq f(x + h)$. Dividing by $h$ gives

$$\frac{f(x + h) - f(x)}{h} \leq 0.$$

Similarly, for $h < 0$ we obtain

$$\frac{f(x + h) - f(x)}{h} \geq 0.$$

Since $f$ is differentiable at $x$, the limit of this difference quotient as $h \to 0$ exists and must be equal from both sides, thus $f'(x) = 0$. □

**Remark 11.** If $x_0$ is an interior point where the first derivative exists and is nonzero, then $x_0$ cannot be a local extremum, by Fermat's theorem. Therefore, to find local extrema among interior points, we only need to check:
(i) Interior points where the first derivative vanishes (Figure 8.1).
(ii) Interior points where the derivative does not exist (Figure 8.2).

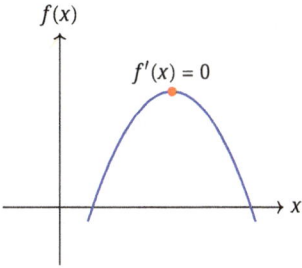

Figure 8.1: Local extremum where the derivative exists and equals zero, e. g., $f(x) = -(x - 2)^2 + 2$.

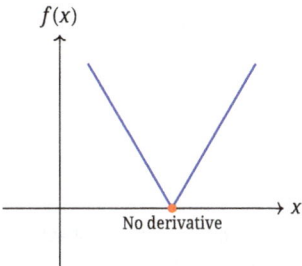

Figure 8.2: Local extremum at a point where the derivative does not exist, e. g., $f(x) = |x - 2|$.

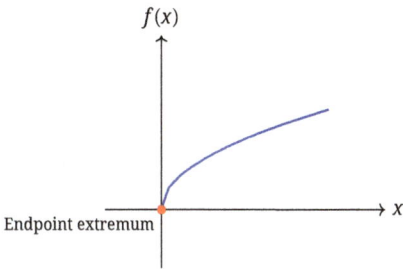

Figure 8.3: Local extremum occurring at an endpoint of a closed interval, e. g., $f(x) = \sqrt{x}, x \geq 0$.

Of course, any point of the domain which is not an interior point is a local extrema point (Figure 8.3).

**Definition 8.2** (Critical point). A **critical point** of a function $f$ is an interior point $x$ such that $f'(x) = 0$. The number $f(x)$ is called a **critical value** of $f$.

Often, "critical points" are defined as interior points where the first derivative vanishes or does not exist.

The following theorem, known as Rolle's theorem, is a fundamental result connecting continuity, differentiability, and the existence of horizontal tangents, providing a powerful analytical tool often utilized to establish broader results in calculus and mathematical analysis.

**Theorem 8.2** (Rolle's theorem, Figure 8.4). *Suppose $f$ is continuous on $[a, b]$, differentiable on $(a, b)$, and $f(a) = f(b)$. Then there exists $x \in (a, b)$ such that $f'(x) = 0$.*

*Proof.* The continuity of $f$ on $[a, b]$ guarantees that $f$ attains a maximum and a minimum on $[a, b]$. If the maximum (or minimum) is achieved at an interior point, then by Fermat's theorem $f'(x) = 0$. If both are attained at endpoints, then $f$ is constant, so $f'(x) = 0$ everywhere. □

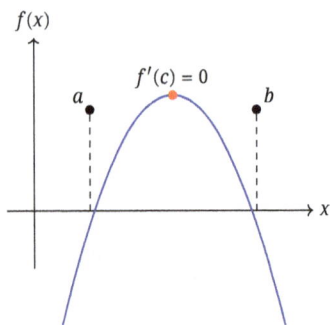

Figure 8.4: Illustration of Rolle's theorem: if a differentiable function satisfies $f(a) = f(b)$, there exists at least one $c \in (a, b)$ with $f'(c) = 0$.

**Example 8.1** (Local extrema including a non-differentiable point). Consider the function

$$f(x) = |x^2 - 1| + 2x.$$

Find all critical points and classify any local extrema. In particular, analyze the behavior at points where $f$ may not be differentiable.

*Solution.* The expression $|x^2 - 1|$ changes definition at $x = \pm 1$, since $x^2 - 1 = 0$ there. So we define $f$ piecewise:

$$f(x) = \begin{cases} x^2 - 1 + 2x, & \text{if } x \leq -1 \text{ or } x \geq 1, \\ -(x^2 - 1) + 2x = -x^2 + 2x + 1, & \text{if } -1 < x < 1. \end{cases}$$

So

$$f(x) = \begin{cases} x^2 + 2x - 1, & x \leq -1, \\ -x^2 + 2x + 1, & -1 < x < 1, \\ x^2 + 2x - 1, & x \geq 1. \end{cases}$$

**Step 1: Check continuity at $x = \pm 1$:**
We verify quickly that $f$ is continuous at $x = \pm 1$:

- $f(-1) = (-1)^2 + 2(-1) - 1 = 1 - 2 - 1 = -2.$
- $\lim_{x \to -1^+} f(x) = -(-1)^2 + 2(-1) + 1 = -1 - 2 + 1 = -2.$
- $\lim_{x \to -1^-} f(x) = (-1)^2 + 2(-1) - 1 = -2.$

Similarly at $x = 1$: $f(1) = 1 + 2 - 1 = 2$, and limit from both sides is 2. So $f$ is continuous everywhere.

**Step 2: Compute derivatives in each interval:**
- For $x < -1$: $f(x) = x^2 + 2x - 1 \Rightarrow f'(x) = 2x + 2$.
- For $-1 < x < 1$: $f(x) = -x^2 + 2x + 1 \Rightarrow f'(x) = -2x + 2$.
- For $x > 1$: $f(x) = x^2 + 2x - 1 \Rightarrow f'(x) = 2x + 2$.

**Step 3: Find critical points:**
Critical points occur where $f'(x) = 0$ or $f'$ does not exist.
1. **Where $f'$ does not exist:**
   - At $x = -1$ and $x = 1$, the function involves absolute value, so we check differentiability.
   - At $x = -1$:
     Left derivative: $\lim_{h \to 0^-} \frac{f(-1+h)-f(-1)}{h}$, $h < 0$: $f'_-(-1) = 2(-1) + 2 = 0$.
     Right derivative: $h > 0$, use middle piece: $f'_+(-1) = -2(-1) + 2 = 2 + 2 = 4$.
     Since $f'_-(-1) \neq f'_+(-1)$, $f'(-1)$ does not exist.
   - At $x = 1$:
     Left derivative: $f'_-(1) = -2(1) + 2 = 0$.
     Right derivative: $f'_+(1) = 2(1) + 2 = 4$. Again, not equal $> f'(1)$ does not exist.
     So $x = -1$ and $x = 1$ are critical points.
2. **Where $f'(x) = 0$:**
   - For $x < -1$: $2x + 2 = 0 \Rightarrow x = -1$, but $x = -1$ is not in $(-\infty, -1) >$ not included.
   - For $-1 < x < 1$: $-2x + 2 = 0 \Rightarrow x = 1$, but $x = 1$ not in $(-1, 1) >$ not included.
   - For $x > 1$: $2x + 2 = 0 \Rightarrow x = -1$, not in $(1, \infty)$.

So the only critical points are $x = -1$ and $x = 1$.

**Step 4: Analyze the sign of $f'$ near critical points:**
Even though $f$ is not differentiable at $x = \pm1$, we can analyze the sign of $f'$ in the intervals:
- On $(-\infty, -1)$: $f'(x) = 2x + 2$. For $x < -1$, say $x = -2$: $f'(-2) = -4 + 2 = -2 < 0 >$ decreasing.
- On $(-1, 1)$: $f'(x) = -2x + 2$. For $x = 0$: $f'(0) = 2 > 0 >$ increasing.
- On $(1, \infty)$: $f'(x) = 2x + 2$. For $x = 2$: $f'(2) = 6 > 0 >$ increasing.

**Step 5: Classify critical points using the definition of local extrema:**
- At $x = -1$:
  - $f$ is decreasing on $(-\infty, -1)$ and increasing on $(-1, 1)$.
  - So $f(x) \geq f(-1)$ for all $x$ near $-1$.
  - $f(-1) = (-1)^2 + 2(-1) - 1 = 1 - 2 - 1 = -2$.
    Therefore, by definition, $f$ has a **local minimum** at $x = -1$, even though $f'(-1)$ does not exist.

– **At** $x = 1$:
  – $f$ is increasing on $(-1, 1)$ and also increasing on $(1, \infty)$.
  – No change in sign of derivative > no local extremum.
  – Check values: $f(1) = 1 + 2 - 1 = 2$.
    For $x = 0.9$: $f(0.9) = -(0.81) + 1.8 + 1 = 1.99 < 2$,
    for $x = 1.1$: $f(1.1) = (1.21) + 2.2 - 1 = 2.41 > 2$.
    But since $f$ is increasing through $x = 1$, no local max or min.

**Final answer:**
– $x = -1$: **local minimum**, $f(-1) = -2$, and $f$ is not differentiable here.
– $x = 1$: **not a local extremum** despite being a critical point.

**Graph of** $f(x) = |x^2 - 1| + 2x$:

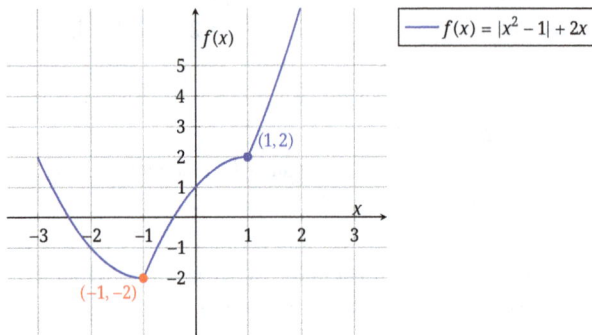

**Conclusion:**
This example shows that:
– A function can have a **local extremum at a point where it is not differentiable**.
– Critical points include both where $f'(x) = 0$ and where $f'(x)$ does not exist.
– The **definition of local extremum** (based on function values) must be used when derivatives fail. □

**Example 8.2** (Critical points and local extrema). Find all critical points of the function $f(x) = x^3 - 3x$ and classify them using Fermat's theorem.

*Solution.*
**Compute** $f'$:
To find the critical points, we first compute the first derivative of $f$:

$$f'(x) = \frac{d}{dx}(x^3 - 3x) = 3x^2 - 3.$$

**Find the roots of** $f'$:

Set the derivative equal to zero and solve

$$3x^2 - 3 = 0 \quad \Rightarrow \quad x^2 = 1 \quad \Rightarrow \quad x = \pm 1.$$

So, the critical points are at

$$x = -1 \quad \text{and} \quad x = 1.$$

Fermat's theorem states that if $f$ has a local extremum at a point $c$, and $f$ is differentiable at $c$, then $f'(c) = 0$. We now analyze the sign of $f'(x)$ around the critical points to classify them.

**Interval test:**

$$f'(x) = 3x^2 - 3 = 3(x - 1)(x + 1).$$

Since $f$ is everywhere continuous, we are able to get the following conclusions:
- On $(-\infty, -1)$: Choose $x = -2 \Rightarrow f'(-2) = 3(4 - 1) = 9 > 0$.
- On $(-1, 1)$: Choose $x = 0 \Rightarrow f'(0) = -3 < 0$.
- On $(1, \infty)$: Choose $x = 2 \Rightarrow f'(2) = 3(4 - 1) = 9 > 0$.

**Conclusion:**
- At $x = -1, f'$ changes from positive to negative: local **maximum**.
- At $x = 1, f'$ changes from negative to positive: local **minimum**.

**Step 4: Compute function values at critical points:**

$$f(-1) = (-1)^3 - 3(-1) = -1 + 3 = 2,$$
$$f(1) = (1)^3 - 3(1) = 1 - 3 = -2.$$

**Final answer:**
- $x = -1$: local maximum, $f(-1) = 2$.
- $x = 1$: local minimum, $f(1) = -2$.

We plot the function and highlight the critical points found:

$$f(x) = x^3 - 3x$$

with critical points at $x = -1$ (local maximum) and $x = 1$ (local minimum).

**Observation:**
- The curve rises to a local maximum at $x = -1$, where $f(-1) = 2$.
- It then falls to a local minimum at $x = 1$, where $f(1) = -2$.
- The function is increasing on $(-\infty, -1)$ and $(1, \infty)$, and decreasing on $(-1, 1)$. $\qquad \square$

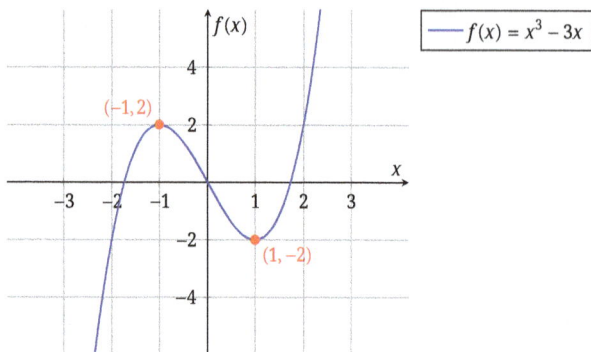

As one can see, the above procedure does not need the existence of the second derivative in order to characterize the critical points. However, if the second derivative exists, one can use the following theorem instead.

**Theorem 8.3** (Second derivative test). *Suppose $f'(a) = 0$. If $f''(a) > 0$, then $f$ has a local minimum at $a$; if $f''(a) < 0$, then $f$ has a local maximum at $a$.*

*Proof.* By the definition of the derivative,

$$f''(a) = \lim_{h \to 0} \frac{f'(a+h) - f'(a)}{h}.$$

Since $f'(a) = 0$, we have

$$f''(a) = \lim_{h \to 0} \frac{f'(a+h)}{h}.$$

If $f''(a) > 0$, then for small $h$, $f'(a + h)$ has the same sign as $h$, so $f$ increases after $a$ and decreases before $a$; thus, $f$ has a local minimum at $a$. The argument for $f''(a) < 0$ (local maximum) is similar. □

**Remark 12.** If $f'(a) = 0$ and $f''(a) = 0$, no conclusion can be drawn from the second derivative alone; further analysis is needed. For example, $f(x) = x^3$ and $g(x) = x^4$ (Figure 8.5) both have $f'(0) = g'(0) = f''(0) = g''(0) = 0$. For $f(x) = x^3$, $x = 0$ is not a local minimum or maximum; for $g(x) = x^4$, $x = 0$ is a local minimum.

**Example 8.3** (Application of Rolle's theorem). Verify Rolle's theorem for the function $f(x) = \sin x$ on the interval $[0, \pi]$.

*Solution.*
**Step 1: Conditions of Rolle's theorem:**
Let $f(x) = \sin x$, defined on the closed interval $[0, \pi]$.

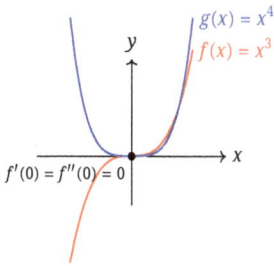

Figure 8.5: Illustration: At $x = 0$, $f'(0) = f''(0) = 0$ for both $f(x) = x^3$ and $g(x) = x^4$. Yet $f(x) = x^3$ has no local extremum at 0, while $g(x) = x^4$ has a local minimum at 0. Thus, no conclusion can be drawn from the second derivative alone when it equals zero.

Rolle's theorem states that if:
1. $f$ is continuous on the closed interval $[a, b]$,
2. $f$ is differentiable on the open interval $(a, b)$,
3. $f(a) = f(b)$,

then there exists at least one $c \in (a, b)$ such that

$$f'(c) = 0.$$

We check these conditions for $f(x) = \sin x$ on $[0, \pi]$:
- **Continuity:** $\sin x$ is continuous on $\mathbb{R}$, so it is continuous on $[0, \pi]$.
- **Differentiability:** $\sin x$ is differentiable on $\mathbb{R}$, so it is differentiable on $(0, \pi)$.
- **Equal endpoints:**

$$f(0) = \sin 0 = 0, \quad f(\pi) = \sin \pi = 0 \quad \Rightarrow \quad f(0) = f(\pi).$$

All conditions of Rolle's theorem are satisfied.

**Step 2: Find $c \in (0, \pi)$ such that $f'(c) = 0$:**
Compute the derivative

$$f'(x) = \cos x.$$

Set the derivative equal to zero

$$\cos x = 0 \quad \Rightarrow \quad x = \frac{\pi}{2}.$$

Since $\frac{\pi}{2} \in (0, \pi)$, we conclude

$$f'\left(\frac{\pi}{2}\right) = \cos\left(\frac{\pi}{2}\right) = 0.$$

**Conclusion:**
Rolle's theorem is verified for $f(x) = \sin x$ on $[0, \pi]$. There exists $c = \frac{\pi}{2} \in (0, \pi)$ such that

$$f'(c) = 0.$$

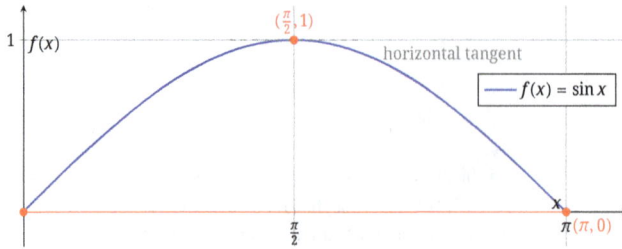

**Conclusion:** The function starts and ends at the same height, and at $x = \frac{\pi}{2}$, the derivative is zero, as shown by the horizontal tangent. This confirms the existence of a point satisfying Rolle's theorem. □

A consequence of Rolle's theorem is the following mean value theorem.

**Theorem 8.4** (Mean value theorem (Figure 8.6)). *Suppose $f$ is continuous on $[a,b]$ and differentiable on $(a,b)$. Then there exists $x \in (a,b)$ such that*

$$f'(x) = \frac{f(b) - f(a)}{b - a}.$$

*Proof.* Define $h(x) = f(x) - (\frac{f(b)-f(a)}{b-a})(x - a)$. The function $h$ is continuous on $[a,b]$ and differentiable on $(a,b)$, and $h(a) = h(b) = f(a)$. By Rolle's theorem, there exists $x \in (a,b)$ such that $h'(x) = 0$, i.e., $f'(x) = \frac{f(b)-f(a)}{b-a}$. □

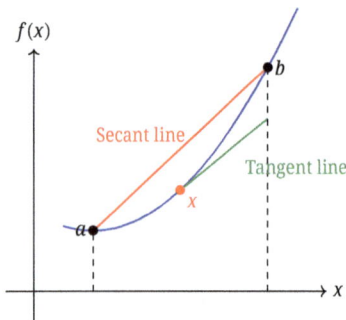

**Figure 8.6:** Mean value theorem: The tangent line at some point $x \in (a,b)$ is parallel to the secant line connecting $(a, f(a))$ and $(b, f(b))$.

The following corollary establishes a fundamental relationship between functions having identical derivatives, revealing that such functions differ by no more than a constant, a result crucial for understanding indefinite integrals and antiderivatives.

**Corollary 8.1** (Functions with equal derivatives differ by a constant). *If $f, g$ are defined on the same interval and $f'(x) = g'(x)$ for all $x$ in the interval, then there exists $c \in \mathbb{R}$ such that $f(x) = g(x) + c$.*

*Proof.* First, if $f'(x) = 0$ for every $x$ in the domain, then $f$ is constant. Indeed, for any $a < b$ in the domain, the mean value theorem gives $x \in (a, b)$ such that $f'(x) = \frac{f(b)-f(a)}{b-a}$. But $f'(x) = 0$ everywhere, so $f(a) = f(b)$. Hence $f$ is constant. For the general case, set $h(x) = f(x) - g(x)$. Then $h'(x) = 0$, and so $h$ is constant. □

## 8.2 Examples

**Example 8.4** (Inequality with parameters). Let $x \in [0, 1]$ and $a \in [0, 1]$. Prove that

$$1 + xa \geq (1 + a)^x.$$

*Solution.* If $a = 0$, the inequality is obvious. Now suppose $a \in (0, 1]$. Consider the function $f(x) = 1 + xa - (1+a)^x$. Its first derivative is $f'(x) = a - (1+a)^x \ln(1+a)$. The derivative vanishes at

$$x^* = \frac{\ln a - \ln \ln(1 + a)}{\ln(1 + a)},$$

which is the unique critical point and satisfies $x^* \in [0, 1]$ (can be shown by analyzing $g(a) = \ln a - \ln \ln(1+a)$ and $h(a) = (1+a) \ln(1+a) - a$ for $a \in (0, 1]$). The second derivative is $f''(x) = -(1 + a)^x \ln^2(1 + a)$, which is always negative when $a \in (0, 1]$, so $f$ has a local maximum at $x^*$. Since $f(0) = f(1) = 0$, we must have $f(x^*) \geq 0$. As there is no local minimum in $(0, 1)$, it follows that $f(x) \geq 0$ for all $x \in [0, 1]$. Otherwise, a value $\xi$ with $f(\xi) < 0$ would create a contradiction as explained in detail above. □

**Example 8.5** (Mean value theorem application). Apply the mean value theorem to the function $f(x) = x^2$ on the interval $[1, 3]$. Find the value $c \in (1, 3)$ guaranteed by the theorem.

*Solution.*
**Step 1: Check conditions of the mean value theorem:**
The mean value theorem (MVT) states that if a function $f$ satisfies:
- It is continuous on the closed interval $[a, b]$,
- It is differentiable on the open interval $(a, b)$,

then there exists at least one number $c \in (a, b)$ such that

$$f'(c) = \frac{f(b) - f(a)}{b - a}.$$

For $f(x) = x^2$ on $[1, 3]$,
- $f$ is a polynomial, so it is continuous and differentiable everywhere, including on $[1, 3]$ and $(1, 3)$.

**Step 2: Compute the average rate of change:**

$$\frac{f(3) - f(1)}{3 - 1} = \frac{3^2 - 1^2}{2} = \frac{9 - 1}{2} = \frac{8}{2} = 4.$$

**Step 3: Find** $c \in (1, 3)$ **such that** $f'(c) = 4$:
We compute the derivative

$$f'(x) = \frac{d}{dx}(x^2) = 2x.$$

Now solve

$$2c = 4 \quad \Rightarrow \quad c = 2.$$

**Conclusion:**
There exists $c = 2 \in (1, 3)$ such that

$$f'(c) = 4 = \frac{f(3) - f(1)}{3 - 1}.$$

This confirms the mean value theorem for $f(x) = x^2$ on the interval $[1, 3]$.

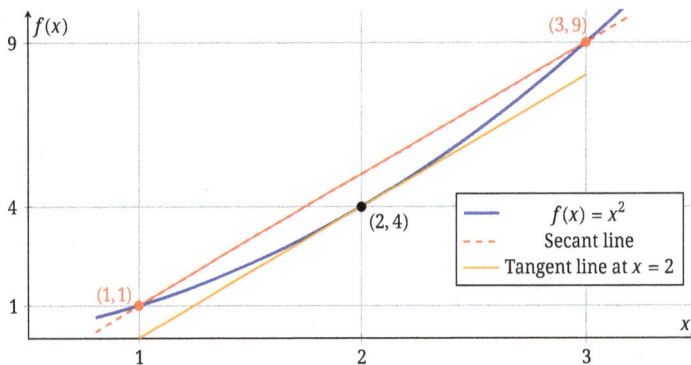

**Explanation:**
- The red dashed line is the secant line from $x = 1$ to $x = 3$ with slope 4.
- The orange tangent line at $x = 2$ has the same slope, confirming the mean value theorem.
- The point $(2, 4)$ is where the tangent touches the curve. □

**Example 8.6.** Find all the local extrema of the function $f(x) = x^{\frac{1}{x}}$ for $x > 0$.

*Solution.* We study the local extrema of the function

$$f(x) = x^{1/x}, \quad x > 0.$$

**First derivative and critical point**

As before, we have

$$f(x) = x^{1/x} = e^{\frac{\ln x}{x}}.$$

Let $g(x) = \frac{\ln x}{x}$. Then

$$f'(x) = e^{g(x)} \cdot g'(x) = x^{1/x} \cdot \frac{1 - \ln x}{x^2}.$$

Critical points occur when $1 - \ln x = 0 \implies x = e$.

**Second derivative**

We compute $f''(x)$:

Recall

$$f'(x) = x^{1/x} \cdot \frac{1 - \ln x}{x^2}.$$

Let us set

$$u(x) = x^{1/x}, \quad v(x) = \frac{1 - \ln x}{x^2}.$$

Then

$$f''(x) = u'(x)v(x) + u(x)v'(x).$$

We already have

$$u(x) = x^{1/x}, \quad u'(x) = x^{1/x} \cdot \frac{1 - \ln x}{x^2}.$$

Now, compute $v'(x)$

$$v(x) = \frac{1 - \ln x}{x^2}.$$

Set $a(x) = 1 - \ln x$, $b(x) = x^2$, then

$$v'(x) = \frac{a'(x)b(x) - a(x)b'(x)}{[b(x)]^2}.$$

Here, $a'(x) = -\frac{1}{x}$, $b(x) = x^2$, $b'(x) = 2x$, so

$$v'(x) = \frac{(-\frac{1}{x})x^2 - (1 - \ln x) \cdot 2x}{x^4}$$

$$= \frac{-x - 2x(1 - \ln x)}{x^4}$$

$$= \frac{-x - 2x + 2x \ln x}{x^4}$$

$$= \frac{-3x + 2x \ln x}{x^4}$$

$$= \frac{x(2 \ln x - 3)}{x^4}$$

$$= \frac{2 \ln x - 3}{x^3}.$$

Therefore,

$$f''(x) = \left[ x^{1/x} \cdot \frac{1 - \ln x}{x^2} \right] \cdot \frac{1 - \ln x}{x^2} + x^{1/x} \cdot \frac{2 \ln x - 3}{x^3}$$

$$= x^{1/x} \left( \frac{(1 - \ln x)^2}{x^4} + \frac{2 \ln x - 3}{x^3} \right).$$

**Second derivative at $x = e$**

Recall $\ln e = 1$, so at $x = e$,

$$1 - \ln e = 0, \quad 2 \ln e - 3 = 2 \cdot 1 - 3 = -1.$$

Therefore,

$$f''(e) = e^{1/e} \left( \frac{0^2}{e^4} + \frac{-1}{e^3} \right) = -\frac{e^{1/e}}{e^3} < 0.$$

Thus, $f''(e) < 0$, so $f$ has a **local maximum** at $x = e$.

**Summary**
- The function has a unique local (and global) maximum at $x = e$.
- $f(e) = e^{1/e} \approx 1.4447$.
- $f''(e) < 0$, so the maximum is strict.
- $f(x) \to 0$ as $x \to 0^+$ and as $x \to +\infty$.

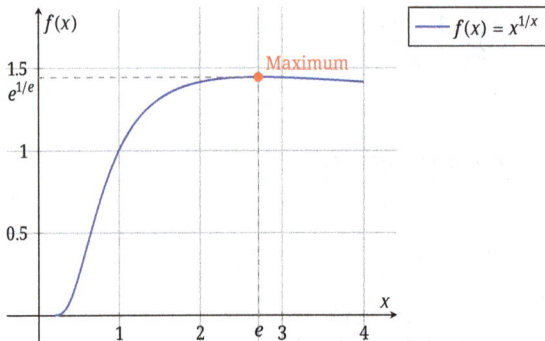

□

**Example 8.7.** Find all the local extrema of the function $f(x) = xe^{-x}$ for $x > 0$.

*Solution.* We study the local extrema of the function

$$f(x) = x\,e^{-x}, \quad x > 0.$$

### First derivative and critical point
Compute the derivative using the product rule

$$f'(x) = e^{-x} + x \cdot (-e^{-x}) = e^{-x} - xe^{-x} = e^{-x}(1 - x).$$

Set $f'(x) = 0 \implies 1 - x = 0 \implies x = 1$.

### Second derivative

$$\begin{aligned}
f''(x) &= \frac{d}{dx}\left(e^{-x}(1 - x)\right) \\
&= (-e^{-x})(1 - x) + e^{-x}(-1) \\
&= -e^{-x}(1 - x) - e^{-x} \\
&= -e^{-x}(1 - x + 1) = -e^{-x}(2 - x).
\end{aligned}$$

At $x = 1$,

$$f''(1) = -e^{-1}(2 - 1) = -\frac{1}{e} < 0.$$

So $f$ has a **local maximum** at $x = 1$.

### Summary
- The function has a unique local (and global) maximum at $x = 1$.
- $f(1) = 1 \cdot e^{-1} = \frac{1}{e}$.
- $f''(1) < 0$, so the maximum is strict.
- $f(x) \to 0$ as $x \to 0^+$ and as $x \to +\infty$.

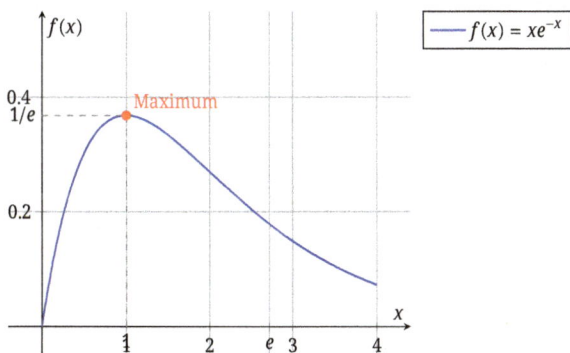

**Example 8.8.** Find all the local extrema of the function $f(x) = \frac{\ln x}{x}$ for $x > 0$.

*Solution.*

**First derivative and critical point**

Compute the derivative

$$f'(x) = \frac{1 \cdot x - \ln x \cdot 1}{x^2} = \frac{1 - \ln x}{x^2}.$$

Set $f'(x) = 0 \implies 1 - \ln x = 0 \implies x = e$.

**Second derivative**

$$f''(x) = \frac{-x^2 \cdot \frac{1}{x} - (1 - \ln x) \cdot 2x}{x^4}$$

$$= \frac{-x - 2x(1 - \ln x)}{x^4}$$

$$= \frac{-x - 2x + 2x \ln x}{x^4}$$

$$= \frac{2x \ln x - 3x}{x^4}$$

$$= \frac{2 \ln x - 3}{x^3}.$$

At $x = e$,

$$f''(e) = \frac{2 \cdot 1 - 3}{e^3} = \frac{-1}{e^3} < 0.$$

Thus, $f$ has a **local maximum** at $x = e$.

**Summary**

- The function has a unique local (and global) maximum at $x = e$.
- $f(e) = \frac{1}{e}$.
- $f''(e) < 0$, so the maximum is strict.
- $f(x) \to 0$ as $x \to 0^+$ and as $x \to +\infty$.

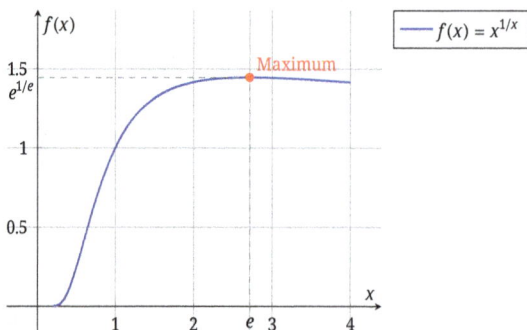

**Example 8.9.** Use the second derivative test to classify the critical points of $f(x) = x^4 - 4x^2$.

*Solution.*

$$f'(x) = 4x^3 - 8x = 4x(x^2 - 2) \quad \Rightarrow \quad f'(x) = 0 \quad \text{when } x = 0, \pm\sqrt{2}.$$

**Step 2: Second derivative**

$$f''(x) = 12x^2 - 8.$$

**Step 3: Classification**

| $x$ | $f''(x)$ | Conclusion | $f(x)$ |
|---|---|---|---|
| $-\sqrt{2}$ | $f''(-\sqrt{2}) = 16 > 0$ | Local minimum | $-4$ |
| $0$ | $f''(0) = -8 < 0$ | Local maximum | $0$ |
| $\sqrt{2}$ | $f''(\sqrt{2}) = 16 > 0$ | Local minimum | $-4$ |

**Conclusion**
- Local minimum at $x = \pm\sqrt{2}$ with value $f = -4$.
- Local maximum at $x = 0$ with value $f = 0$.

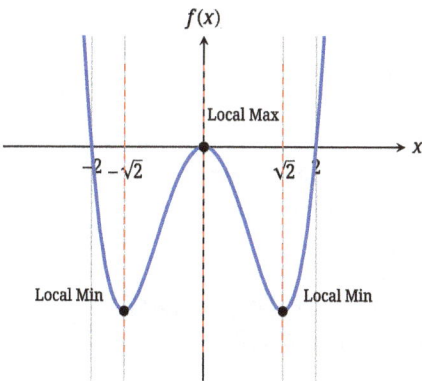

## 8.3 Python and AI

**AI Request**
Write a Python code that computes all the local extrema of a function. I need also the graph of the function.

**AI Request**
Compute all the local extrema of $x^{\frac{1}{x}}$ step-by-step and give me the Latex code.

## 8.4 Exercises

### True or false

Determine whether each statement is **true** or **false**. Justify your answer.

1. If $f'(a) = 0$, then $f$ has a local extremum at $a$.

   **Answer:** _____

2. A function can have a local extremum at a point where it is not differentiable.

   **Answer:** _____

3. If $f$ is continuous on $[a, b]$, it always attains both a maximum and a minimum on $[a, b]$.

   **Answer:** _____

4. The second derivative test always gives a conclusive answer for all critical points.

   **Answer:** _____

5. If $f$ has a local maximum at $x_0$ and $f''(x_0) < 0$, then $f'(x_0) = 0$.

   **Answer:** _____

### Multiple choice questions

Choose the correct option (A, B, C, or D) for each question.

1. At which points do we usually check for local extrema of a differentiable function on $[a, b]$?

   A. Points where $f'(x) = 0$    B. Points where $f'(x)$ does not exist

   C. Endpoints $a$, $b$            D. All of the above

   **Answer:** _____

2. If $f(x) = |x|$, which of the following is true?

   A. $x = 0$ is a local minimum         B. $f$ is differentiable at $x = 0$

   C. $f$ has a local maximum at $x = 0$    D. $f(x)$ has no extrema

   **Answer:** _____

3. Which statement about Rolle's theorem is **not** true?

   A. $f(a) = f(b)$ is required           B. $f$ must be differentiable on $[a, b]$

   C. $f$ must be continuous on $[a, b]$    D. There is some $c \in (a, b)$ with $f'(c) = 0$

   **Answer:** _____

4. For $f(x) = x^4$, what can be said about the critical point at $x = 0$?

   A. Local maximum    B. Local minimum

   C. Neither            D. Not defined

   **Answer:** _____

5. If $f$ is continuous on $[a, b]$ and differentiable on $(a, b)$, then the mean value theorem says:

   A. $f'(c) = \frac{f(b) - f(a)}{b - a}$ for some $c \in (a, b)$    B. $f(a) = f(b)$

   C. $f'(a) = f'(b)$                   D. $f$ is increasing on $[a, b]$

   **Answer:** _____

## Matching exercise

Match each concept with its appropriate description.

1. Critical point                                     A. $f$ not differentiable at this point
2. Local minimum for $f(x) = x^2$      B. The point where $f'(x) = 0$
3. Corner point (e. g., $|x|$ at 0)         C. $x = 0$
4. Fermat's theorem                            D. $f$ has a local extremum and $f'$ exists $\implies f'(x) = 0$
5. Second derivative test                     E. Uses $f''(x)$ to classify extrema

**Answers: 1.** _____   **2.** _____   **3.** _____   **4.** _____   **5.** _____

## Fill in the blanks

Fill in each blank with the most appropriate word or expression.

1.  A point where the derivative of a function is zero or does not exist is called a

    _____.

    **Answer:** _____

2.  The function $f(x) = |x|$ has a _____ at $x = 0$.

    **Answer:** _____

3.  If $f''(x_0) > 0$ and $f'(x_0) = 0$, then $f$ has a local _____ at $x_0$.

    **Answer:** _____

4.  According to the extreme value theorem, a continuous function on a closed interval

    attains a _____ and a _____.

    **Answer:** _____

5.  Rolle's theorem requires that $f$ be continuous on $[a, b]$ and differentiable on

    _____.

    **Answer:** _____

## Theoretical and computational problems

Solve each problem step-by-step.

1.  Find all the local extrema of $f(x) = x^3 - 6x^2 + 9x + 1$ and classify them using the second derivative test.
2.  Prove that $f(x) = |x|$ has a global minimum at $x = 0$.
3.  Let $f(x) = x^4 - 2x^2$. Find and classify all local extrema.
4.  For $f(x) = \sin x$ on $[0, 2\pi]$, find all points where $f$ attains its absolute maximum and minimum.
5.  Show, using the mean value theorem, that for any $x > 0$, $\ln(x) < x - 1$ if $x \neq 1$.

## Python-based exercises

Use Python (sympy, numpy, matplotlib) to solve the following:

1.  Use sympy to compute and solve $f'(x) = 0$ for $f(x) = x^3 - 3x + 2$.
2.  Plot $f(x) = |x|$ and $f(x) = x^2$ on $[-2, 2]$ and visually compare their minima.
3.  Numerically estimate where $f(x) = x^3 - 6x^2 + 9x$ has a horizontal tangent.
4.  Use sympy to verify Fermat's theorem for $f(x) = x^4$ at $x = 0$.

# 9 Derivatives and monotonicity–convexity

## Contents

Understanding how derivatives reveal the behavior of functions is central in calculus. The sign of the first derivative tells us where a function is increasing or decreasing, while the second derivative describes how a function curves, i. e., whether it is convex or concave.

In this chapter, we define monotonicity and convexity, and show how to use derivatives to analyze them. We present key theorems linking derivatives with increasing/decreasing behavior and curvature, and introduce inflection points, where the graph changes concavity.

Through definitions, theorems, and worked examples, you will develop the tools to interpret and predict the shape and growth of functions.

## 9.1 Introduction

The relationship between derivatives and the shape of functions is a fundamental theme in calculus. By examining the sign and behavior of a function's derivatives, we gain powerful insight into how the function changes, where it increases or decreases, and how it bends, i. e., whether it is convex (curving upwards) or concave (curving downwards).

In this chapter, we systematically explore the interplay between derivatives, monotonicity, and convexity. We begin by reviewing the precise definitions of strictly increasing and strictly decreasing functions, and show how the first derivative determines monotonicity: a positive derivative signals a function that increases, while a negative derivative signals a decreasing trend. Using the mean value theorem, we rigorously establish the connection between the sign of the derivative and the monotonic behavior of a function.

Next, we introduce the concepts of convexity and concavity, both geometrically and analytically. The second derivative emerges as a key tool: if the second derivative is positive, the graph is convex; if negative, concave. We prove these characterizations and discuss their significance, emphasizing the role of inflection points—where the concavity of a function changes.

https://doi.org/10.1515/9783112228289-009

Throughout the chapter, we illustrate these ideas with examples, counterexamples, and graphical representations. You will learn to identify critical points, construct sign charts, and use both the first and second derivatives to analyze and interpret the shape and growth of functions. Special attention is paid to common pitfalls and subtle cases, such as points where a function is convex but not twice differentiable.

By mastering the content of this chapter, you will develop a deeper intuition for how calculus describes the geometry of functions and prepares you for advanced applications in optimization, economics, and scientific modeling.

## 9.2 Increasing/decreasing functions

**Definition 9.1.** A function $f$ is **strictly increasing** (see Figure 9.1), on an interval $I$ if $f(a) < f(b)$ whenever $a < b$ and $a, b \in I$. If the inequality is reversed, $f$ is **strictly decreasing** (see Figure 9.2).

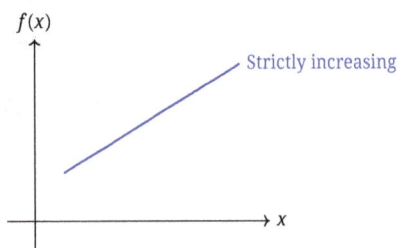

Figure 9.1: Example of a strictly increasing function: $f'(x) > 0$.

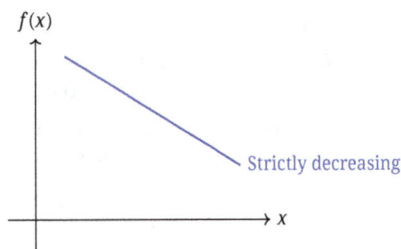

Figure 9.2: Example of a strictly decreasing function: $f'(x) < 0$.

**Corollary 9.1** (Sign of derivative implies monotonicity). *If $f'(x) > 0$ for all $x \in I$, then $f$ is strictly increasing on $I$. If $f'(x) < 0$ for all $x \in I$, then $f$ is strictly decreasing on $I$.*

*Proof.* If $f'(x) > 0$ for all $x$ in an interval, pick any $a < b$ in $I$. The mean value theorem provides $x \in (a, b)$ with $f'(x) = \frac{f(b) - f(a)}{b - a}$. Since $f'(x) > 0$, it follows that $f(b) > f(a)$. The case $f'(x) < 0$ is analogous. $\square$

## 9.3 Convex/concave functions

**Definition 9.2** (Concave/convex functions). A function $f$ is called **convex** (see Figure 9.3) on an interval $I$ if for any $x_1, x_2 \in I$ and any $\lambda \in [0,1]$,

$$f\big(\lambda x_1 + (1-\lambda)x_2\big) \leq \lambda f(x_1) + (1-\lambda)f(x_2).$$

It is called **concave** (see Figure 9.4) on $I$ if for any $x_1, x_2 \in I$ and any $\lambda \in [0,1]$,

$$f\big(\lambda x_1 + (1-\lambda)x_2\big) \geq \lambda f(x_1) + (1-\lambda)f(x_2).$$

If these inequalities are strict for all distinct $x_1, x_2 \in I$ and $\lambda \in (0,1)$, the function is called **strictly convex** or **strictly concave**, respectively.

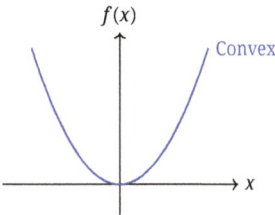

Figure 9.3: Example of a convex function: $f''(x) > 0$.

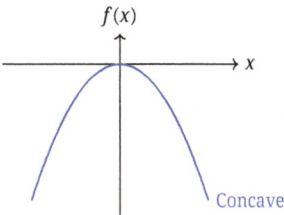

Figure 9.4: Example of a concave function: $f''(x) < 0$.

**Theorem 9.1** (Second derivative and convexity/concavity). *Let $f : (a, b) \to \mathbb{R}$ be a twice differentiable function on the open interval $(a, b)$. Then:*
1. *The function $f$ is **convex** (or concave upward) on $(a, b)$ if and only if $f''(x) \geq 0$ for all $x \in (a, b)$.*
2. *The function $f$ is **concave** (or concave downward) on $(a, b)$ if and only if $f''(x) \leq 0$ for all $x \in (a, b)$.*

*Proof.* We will prove part (1); the proof of part (2) follows similarly by reversing the inequalities. Part (1) requires proving two directions:

**Direction 1 (forward implication): If $f''(x) \geq 0$ for all $x \in (a, b)$, then $f$ is convex on $(a, b)$.**

Suppose that $f''(x) \geq 0$ for all $x \in (a, b)$. This implies that the first derivative $f'(x)$ is **increasing** (nondecreasing) on $(a, b)$.

Let $x_1, x_2 \in (a, b)$ with $x_1 < x_2$, and let $c \in (x_1, x_2)$ be any point in between. We aim to show that the graph of $f$ lies *below or on* the chord connecting $(x_1, f(x_1))$ and $(x_2, f(x_2))$, which is the definition of convexity.

Using the mean value Theorem 8.4 (MVT), there exist points $\xi_1 \in (x_1, c)$ and $\xi_2 \in (c, x_2)$ such that

$$f'(\xi_1) = \frac{f(c) - f(x_1)}{c - x_1},$$

$$f'(\xi_2) = \frac{f(x_2) - f(c)}{x_2 - c}.$$

Since $x_1 < \xi_1 < c < \xi_2 < x_2$, and $f'(x)$ is increasing, we have

$$f'(\xi_1) \le f'(\xi_2).$$

Substituting the expressions from MVT,

$$\frac{f(c) - f(x_1)}{c - x_1} \le \frac{f(x_2) - f(c)}{x_2 - c}.$$

Since $c - x_1 > 0$ and $x_2 - c > 0$, we can cross-multiply

$$(x_2 - c)(f(c) - f(x_1)) \le (c - x_1)(f(x_2) - f(c)).$$

Expand both sides:

$$(x_2 - c)f(c) - (x_2 - c)f(x_1) \le (c - x_1)f(x_2) - (c - x_1)f(c).$$

Rearrange terms to isolate $f(c)$:

$$(x_2 - c)f(c) + (c - x_1)f(c) \le (x_2 - c)f(x_1) + (c - x_1)f(x_2),$$
$$(x_2 - c + c - x_1)f(c) \le (x_2 - c)f(x_1) + (c - x_1)f(x_2),$$
$$(x_2 - x_1)f(c) \le (x_2 - c)f(x_1) + (c - x_1)f(x_2).$$

Divide by $(x_2 - x_1)$ (which is positive):

$$f(c) \le \frac{x_2 - c}{x_2 - x_1}f(x_1) + \frac{c - x_1}{x_2 - x_1}f(x_2).$$

This inequality is the precise condition for $f$ to be convex on $(a, b)$.

**Direction 2 (reverse implication): If $f$ is convex on $(a, b)$, then $f''(x) \ge 0$ for all $x \in (a, b)$.**

Suppose that $f$ is convex on $(a, b)$ and is twice differentiable. We want to show that $f''(x) \ge 0$ for every $x \in (a, b)$.

By the definition of convexity, for any $x_0 \in (a, b)$ and for any $h > 0$ such that $x_0 - h$ and $x_0 + h$ are in $(a, b)$, the following inequality holds:

$$\frac{f(x_0) - f(x_0 - h)}{x_0 - (x_0 - h)} \leq \frac{f(x_0 + h) - f(x_0)}{(x_0 + h) - x_0}.$$

This simplifies to

$$\frac{f(x_0) - f(x_0 - h)}{h} \leq \frac{f(x_0 + h) - f(x_0)}{h}.$$

Let's consider the difference quotient for the second derivative. From the above inequality, we can write

$$f(x_0) - f(x_0 - h) \leq f(x_0 + h) - f(x_0).$$

Rearranging the terms,

$$f(x_0 + h) - 2f(x_0) + f(x_0 - h) \geq 0.$$

Divide by $h^2$ (which is positive):

$$\frac{f(x_0 + h) - 2f(x_0) + f(x_0 - h)}{h^2} \geq 0.$$

Now, we take the limit as $h \to 0$. By the definition of the second derivative,

$$\lim_{h \to 0} \frac{f(x_0 + h) - 2f(x_0) + f(x_0 - h)}{h^2} = f''(x_0).$$

Since the limit of a nonnegative function is nonnegative, we have

$$f''(x_0) \geq 0.$$

This holds for any $x_0 \in (a, b)$, thus $f''(x) \geq 0$ for all $x \in (a, b)$.

For part (2), if $f$ is concave, then $-f$ is convex. By part (1), $(-f)''(x) \geq 0$, which implies $-f''(x) \geq 0$, or $f''(x) \leq 0$. Conversely, if $f''(x) \leq 0$, then $-f''(x) \geq 0$, so $(-f)''(x) \geq 0$. By part (1), $-f$ is convex, which means $f$ is concave.

This completes the proof. □

**Remark 13.** It is important to note that convexity does not necessarily imply the existence of a second derivative. A function may be convex on an interval without being twice differentiable at every point in that interval.

For example, the function $f(x) = |x|$ is convex on $\mathbb{R}$, but it is not differentiable at $x = 0$, and hence its second derivative does not exist at this point.

Therefore, while the condition $f''(x) > 0$ implies strict convexity, the converse does not always hold, i. e., convexity can exist even when $f''(x)$ fails to exist at certain points.

## 9.3.1 Inflection points

**Definition 9.3** (Inflection point). An **inflection point** (see Figure 9.5) is a point on the graph of a function where the concavity changes. This means that the function transitions from being:

- **convex** to **concave**, or
- **concave** to **convex**.

Mathematically, if $f''(x)$ exists in an open interval containing $x = a$, and

$$\text{the sign of } f''(x) \text{ changes at } x = a,$$

then $x = a$ is an inflection point.

**Theorem 9.2** (Condition for inflection point). *If $f$ is a twice differentiable function and $(x_0, f(x_0))$ is an inflection point, then $f''(x_0) = 0$.*

*Proof.* If $f''(x_0) \neq 0$, then by continuity of $f''$, there would be a neighborhood around $x_0$ where $f''(x)$ has the same sign as $f''(x_0)$. This would imply that the concavity does not change at $x_0$, contradicting the definition of an inflection point. Therefore, it must be that $f''(x_0) = 0$. ☐

**Note.** While $f''(x_0) = 0$ is a necessary condition for an inflection point (if $f''$ is continuous), **it is not a sufficient condition**. For an inflection point to exist at $x_0$, the second derivative $f''(x)$ must actually change sign around $x_0$.

**Example.** Consider $f(x) = x^4$. We have $f'(x) = 4x^3$ and $f''(x) = 12x^2$. At $x = 0, f''(0) = 0$. However, for $x < 0, f''(x) > 0$, and for $x > 0, f''(x) > 0$. The concavity does not change at $x = 0$ (the function remains convex). Therefore, $(0,0)$ is not an inflection point for $f(x) = x^4$.

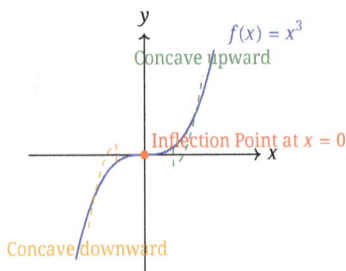

The graph of $f(x) = x^3$ has an inflection point at $x = 0$, where the concavity changes from downward to upward.

**Figure 9.5:** Graph of $f(x) = x^3$ with inflection point at $x = 0$.

## 9.4 Examples

**Example 9.1** (Monotonicity and derivatives). Determine the intervals on which the function $f(x) = x^3 - 3x$ is increasing or decreasing.

*Solution.*

$$f'(x) = 3x^2 - 3 = 3(x - 1)(x + 1),$$
$$f'(x) = 0 \quad \Rightarrow \quad x = -1, \quad x = 1.$$

**Sign table of the first derivative**

| $x$ | $(-\infty, -1)$ | $(-1, 1)$ | $(1, \infty)$ |
|---|---|---|---|
| $f'(x)$ | $+$ | $-$ | $+$ |
| Behavior | Increasing | Decreasing | Increasing |

**Conclusion**
- $f$ is **increasing** on $(-\infty, -1) \cup (1, \infty)$.
- $f$ is **decreasing** on $(-1, 1)$.

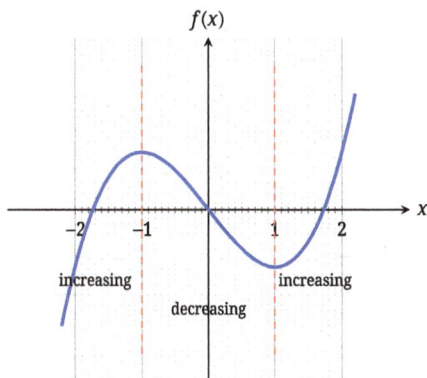

□

**Example 9.2** (Finding inflection points). Find the inflection points of $f(x) = x^3 - 3x$.

*Solution.*
**First and second derivatives**

$$f'(x) = 3x^2 - 3, \quad f''(x) = 6x.$$

Set $f''(x) = 0 \Rightarrow x = 0$, so we test the sign of $f''(x)$ around 0.

## Sign table of the second derivative

| $x$ | $(-\infty, 0)$ | $(0, \infty)$ |
|---|---|---|
| $f''(x)$ | $-$ | $+$ |
| Concavity | Concave (down) | Convex (up) |

## Conclusion

- The function has an **inflection point** at $(0, 0)$.
- $f$ is **concave** on $(-\infty, 0)$ and **convex** on $(0, \infty)$.

## Graphical representation

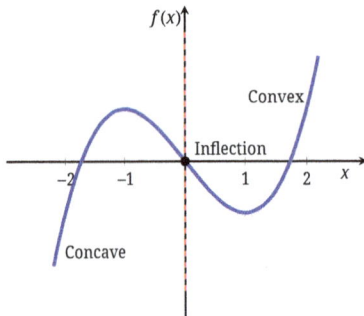

$\square$

## 9.5 Python and AI

**AI Request**

Write a Python code that checks if a given function is convex or concave on a given interval.

**AI Request**

Study the function $f(x) = x^3 - 3x^2 + 2$ and step-by-step compute the intervals of monotonicity and convexity. Give me the Latex code of the solution.

## 9.6 Exercises

### True or false

Decide whether each statement is **true** or **false**. Justify your answer.

1. If $f'(x) > 0$ for all $x$ in an interval, then $f$ is strictly increasing on that interval.
   **Answer:** _____
2. If $f''(x) < 0$ for all $x$ in an interval, then $f$ is convex there.
   **Answer:** _____

3. Every inflection point occurs where $f''(x) = 0$.

    **Answer:** _____

4. If $f$ is concave on an interval, then the graph of $f$ lies below any chord joining two points of its graph on that interval.

    **Answer:** _____

5. If $f$ is strictly increasing, then $f'(x) > 0$ everywhere.

    **Answer:** _____

## Multiple choice questions

Choose the correct option (A, B, C, or D) for each question.

1. Which of the following is the correct criterion for strict convexity on an open interval?

    A. $f'(x) > 0$ for all $x$    B. $f''(x) > 0$ for all $x$

    C. $f(x) > 0$ for all $x$    D. $f''(x) < 0$ for all $x$

    **Answer:** _____

2. If $f''(x)$ changes sign at $x = a$, then $x = a$ is

    A. a local extremum    B. a critical point

    C. an inflection point    D. an endpoint

    **Answer:** _____

3. For which of the following functions is the graph concave upward on $\mathbb{R}$?

    A. $f(x) = x^2$   B. $f(x) = -x^2$

    C. $f(x) = x^3$   D. $f(x) = |x|$

    **Answer:** _____

4. The function $f(x) = x^3$ is

    A. always convex                        B. always concave

    C. has an inflection point at $x = 0$    D. both A and B

    **Answer:** _____

5. Which of the following is a sufficient condition for $f$ to be decreasing on $I$?

    A. $f'(x) > 0$ for all $x \in I$    B. $f'(x) < 0$ for all $x \in I$

    C. $f''(x) > 0$ for all $x \in I$   D. $f(x)$ is continuous on $I$

    **Answer:** _____

## Matching exercise

Match each concept with its correct description.

1. Strictly increasing function    A. $f''(x) > 0$ everywhere
2. Inflection point                 B. $f(a) < f(b)$ whenever $a < b$
3. Convex function                  C. Point where $f''(x)$ changes sign
4. Concave function                 D. $f''(x) < 0$ everywhere
5. Chord lies below graph           E. Concave function

**Answers:** 1. _____    2. _____    3. _____    4. _____    5. _____

## Fill in the blanks

Fill in each blank with the most appropriate word or phrase.

1.  If $f'(x) > 0$ for all $x$ in an interval, then $f$ is _____ on that interval.

    **Answer:** _____

2.  If $f''(x) < 0$ on an interval, $f$ is said to be _____ there.

    **Answer:** _____

3.  A point $x = a$ where the graph of $f$ changes from convex to concave is called an

    _____.

    **Answer:** _____

4.  For a convex function, the line segment joining any two points of its graph lies _____ the graph.

    **Answer:** _____

5.  The condition $f''(x) = 0$ is _____ for an inflection point, but not sufficient.

    **Answer:** _____

## Theoretical and computational problems

Solve each problem step-by-step.

1.  Find the intervals of monotonicity and concavity for $f(x) = x^3 - 3x^2 + 2$.
2.  Prove that $f(x) = e^x$ is strictly increasing and strictly convex on $\mathbb{R}$.
3.  Let $f(x) = x^4 - 2x^2$. Find and classify all inflection points.
4.  Find the intervals where $f(x) = \ln x$ is concave or convex.
5.  Show that if $f''(x) > 0$ on $(a, b)$, then $f'$ is strictly increasing on $(a, b)$.

## Python-based exercises

Use Python (sympy, numpy, matplotlib) to solve the following:

1.  Plot $f(x) = x^4 - 2x^2$ and use sympy to find numerically the inflection points.
2.  Write a Python function that tests if a given function is strictly increasing or decreasing on an interval.
3.  For $f(x) = x^3 - 3x$, plot its graph and indicate all intervals where it is convex or concave.
4.  Use sympy to check for convexity/concavity of $f(x) = \ln x$ on $(0, \infty)$.
5.  Numerically estimate the point of inflection of $f(x) = \tanh x$.

# 10 Inverse functions

**Contents**

Inverse functions allow us to "reverse" the effect of a function and solve equations in terms of known operations. Invertibility is closely tied to one-to-one behavior and monotonicity, ensuring that each output corresponds to a unique input.

In this chapter, we define when a function admits an inverse and explain the importance of continuity and strict monotonicity. We present the inverse function theorem, showing how to compute the derivative of an inverse function.

With motivating examples and practical methods, you will learn to analyze, differentiate, and interpret inverse functions in mathematics and beyond.

## 10.1 Introduction

Inverse functions play a central role in mathematics, providing a systematic way to "undo" the action of a function and to solve equations in terms of known operations. From solving algebraic equations and logarithms, to trigonometric and exponential inverses, the concept of invertibility underlies much of modern analysis.

In this chapter, we explore the foundational properties and theorems of inverse functions. We begin by clarifying the relationship between strict monotonicity, injectivity (one-to-one-ness), and the existence of an inverse function. The interplay between continuity, monotonicity, and invertibility on intervals is developed in detail, highlighting why the structure of the domain matters.

The differentiability of inverse functions is another key topic: we present and prove the inverse function theorem, which allows us to compute the derivative of an inverse in terms of the derivative of the original function. Careful examples—including linear, quadratic, exponential, and trigonometric functions—demonstrate how these abstract results translate into effective computational techniques.

Special attention is paid to graphical and geometric interpretations, such as the reflection property of function graphs across the line $y = x$. Throughout, we emphasize both the necessity of the various hypotheses and common pitfalls, such as failure of injectivity or continuity.

By the end of this chapter, you will have mastered the essential theory and tools for working with inverse functions and their derivatives, equipping you to analyze, compute, and interpret inverses in a wide range of mathematical and applied settings.

https://doi.org/10.1515/9783112228289-010

## 10.2 Basic results

The following theorem clarifies the fundamental interplay between monotonicity, continuity, and invertibility of functions, showing that strictly monotonic functions possess continuous inverses, thus providing critical insights into the structure and behavior of inverse functions.

**Theorem 10.1.** *If a function f is strictly monotonic on its domain, then it is one-to-one there. Its inverse $f^{-1}$ is also strictly monotonic, of the same monotonicity type. Moreover, if f is one-to-one and continuous on $I = [a, b]$, then f is strictly monotonic and $f^{-1}$ is continuous on its domain.*

*Proof.* We will assume that $f$ is strictly increasing; the proof for strictly decreasing is similar. The function $f$ is one-to-one because, for any $x, y$ in its domain with $x < y$, we have $f(x) < f(y)$, so it is one-to-one and its inverse is defined. We will show that its inverse is also strictly increasing. Suppose not; then we could find $x, y$ in the domain of $f^{-1}$ with $x < y$ and $f^{-1}(x) \geq f^{-1}(y)$. But since $f$ is strictly increasing, we would have

$$f(f^{-1}(x)) \geq f(f^{-1}(y)),$$

which means $x \geq y$, a contradiction.

Next, we show that if $f$ is one-to-one and continuous on $I = [a, b]$, then it is strictly monotonic on $I$. Suppose $f(a) < f(b)$ and pick any $a < x < b$. We claim that necessarily $f(a) < f(x) < f(b)$. Since $f$ is one-to-one, it cannot be that $f(x) = f(a)$ or $f(x) = f(b)$. The case $f(x) < f(a)$ is also impossible, because by the intermediate value theorem, there would exist $x^*$ with $f(x^*) = f(a)$. Similarly, for $f(x) > f(b)$. Now, pick any $y$ with $a < x < y < b$. Then $f(x) < f(b)$, so working in the interval $[x, b]$ as above, we have $f(x) < f(y) < f(b)$, which means $f$ is strictly increasing. The case $f(a) > f(b)$ is handled similarly, showing $f$ is strictly decreasing.

Now, we prove that the inverse $f^{-1}$ is continuous when $f$ is one-to-one and continuous. We have already shown that $f$ is strictly monotonic, so we may assume it is strictly increasing. Define the function $F(x) = f(x + \varepsilon) - f(x)$ for some $\varepsilon > 0$. Since $F$ is continuous, it attains its minimum, say $\delta > 0$, at some point in $[a, b - \varepsilon]$. If we choose two points $x_1, x_2$ separated by at least $\varepsilon$, i. e., without loss of generality $x_2 \geq x_1 + \varepsilon$, then

$$|y_1 - y_2| = |f(x_1) - f(x_2)| = y_2 - y_1 = f(x_2) - f(x_1) \geq f(x_1 + \varepsilon) - f(x_1) \geq \delta,$$

because $f$ is strictly increasing (see the definition of $F$). So, in general, if $|x_1 - x_2| \geq \varepsilon$, then

$$|y_2 - y_1| = |f(x_2) - f(x_1)| \geq \delta.$$

Now, if we choose $y_1, y_2$ so that $|y_1 - y_2| < \delta$, then also $|f(x_1) - f(x_2)| < \delta$, which forces $|x_1 - x_2| < \varepsilon$; otherwise, we get a contradiction. But $|x_1 - x_2| = |f^{-1}(y_1) - f^{-1}(y_2)| < \varepsilon$. □

**Remark 14.** The assumption that the domain is an interval and not just a subset of $\mathbb{R}$ is essential. Consider the function $f : [0,1] \cup [2,3] \to \mathbb{R}$ defined by

$$f(x) = \begin{cases} x, & \text{if } x \in [0,1], \\ x-1, & \text{if } x \in [2,3]. \end{cases}$$

The domain of $f$ is $[0,1] \cup [2,3]$, which is not an interval. The function $f$ is one-to-one and continuous on its domain, but it is not strictly monotonic.

- **One-to-one:** For any $x_1, x_2 \in [0,1] \cup [2,3]$, if $f(x_1) = f(x_2)$, then $x_1 = x_2$. This is because $f(x) = x$ on $[0,1]$ and $f(x) = x-1$ on $[2,3]$, so the images of $[0,1]$ and $[2,3]$ under $f$ do not overlap.
- **Continuous:** The function $f$ is continuous on both $[0,1]$ and $[2,3]$, and since these intervals are disjoint, $f$ is continuous on its entire domain $[0,1] \cup [2,3]$.
- **Not strictly monotonic:** The function $f$ is increasing on each of the intervals $[0,1]$ and $[2,3]$, but it is not strictly monotonic overall. For example, $f(1) = 1$ and $f(2) = 1$, so there is no consistent monotonic behavior across the entire domain.

This example demonstrates that the assumption in the theorem that the domain is an interval is essential. Without this assumption, a function can be one-to-one and continuous but fail to be strictly monotonic.

The following theorem establishes a fundamental formula linking the derivatives of a function and its inverse, providing a powerful tool for analyzing and computing derivatives of inverse functions across calculus and analysis.

**Theorem 10.2.** *Let $f$ be a continuous and one-to-one function defined on an interval $I$, and suppose that $f$ is differentiable at $f^{-1}(b)$ with $f'(f^{-1}(b)) \neq 0$. Then $f^{-1}$ is differentiable at $b$ and*

$$\left(f^{-1}\right)'(b) = \frac{1}{f'(f^{-1}(b))}.$$

*Proof.* Let $b = f(a)$. Then

$$\lim_{h \to 0} \frac{f^{-1}(b+h) - f^{-1}(b)}{h} = \lim_{h \to 0} \frac{f^{-1}(b+h) - a}{h}.$$

For $k$ such that $b + h = f(a+k)$, we have

$$\lim_{h \to 0} \frac{f^{-1}(b+h) - a}{h} = \lim_{h \to 0} \frac{f^{-1}(f(a+k)) - a}{h}$$

$$= \lim_{h \to 0} \frac{k}{f(a+k) - f(a)}.$$

But $k = f^{-1}(b+h) - f^{-1}(b)$ and by the continuity of $f^{-1}$ at $b$, we have $k \to 0$ as $h \to 0$. Since

$$\lim_{k \to 0} \frac{f(a+k) - f(a)}{k} = f'(a) = f'(f^{-1}(b)) \neq 0,$$

we obtain

$$(f^{-1})'(b) = \frac{1}{f'(f^{-1}(b))}. \qquad \square$$

## 10.3 Examples

**Example 10.1** (Linear inverse and its derivative). Let $f: \mathbb{R} \to \mathbb{R}$ be given by $f(x) = 3x + 2$.
(a) Show that $f$ is a bijection and find $f^{-1}$ explicitly.
(b) Use Theorem 10.2 to compute $(f^{-1})'(b)$ for an arbitrary $b \in \mathbb{R}$.

*Solution.*
(a) The map $f$ is strictly increasing (constant positive slope), hence one-to-one, and its image is all of $\mathbb{R}$, so it is onto. Solving $y = 3x + 2$ for $x$ gives

$$f^{-1}(y) = \frac{y - 2}{3}.$$

(b) Differentiating $f$ first,

$$f'(x) = 3 \neq 0 \quad \text{for all } x.$$

By Theorem 10.2,

$$(f^{-1})'(b) = \frac{1}{f'(f^{-1}(b))} = \frac{1}{3} \quad \text{for every } b \in \mathbb{R}. \qquad \square$$

**Example 10.2** (Derivative of arcsin $x$). Define $f(x) = \sin x$ on the restricted domain $I = [-\frac{\pi}{2}, \frac{\pi}{2}]$.
(a) Explain why $f$ satisfies the hypotheses of Theorem 10.2.
(b) Show that $(\arcsin)'(x) = \frac{1}{\sqrt{1-x^2}}$ for $x \in (-1, 1)$.

*Solution.*
(a) On $I$ the sine function is continuous, strictly increasing, and differentiable with $f'(x) = \cos x$, which never vanishes on the open interval $(-\frac{\pi}{2}, \frac{\pi}{2})$. Hence every condition of Theorem 10.2 is met.
(b) For $b \in (-1, 1)$ we have $f^{-1}(b) = \arcsin b$ and

$$f'(f^{-1}(b)) = \cos(\arcsin b) = \sqrt{1 - b^2}.$$

Therefore

$$(\arcsin)'(b) = \frac{1}{\sqrt{1 - b^2}}. \qquad \square$$

**Example 10.3** (Inverse of a quadratic branch). Consider $f(x) = x^2$ restricted to $I = [0, \infty)$.
(a) Verify that $f$ is one-to-one and continuous on $I$.
(b) Determine $f^{-1}$ and prove it is continuous on its domain.
(c) Compute $(f^{-1})'(b)$ for any $b > 0$.

*Solution.*
(a) On $I$ the function $f(x) = x^2$ is strictly increasing, hence injective, and clearly continuous.
(b) Solving $y = x^2$ with $x \geq 0$ gives $f^{-1}(y) = \sqrt{y}$. The square-root function is continuous
   for $y \geq 0$, so $f^{-1}$ is continuous on its domain $(0, \infty)$ and right continuous at $y = 0$.
(c) Since $f'(x) = 2x$ and $f^{-1}(b) = \sqrt{b}$,

$$(f^{-1})'(b) = \frac{1}{f'(\sqrt{b})} = \frac{1}{2\sqrt{b}}. \qquad \square$$

**Example 10.4** (Inverse implicit differentiation). Let $f$ be differentiable and strictly increasing on an interval $I$ with $f(a) = b$. Assume $f'(a) = 5$. Without writing $f^{-1}$ explicitly, show that $(f^{-1})'(b) = \frac{1}{5}$ using implicit
differentiation.

*Solution.* Set $y = f(x)$ with $x \in I$. Writing the inverse relation as $x = f^{-1}(y)$ and differentiating implicitly,

$$1 = \frac{d}{dy}(f(f^{-1}(y))) = f'(f^{-1}(y))\,(f^{-1})'(y).$$

Evaluating at $y = b$ (so that $f^{-1}(b) = a$) gives $1 = f'(a)\,(f^{-1})'(b) = 5\,(f^{-1})'(b)$, hence
$(f^{-1})'(b) = \frac{1}{5}$. $\qquad \square$

## 10.4  Python and AI

**AI Request**
Write a Python code that computes symbolically the inverse of a function, if it exists. Plot both of them.

## 10.5  Exercises

### True or false
Decide whether each statement is **true** or **false**. Justify your answer.
1. If a function $f$ is strictly decreasing on an interval, then its inverse $f^{-1}$ is strictly
   increasing.
   **Answer:** _____

2. A continuous one-to-one function on a closed interval $[a, b]$ must attain a maximum and a minimum there.

    **Answer:** _____

3. If $f$ is differentiable on an interval and $f'(x) \neq 0$ everywhere, then $f$ admits a differentiable inverse on its image.

    **Answer:** _____

4. For every real-valued function $f$, the graph of $y = f^{-1}(x)$ is the reflection of the graph of $y = f(x)$ across the y-axis.

    **Answer:** _____

5. If $f$ is strictly monotonic but *not* continuous on an interval, an inverse function $f^{-1}$ need not exist.

    **Answer:** _____

## Multiple choice questions

Choose the correct option (A, B, C, or D) for each question.

1. Which of the following guarantees that $f^{-1}$ is continuous on its domain?

    A. $f$ is one-to-one only B. $f$ is continuous only

    C. $f$ is one-to-one and continuous D. $f$ is differentiable

    **Answer:** _____

2. Suppose $f$ is differentiable and satisfies $f'(x) \neq 0$ on an interval. What additional hypothesis ensures $(f^{-1})'(b) = \frac{1}{f'(f^{-1}(b))}$ for all $b$ in the range of $f$?

    A. $f$ is convex B. $f$ is continuous and one-to-one

    C. $f''(x)$ exists D. $f$ is bounded

    **Answer:** _____

3. The function $f(x) = x^3$ (defined on $\mathbb{R}$) is

    A. not one-to-one

    B. one-to-one but not differentiable

    C. one-to-one and its inverse is differentiable everywhere

    D. one-to-one but its inverse fails to be differentiable at $x = 0$

    **Answer:** _____

4. Let $f(x) = e^x$ on $\mathbb{R}$. Then $f^{-1}(x)$ equals

    A. $\ln x$ B. $\log_{10} x$

    C. $x e^x$ D. No inverse exists

    **Answer:** _____

5. If $f^{-1}$ exists and is differentiable at $b$, which of the following must hold?

    A. $f'(f^{-1}(b)) \neq 0$ B. $f'(f^{-1}(b)) = 0$

    C. $f''(f^{-1}(b)) \neq 0$ D. $f$ is even

    **Answer:** _____

## Matching exercise

Match each concept with the correct description.

| 1. Reflection property | A. $f^{-1}$ exists and is continuous |
|---|---|
| 2. Horizontal line test | B. $(f^{-1})'(b) = \frac{1}{f'(f^{-1}(b))}$ |
| 3. Inverse function theorem | C. Graph of $f^{-1}$ obtained by reflecting graph of $f$ across $y = x$ |
| 4. One-to-one and continuous | D. Criterion to check injectivity graphically |
| 5. Differentiability condition | E. Requires $f'(x) \neq 0$ |

**Answers: 1._____  2._____  3._____  4._____  5._____**

## Fill in the blanks

Fill in each blank with the most appropriate word or phrase.

1. A function that never takes the same value twice on its domain is called _____ .

   **Answer:** _____

2. The necessary condition $f'(a) \neq 0$ in the inverse function theorem prevents the derivative from _____ at $a$.

   **Answer:** _____

3. For a strictly monotonic and continuous $f$, the domain of $f^{-1}$ equals the _____ of $f$.

   **Answer:** _____

4. The inverse of $\ln x$ is the _____ function.

   **Answer:** _____

5. If $f^{-1}$ exists and is differentiable, its derivative at $b$ is the reciprocal of the _____ of $f$ at $f^{-1}(b)$.

   **Answer:** _____

## Theoretical and computational problems

Solve each problem step-by-step.

1. For $f(x) = x^3 - 3x$, determine whether an inverse exists on (i) $(-\infty, 0]$ and (ii) $[0, \infty)$. If so, find $f^{-1}$ on each interval and state its monotonicity.

2. Let $g(x) = \sqrt{x + 4}$ with domain $[-4, \infty)$. Verify that $g$ has a differentiable inverse and compute $(g^{-1})'(5)$.

3. Show that $h(x) = x + \sin x$ is strictly increasing on $\mathbb{R}$ and deduce that $h^{-1}$ exists. Prove that $h^{-1}$ is differentiable everywhere.

4. Find all $x$ such that the inverse of $p(x) = x^2 + 1$ (restricted to $[0, \infty)$) has derivative $1/4$.

5. Suppose $f$ is continuously differentiable, $f(2) = 5$, and $f'(2) = 4$. Using a linear approximation for $f^{-1}$, estimate $f^{-1}(5.2)$.

## Python-based exercises

Use Python (`sympy`, `numpy`, `matplotlib`) to solve the following:

1. Plot $y = f(x) = x + \sin x$ on $[-6, 6]$ and use `sympy.nsolve` to approximate $f^{-1}(3)$.

2. Write a Python function is_bijective(f, interval) that tests numerically whether a symbolic function f is one-to-one on a given interval by sampling.

3. For $q(x) = \tanh x$, compute numerically the derivative of $q^{-1}$ at $x = 0.5$ and compare it with the theoretical value from the inverse function theorem.

4. Generate a plot showing both $y = e^x$ and its inverse $y = \ln x$ on suitable domains; verify graphically that they are reflections across $y = x$.

5. For $s(x) = x^3 - 3x$, restrict to $[2, \infty)$ and use sympy to (i) find $s^{-1}$ symbolically, (ii) plot both $s$ and $s^{-1}$ on the same axes, and (iii) mark the point whose $x$-coordinate equals $s^{-1}(10)$.

# 11 L'Hospital's rules

## Contents

L'Hospital's rule provides a powerful tool for evaluating limits that yield indeterminate forms such as $0/0$ or $\infty/\infty$. By connecting the limit of a quotient to the limit of the derivatives, it simplifies the analysis of many otherwise difficult expressions.

This chapter explains the precise conditions needed for the rule to apply and presents its rigorous justification via the Cauchy mean value theorem. Through a range of examples, you will learn how to use L'Hospital's rule in both straightforward and more subtle cases, including repeated application and related logarithmic forms.

You will also see common pitfalls and learn how to distinguish when L'Hospital's rule can and cannot be used.

## 11.1 Introduction

The evaluation of indeterminate limits, such as $0/0$ and $\infty/\infty$, is a central theme in calculus and analysis. Many fundamental problems in mathematics, science, and engineering reduce to computing such limits, especially when direct substitution fails. L'Hospital's rule, named after **Guillaume de l'Hospital**, provides a systematic and rigorous technique for resolving these challenging forms by relating the behavior of a function to that of its derivative.

In this chapter, we build a clear and detailed understanding of L'Hospital's rules and their rigorous foundation in the Cauchy mean value theorem. We present the conditions under which L'Hospital's rule applies, clarify its limitations, and discuss its use in both the $0/0$ and $\infty/\infty$ cases. Special attention is given to the hypotheses—such as differentiability and nonvanishing denominators—that are essential for valid application.

Through carefully chosen examples, we demonstrate step-by-step how to apply L'Hospital's rule to a variety of classical limits, including those involving exponential, logarithmic, and trigonometric functions. We also address more subtle indeterminate forms (like $1^{\infty}$) and show how logarithmic manipulation and repeated application of the rule can resolve complex expressions.

By the end of this chapter, you will be able to recognize when and how to use L'Hospital's rule confidently, appreciate its theoretical underpinnings, and avoid common mistakes in its application. Mastery of these techniques is essential for further study in analysis, differential equations, and applied mathematics.

https://doi.org/10.1515/9783112228289-011

**Theorem 11.1** (Cauchy mean value theorem). *If $f$ and $g$ are continuous on $[a, b]$ and differentiable on $(a, b)$, then there exists $x \in (a, b)$ such that*

$$\left(f(b) - f(a)\right)g'(x) = \left(g(b) - g(a)\right)f'(x).$$

*Proof.* Define $h(x) = f(x)(g(b) - g(a)) - g(x)(f(b) - f(a))$. The function $h$ is continuous on $[a, b]$ and differentiable on $(a, b)$, and $h(a) = f(a)g(b) - g(a)f(b) = h(b)$. By Rolle's theorem, there exists $x \in (a, b)$ such that $h'(x) = 0$, from which the required identity follows. □

**Theorem 11.2** (Derivative limit at zero over zero). *Suppose for functions $f$, $g$ that*

$$\lim_{x \to a} f(x) = \lim_{x \to a} g(x) = 0,$$

*and that $f'$, $g'$ exist at $a$ with $g'(a) \neq 0$. Then*

$$\lim_{x \to a} \frac{f(x)}{g(x)} = \frac{f'(a)}{g'(a)}.$$

*Proof.* By continuity at $a$, $f(a) = g(a) = 0$, so

$$\lim_{x \to a} \frac{f(x)}{g(x)} = \lim_{x \to a} \frac{\frac{f(x) - f(a)}{x - a}}{\frac{g(x) - g(a)}{x - a}} = \frac{f'(a)}{g'(a)}. \qquad □$$

The following theorems, collectively known as L'Hospital's rule, provide essential tools for evaluating limits involving indeterminate forms of type $0/0$ and $\infty/\infty$. These results simplify the computation of otherwise challenging limits by reducing them to derivatives, making them fundamental techniques in analysis and calculus.

**Theorem 11.3** (L'Hospital's rule: 0/0 form). *Let $f$, $g$ be differentiable on $(a, b)$, with $a, b \in \mathbb{R} \cup \{-\infty, +\infty\}$, and $g'(x) \neq 0$ for every $x \in (a, b)$. If*

$$\lim_{x \to a^+} f(x) = \lim_{x \to a^+} g(x) = 0,$$

*then*

$$\lim_{x \to a^+} \frac{f(x)}{g(x)} = \lim_{x \to a^+} \frac{f'(x)}{g'(x)}$$

*whenever the right limit exists (finite or infinite). If the right-hand limit does not exist, we cannot conclude about the existence of the left-hand limit.*

*Proof.* We distinguish between the cases $a \in \mathbb{R}$ and $a = \pm\infty$. For $a \in \mathbb{R}$, define

$$F(x) = \begin{cases} f(x), & x \in (a, b) \\ 0, & x = a \end{cases} \quad \text{and} \quad G(x) = \begin{cases} g(x), & x \in (a, b) \\ 0, & x = a \end{cases}$$

so $F$, $G$ are continuous on $[a, b)$ and differentiable on $(a, b)$. By the Cauchy mean value Theorem 11.1, for $x \in (a, b)$, there is $\xi \in (a, x)$ such that

$$\frac{f(x)}{g(x)} = \frac{F(x) - F(a)}{G(x) - G(a)} = \frac{F'(\xi)(x - a)}{G'(\xi)(x - a)} = \frac{f'(\xi)}{g'(\xi)}.$$

Taking the limit as $x \to a^+$ yields

$$\lim_{x \to a^+} \frac{f(x)}{g(x)} = \lim_{x \to a^+} \frac{f'(\xi)}{g'(\xi)} = \lim_{\xi \to a^+} \frac{f'(\xi)}{g'(\xi)} = \lim_{x \to a^+} \frac{f'(x)}{g'(x)}.$$

For $a = +\infty$ (or $-\infty$), substitute $t = 1/x$ and reduce to the previous case. □

---

**Theorem 11.4** (L'Hospital's rule: $\infty/\infty$ form). *Let $f$, $g$ be differentiable on $(a, b)$, with $a$, $b \in \mathbb{R} \cup \{-\infty, +\infty\}$ and $g'(x) \neq 0$ for every $x \in (a, b)$. If $\lim_{x \to a^+} f(x) = \lim_{x \to a^+} g(x) = +\infty$ (or $-\infty$), then*

$$\lim_{x \to a^+} \frac{f(x)}{g(x)} = \lim_{x \to a^+} \frac{f'(x)}{g'(x)}$$

*whenever the right limit exists (finite or infinite). Otherwise, nothing can be concluded about the left limit.*

---

*Proof.* Apply the Cauchy mean value Theorem 11.1 on $(a, a + h)$ and for $c$ with $a < x < c < a + h$, get

$$\frac{f(x) - f(c)}{g(x) - g(c)} = \frac{f(x)}{g(x)} \frac{1 - f(c)/f(x)}{1 - g(c)/g(x)} = \frac{f'(\xi)}{g'(\xi)}$$

for some $\xi$ with $a < x < \xi < c$. Thus,

$$\frac{f(x)}{g(x)} = \frac{f'(\xi)}{g'(\xi)} \frac{1 - g(c)/g(x)}{1 - f(c)/f(x)}.$$

If $\lim_{x \to a^+} \frac{f'(\xi)}{g'(\xi)} = A \in \mathbb{R}$, choose $c$ close to $a$ so that $|\frac{f'(\xi)}{g'(\xi)} - A| < \varepsilon$ for given $\varepsilon > 0$. As $x \to a^+$, $\frac{1 - g(c)/g(x)}{1 - f(c)/f(x)} \to 1$. Hence, $\frac{f(x)}{g(x)} \to \lim_{x \to a^+} \frac{f'(x)}{g'(x)}$.

If $\lim_{x \to a^+} \frac{f'(x)}{g'(x)} = +\infty$, the argument is similar. The case $-\infty$ is analogous by considering $-\frac{f(x)}{g(x)}$. □

## 11.2 Examples

---

**Example 11.1** (Application of the Cauchy mean value theorem). Apply the Cauchy mean value theorem to the functions $f(x) = x^2$ and $g(x) = x^3$ on the interval $[1, 2]$. Find a point $c \in (1, 2)$ such that

$$\big(f(b) - f(a)\big)g'(c) = \big(g(b) - g(a)\big)f'(c).$$

*Solution.* We are given

$$f(x) = x^2, \quad g(x) = x^3.$$

Compute values at endpoints
- $f(1) = 1, f(2) = 4.$
- $g(1) = 1, g(2) = 8.$

Compute derivatives
- $f'(x) = 2x, g'(x) = 3x^2.$

Apply the formula

$$(f(2) - f(1))g'(c) = (g(2) - g(1))f'(c) \quad \Rightarrow \quad (4 - 1)(3c^2) = (8 - 1)(2c)$$
$$\Rightarrow \quad 3 \cdot 3c^2 = 7 \cdot 2c \quad \Rightarrow \quad 9c^2 = 14c.$$

Solve

$$9c^2 - 14c = 0 \quad \Rightarrow \quad c(9c - 14) = 0 \quad \Rightarrow \quad c = 0 \text{ or } c = \frac{14}{9}.$$

Since $c \in (1, 2)$, we take

$$c = \frac{14}{9} \approx 1.56.$$

Thus, there exists $c \in (1, 2)$ satisfying the conclusion of the Cauchy mean value theorem. □

**Example 11.2** (L'Hospital's rule for $\frac{0}{0}$). Evaluate the limit

$$\lim_{x \to 0} \frac{\sin x}{x}$$

using L'Hospital's rule.

*Solution.* As $x \to 0$, both numerator and denominator approach zero

$$\lim_{x \to 0} \sin x = 0, \quad \lim_{x \to 0} x = 0.$$

So we apply L'Hospital's rule

$$\lim_{x \to 0} \frac{\sin x}{x} = \lim_{x \to 0} \frac{\cos x}{1} = \cos(0) = 1.$$

Hence,

$$\lim_{x \to 0} \frac{\sin x}{x} = 1.$$

☐

**Example 11.3** (L'Hospital's rule for $\frac{\infty}{\infty}$). Evaluate the limit

$$\lim_{x \to \infty} \frac{x^2}{e^x}$$

using L'Hospital's rule.

*Solution.* As $x \to \infty$, both numerator and denominator tend to infinity

$$\lim_{x \to \infty} x^2 = \infty, \quad \lim_{x \to \infty} e^x = \infty.$$

Apply L'Hospital's rule once

$$\lim_{x \to \infty} \frac{x^2}{e^x} = \lim_{x \to \infty} \frac{2x}{e^x}.$$

Still in $\frac{\infty}{\infty}$ form; apply again

$$= \lim_{x \to \infty} \frac{2}{e^x} = 0.$$

Therefore

$$\lim_{x \to \infty} \frac{x^2}{e^x} = 0.$$

☐

**Example 11.4** (Repeated application of L'Hospital's rule). Evaluate

$$\lim_{x \to 0} \frac{e^x - 1 - x}{x^2}.$$

*Solution.* First, check the form

$$\lim_{x \to 0} (e^x - 1 - x) = 0, \quad \lim_{x \to 0} x^2 = 0 \Rightarrow \frac{0}{0}.$$

Apply L'Hospital's rule

$$\lim_{x \to 0} \frac{e^x - 1 - x}{x^2} = \lim_{x \to 0} \frac{e^x - 1}{2x}.$$

Still $\frac{0}{0}$, so apply again

$$= \lim_{x \to 0} \frac{e^x}{2} = \frac{e^0}{2} = \frac{1}{2}.$$

Thus

$$\lim_{x \to 0} \frac{e^x - 1 - x}{x^2} = \frac{1}{2}. \qquad \square$$

**Example 11.5** (L'Hospital's rule: logarithmic limit). Evaluate the limit

$$\lim_{x \to 0^+} \frac{\ln(1 + x)}{x}.$$

*Solution.* As $x \to 0^+$, both numerator and denominator approach zero

$$\ln(1 + x) \to 0, \quad x \to 0.$$

Thus, we have a $\frac{0}{0}$ indeterminate form.
    Apply L'Hospital's rule

$$\lim_{x \to 0^+} \frac{\ln(1 + x)}{x} = \lim_{x \to 0^+} \frac{1/(1 + x)}{1} = \lim_{x \to 0^+} \frac{1}{1 + x} = 1.$$

Therefore,

$$\boxed{1}$$

is the value of the limit. $\qquad \square$

**Example 11.6** (L'Hospital's rule: power functions). Compute the limit

$$\lim_{x \to 0^+} \frac{1 - \cos x}{x^2}.$$

*Solution.* As $x \to 0^+$, $1 - \cos x \to 0$ and $x^2 \to 0$, so the limit is of the $\frac{0}{0}$ form.
    Apply L'Hospital's rule

$$\lim_{x \to 0^+} \frac{1 - \cos x}{x^2} = \lim_{x \to 0^+} \frac{\sin x}{2x}.$$

Still $\frac{0}{0}$, so apply again

$$= \lim_{x \to 0^+} \frac{\cos x}{2} = \frac{1}{2}.$$

So,

$$\boxed{\dfrac{1}{2}}$$

is the value of the limit. □

**Example 11.7** (L'Hospital's rule: exponential minus polynomial). Evaluate

$$\lim_{x \to \infty} \frac{e^x}{x^n}$$

for any positive integer $n$.

*Solution.* As $x \to \infty$, $e^x \to \infty$ and $x^n \to \infty$ $(n \geq 1)$, so we have an $\frac{\infty}{\infty}$ form.
Apply L'Hospital's rule $n$ times

$$\lim_{x \to \infty} \frac{e^x}{x^n} = \lim_{x \to \infty} \frac{e^x}{nx^{n-1}} = \cdots = \lim_{x \to \infty} \frac{e^x}{n!} = \infty.$$

Thus, the exponential grows much faster than any power, so

$$\boxed{\infty}$$

is the value of the limit. □

**Example 11.8** (Indeterminate form $1^\infty$ via L'Hospital's rule). Evaluate

$$\lim_{x \to 0^+} (1 + 2x)^{1/x}.$$

*Solution.* We can rewrite the limit in exponential form by expressing the power as an exponent:

$$\lim_{x \to 0^+} (1 + 2x)^{1/x} = \lim_{x \to 0^+} \exp\left( \frac{1}{x} \ln(1 + 2x) \right).$$

Now, compute the inner limit

$$\lim_{x \to 0^+} \frac{1}{x} \ln(1 + 2x).$$

This is an indeterminate form $\frac{0}{0}$ if we let $y = 2x$, but instead, we can proceed directly or use Taylor expansion

$$\ln(1 + 2x) = 2x - 2x^2 + O(x^3).$$

So,

$$\frac{1}{x} \ln(1 + 2x) = \frac{1}{x}(2x - 2x^2 + O(x^3)) = 2 - 2x + O(x^2).$$

Therefore,

$$\lim_{x \to 0^+} \frac{1}{x} \ln(1 + 2x) = 2.$$

So the original limit is

$$\exp\left( \lim_{x \to 0^+} \frac{1}{x} \ln(1 + 2x) \right) = \exp(2).$$

**Final answer:**

$$\boxed{\lim_{x \to 0^+} (1 + 2x)^{1/x} = \exp(2)}$$  ☐

## 11.3 Python and AI

**AI Request**

Compute the

$$\lim_{x \to \infty} \frac{e^x}{x^n}$$

for any positive integer $n$. Give me the solution step-by-step in Latex.

**AI Request**

Provide any historical information you know about Guillaume de l'Hospital.

## 11.4 Exercises

### True or false

Decide whether each statement is **true** or **false**. Justify your answer.

1. L'Hospital's rule can be applied to any indeterminate form.
2. If $\lim_{x \to a} f(x) = \lim_{x \to a} g(x) = 0$ and $g'(x) \neq 0$ near $a$, then $\lim_{x \to a} \frac{f(x)}{g(x)} = \lim_{x \to a} \frac{f'(x)}{g'(x)}$ always exists.
3. If $f(x)$ and $g(x)$ are differentiable and $g'(x) \neq 0$, then L'Hospital's rule can always be applied for $x \to a$.
4. The limit $\lim_{x \to \infty} \frac{\ln x}{x}$ can be evaluated using L'Hospital's rule.
5. For the indeterminate form 0/0, the Cauchy mean value theorem underlies L'Hospital's rule.

### Multiple choice questions

Choose the correct option (A, B, C, or D) for each question.

1. Which of the following is not an indeterminate form to which L'Hospital's rule applies directly?

    A. $0/0$     B. $\infty/\infty$     C. $1^{\infty}$     D. $0 \cdot \infty$

    **Answer:** _____

2. What is the first step in evaluating $\lim_{x \to 0} \frac{e^x - 1}{\sin x}$ using L'Hospital's rule?

    A. Differentiate numerator and denominator

    B. Substitute $x = 0$ directly

    C. Rewrite in terms of $\tan x$

    D. Multiply numerator and denominator by $x$

    **Answer:** _____

3. Which limit is of the form $\infty/\infty$?

    A. $\lim_{x \to 0^+} \frac{\ln x}{x}$

    B. $\lim_{x \to \infty} \frac{x^3}{e^x}$

    C. $\lim_{x \to 0} \frac{\sin x}{x}$

    D. $\lim_{x \to 0^+} x \ln x$

    **Answer:** _____

4. To evaluate $\lim_{x \to 0^+} x^x$, which technique is most helpful?

    A. Apply L'Hospital's rule directly

    B. Take logarithms and rewrite as an exponential

    C. Use a Taylor expansion

    D. Apply the mean value theorem

    **Answer:** _____

5. The Cauchy mean value theorem is a generalization of

    A. Fermat's theorem

    B. Rolle's theorem

    C. The mean value theorem

    D. L'Hospital's rule

    **Answer:** _____

## Matching exercise

Match each limit to the correct answer.

1. $\lim_{x \to 0} \frac{\tan x}{x}$     A. 0

2. $\lim_{x \to \infty} \frac{\ln x}{x}$     B. 1

3. $\lim_{x \to 0^+} x \ln x$     C. $+\infty$

4. $\lim_{x \to 0} \frac{e^x - 1}{x}$     D. $-\infty$

5. $\lim_{x \to \infty} \frac{e^x}{x^2}$     E. 1

**Answers: 1.** _____    **2.** _____    **3.** _____    **4.** _____    **5.** _____

## Fill in the blanks

Complete each statement with the most appropriate word or phrase.

1.  L'Hospital's rule can be used for the indeterminate forms _____ and
    _____.

2.  To evaluate a limit of the type $1^\infty$, it is helpful to first take the _____
    of both sides.

3.  If both numerator and denominator approach infinity, the limit is said to be of the
    form _____.

4.  The fundamental tool used in proving L'Hospital's rule is the _____
    mean value theorem.

5.  The rule requires that the derivative of the denominator does not become _____
    near the limiting point.

## Theoretical/computational problems

Solve step-by-step.

1.  Use L'Hospital's rule to evaluate $\lim_{x \to 0} \frac{\ln(1+x)}{x}$.

2.  Show that $\lim_{x \to \infty} \frac{\ln x}{x} = 0$ using L'Hospital's rule.

3.  Evaluate $\lim_{x \to 0^+} x^x$.

4.  For which values of $n$ does $\lim_{x \to \infty} \frac{x^n}{e^x} = 0$?

5.  Use L'Hospital's rule to compute $\lim_{x \to 0} \frac{\tan x - x}{x^3}$.

## Python-based exercises

Use sympy or numpy to check or illustrate the following limits numerically or symbolically:

1.  Numerically estimate $\lim_{x \to 0} \frac{\sin x}{x}$ and compare to the theoretical value.

2.  Use sympy to symbolically evaluate $\lim_{x \to 0} \frac{e^x - 1 - x}{x^2}$.

3.  Plot $f(x) = \frac{x^2}{e^x}$ for $x \in [0, 10]$ and describe its end behavior.

4.  Numerically investigate the behavior of $x^x$ as $x \to 0^+$ and as $x \to 1$.

5.  Write a function to compute $\lim_{x \to a} \frac{f(x)}{g(x)}$ numerically and test it for several pairs
    $(f, g)$.

# 12 Asymptotes and asymptotic expansions

**Contents**

Asymptotes describe the long-term behavior of functions and are crucial in understanding graphs, limits, and the qualitative features of many models. This chapter introduces the main types of asymptotes: vertical, horizontal, oblique (slant), and polynomial.

We provide rigorous definitions, illustrate each type with examples and graphs, and show how to determine them for a variety of functions, including rational and transcendental cases. The methods include algebraic simplification, limit calculations, and polynomial division.

You will also see how polynomial and asymptotic approximations are used to estimate functions for large values of the variable, a technique essential in analysis and applied mathematics.

## 12.1 Introduction

The concept of asymptotes lies at the heart of mathematical analysis and the study of functions. Asymptotes provide a precise language to describe how functions behave as the variable approaches infinity or a critical point where the function becomes unbounded. They play a central role not only in pure mathematics—such as calculus, algebra, and real analysis—but also in applied fields, including physics, engineering, and economics, where the long-term or extreme behavior of a model is often crucial.

This chapter offers a systematic study of asymptotes, beginning with the rigorous definitions of vertical, horizontal, and oblique (slant) asymptotes. We clarify how these different types arise for a wide range of functions and discuss the significance of each: vertical asymptotes are connected with singularities or points of discontinuity, while horizontal and oblique asymptotes capture end behavior at infinity.

A special emphasis is placed on polynomial asymptotes—generalizations of horizontal and oblique asymptotes—which provide the best possible polynomial approximation to a function at infinity. We show how to compute these polynomial asymptotes, interpret their coefficients, and relate them to the algebraic structure of the function.

In addition, we explore practical techniques for finding asymptotes: algebraic manipulation, polynomial division, limit calculations, and the use of L'Hospital's rule. Each

https://doi.org/10.1515/9783112228289-012

method is illustrated with carefully chosen examples, accompanied by clear graphs that visualize how functions approach their asymptotes from different directions.

We also introduce the notion of asymptotic expansions, which extend the idea of asymptotes by expressing functions as infinite series that approximate their behavior in a certain limit. This prepares the ground for more advanced topics in analysis, such as Taylor expansions and the method of dominant balance.

By mastering the material in this chapter, you will be able to analyze the asymptotic behavior of a broad class of functions, distinguish between different types of asymptotes, and use asymptotic approximations as powerful tools in both theoretical and applied contexts.

**Definition 12.1** (Vertical asymptote (Figure 12.1)). A line $x = a$ is called a *vertical asymptote* of the real function $f$ if at least one of the one-sided limits

$$\lim_{x \to a^-} f(x), \quad \lim_{x \to a^+} f(x)$$

is infinite (i. e., $+\infty$ or $-\infty$).

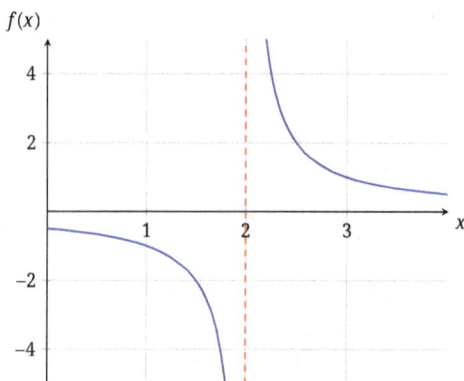

**Figure 12.1:** Graph of $f(x) = \frac{1}{x-2}$ with a **vertical asymptote** at $x = 2$. The function approaches infinity as $x$ approaches 2 from either side.

**Definition 12.2** (Horizontal asymptote (Figure 12.2)). A line $y = L$ is called a *horizontal asymptote* of the real function $f$ if

$$\lim_{x \to +\infty} f(x) = L \quad \text{or} \quad \lim_{x \to -\infty} f(x) = L.$$

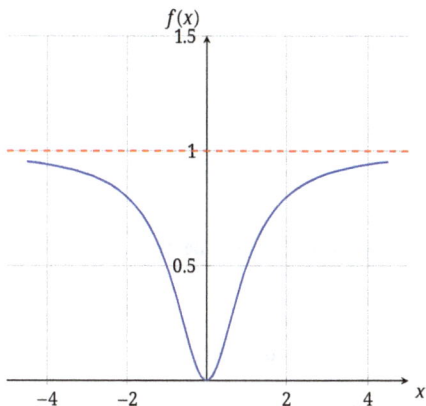

**Figure 12.2:** Graph of $f(x) = \frac{x^2}{x^2+1}$ with a **horizontal asymptote** at $y = 1$. The function approaches 1 as $x$ tends to positive or negative infinity.

**Definition 12.3** (Oblique (slant) asymptote (Figure 12.3)). A line

$$y = mx + b$$

is called an *oblique asymptote* of $f$ as $x \to \pm\infty$ if

$$\lim_{x\to\pm\infty}\left[f(x) - (mx + b)\right] = 0.$$

Equivalently,

$$m = \lim_{x\to\pm\infty} \frac{f(x)}{x}, \quad b = \lim_{x\to\pm\infty}\left[f(x) - mx\right].$$

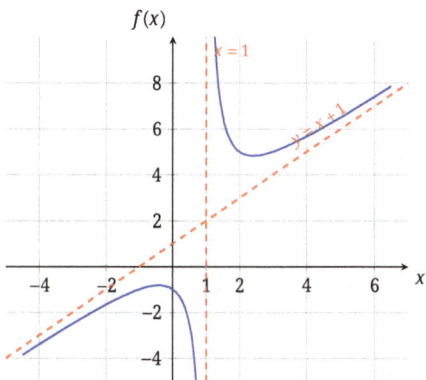

**Figure 12.3:** Graph of $f(x) = \frac{x^2+1}{x-1}$ with its oblique asymptote $y = x + 1$ and vertical asymptote $x = 1$.

**Definition 12.4** (Polynomial asymptote). A polynomial

$$p_n(x) = a_n x^n + a_{n-1}x^{n-1} + \cdots + a_0$$

is called a *polynomial asymptote of degree $n$* of $f$ at infinity if

$$\lim_{x \to \pm\infty} [f(x) - p_n(x)] = 0.$$

In particular,

$$a_n = \lim_{x \to \pm\infty} \frac{f(x)}{x^n}, \quad a_{n-1} = \lim_{x \to \pm\infty} \frac{f(x) - a_n x^n}{x^{n-1}}, \dots, a_0 = \lim_{x \to \pm\infty} [f(x) - a_n x^n - \dots - a_1 x].$$

**Example 12.1** (The function $e^x$ grows faster than any polynomial). Determine whether the function $f(x) = e^x$ has a polynomial asymptote as $x \to \infty$ or $x \to -\infty$.

*Solution.* We analyze the behavior of $f(x) = e^x$ at both ends of the real line.

**As $x \to \infty$: no polynomial asymptote exists**

Suppose, for contradiction, that there exists a polynomial $p_n(x)$ of degree $n$ such that

$$\lim_{x \to \infty} (e^x - p_n(x)) = 0.$$

Then $e^x = p_n(x) + o(1)$, so $e^x$ would grow at most like a polynomial. But it is well known that

$$\lim_{x \to \infty} \frac{e^x}{x^n} = \infty \quad \text{for every } n \in \mathbb{N}.$$

This means $e^x$ grows faster than any polynomial.

**As $x \to -\infty$: the zero polynomial is an asymptote**

We have

$$\lim_{x \to -\infty} e^x = 0.$$

Consider the constant polynomial $p(x) = 0$. Then

$$\lim_{x \to -\infty} (e^x - 0) = 0.$$

By definition, this means $p(x) = 0$ is a polynomial asymptote of $f(x) = e^x$ as $x \to -\infty$.

**Conclusion**

- As $x \to +\infty$: $e^x$ grows faster than any polynomial.
- As $x \to -\infty$: $e^x \to 0$, and $\lim_{x \to -\infty}(e^x - 0) = 0$. The **zero polynomial** $p(x) = 0$ is a polynomial asymptote.
- At $+\infty$, polynomial asymptotes require sub-exponential growth.

**Graph of $f(x) = e^x$**

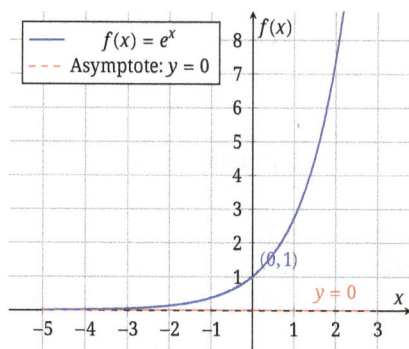

**Example 12.2** (The logarithm grows slower than any positive power). Show that for any $p > 0$,

$$\lim_{x \to \infty} \frac{\ln x}{x^p} = 0.$$

This implies that $\ln x$ grows slower than any positive power of $x$, and hence slower than any nonconstant polynomial.

*Solution.* Let $p > 0$ be given. We want to evaluate

$$\lim_{x \to \infty} \frac{\ln x}{x^p}.$$

This is an indeterminate form of type $\frac{\infty}{\infty}$, so we apply **L'Hospital's rule**. Differentiate numerator and denominator with respect to $x$

$$\frac{d}{dx}(\ln x) = \frac{1}{x}, \quad \frac{d}{dx}(x^p) = px^{p-1}.$$

Then

$$\lim_{x \to \infty} \frac{\ln x}{x^p} = \lim_{x \to \infty} \frac{1/x}{px^{p-1}} = \lim_{x \to \infty} \frac{1}{px \cdot x^{p-1}} = \lim_{x \to \infty} \frac{1}{px^p} = 0.$$

Since $p > 0$, $x^p \to \infty$, so the denominator goes to infinity, and the whole expression tends to 0.

Therefore,

$$\lim_{x \to \infty} \frac{\ln x}{x^p} = 0 \quad \text{for all } p > 0.$$

**Implication for polynomials**

Let $q(x) = a_n x^n + \cdots + a_0$ be a polynomial of degree $n \geq 1$ with $a_n > 0$. Then as $x \to \infty$, $q(x) \sim a_n x^n$, so

$$\lim_{x \to \infty} \frac{\ln x}{q(x)} = \lim_{x \to \infty} \frac{\ln x}{a_n x^n + \cdots} = 0$$

since $n > 0$. Thus, $\ln x$ grows slower than any nonconstant polynomial.

This result shows that:

- Despite $\ln x \to \infty$ as $x \to \infty$, it does so **extremely slowly**.
- For example, even $x^{0.1}$ will eventually surpass $\ln x$, no matter how large $x$ is.
- This is in sharp contrast to exponential functions, which grow faster than any polynomial—here, the logarithm is at the opposite extreme.

**Graphical comparison**

Below we plot $f(x) = \ln x$, $g(x) = x^{0.2}$, and $h(x) = x^{0.5}$ for $x \in (0, 100]$. Although $\ln x$ starts higher, the power functions eventually dominate.

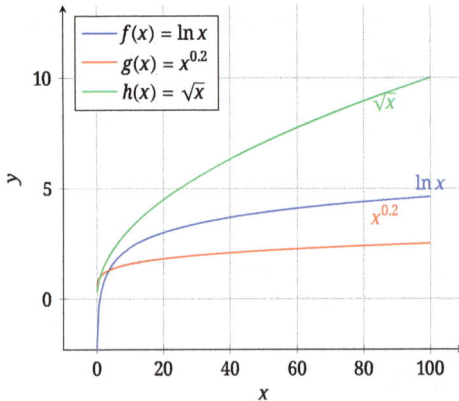

**Numerical illustration**

| $x$ | $\ln x$ | $x^{0.1}$ | $x^{0.2}$ |
|---|---|---|---|
| 10 | 2.30 | 1.26 | 1.58 |
| 100 | 4.61 | 1.58 | 2.51 |
| $10^4$ | 9.21 | 2.51 | 6.31 |
| $10^6$ | 13.8 | 3.98 | 15.8 |
| $10^9$ | 20.7 | 6.31 | 39.8 |

Even though $\ln x$ grows without bound, it is overtaken by $x^{0.2}$ around $x \approx 10^6$, and the gap widens rapidly. □

**Remark 15** (Practical use of polynomial asymptotes). Finding the polynomial asymptotes of a function one can approximate the value of the function for big enough $|x|$ by computing the value of the corresponding asymptote.

## 12.2 Examples

**Example 12.3** (Vertical asymptote). Find all vertical asymptotes of the function

$$f(x) = \frac{x^2 + 1}{x - 2}.$$

*Solution.*
**Step 1: Denominator zero**
The function is undefined when the denominator is zero

$$x - 2 = 0 \quad \Rightarrow \quad x = 2.$$

This is a candidate for a vertical asymptote.

**Step 2: Limit behavior near $x = 2$**

$$\lim_{x \to 2^-} \frac{x^2 + 1}{x - 2} = -\infty, \quad \lim_{x \to 2^+} \frac{x^2 + 1}{x - 2} = +\infty.$$

Since the limit diverges, there is a vertical asymptote at

$$\boxed{x = 2.}$$

$\square$

**Example 12.4** (Horizontal asymptote). Find the horizontal asymptote(s) of the function

$$f(x) = \frac{x^2}{x^2 + 1}.$$

*Solution.*
**Step 1: Degree comparison**

$$\text{Degree of numerator} = 2, \quad \text{Degree of denominator} = 2$$
$$\Rightarrow \quad \text{Horizontal asymptote: } y = \frac{1}{1} = 1.$$

**Step 2: Limit at infinity**

$$\lim_{x \to \pm\infty} \frac{x^2}{x^2 + 1} = \frac{1}{1 + \frac{1}{x^2}} \to 1.$$

**Conclusion**

The function has a single horizontal asymptote at

$$\boxed{y = 1.}$$ ☐

**Example 12.5** (Oblique (slant) asymptote).  Find the oblique asymptote of

$$f(x) = \frac{x^2 + 1}{x - 1}.$$

*Solution.*  We perform polynomial long division or use limits.

Let's divide

$$\frac{x^2 + 1}{x - 1} = x + 1 + \frac{2}{x - 1}.$$

As $x \to \pm\infty$, the term $\frac{2}{x-1} \to 0$, so

$$f(x) \approx x + 1.$$

Therefore, the line $y = x + 1$ is an oblique asymptote of $f(x)$. ☐

**Example 12.6** (Polynomial asymptote).  Find a polynomial asymptote of degree 2 for the function

$$f(x) = \frac{x^3 + x + 1}{x}.$$

*Solution.*  First simplify

$$f(x) = \frac{x^3 + x + 1}{x} = x^2 + 1 + \frac{1}{x}.$$

As $x \to \pm\infty$, $\frac{1}{x} \to 0$, so

$$f(x) \approx x^2 + 1 \quad \text{for big enough } x.$$

Thus, the polynomial asymptote of degree 2 is

$$p(x) = x^2 + 1.$$ ☐

**Example 12.7** (Using limits to find asymptotes).  Find the oblique asymptote of

$$f(x) = \frac{2x^2 + 3x + 1}{x + 1}.$$

*Solution.* Use the formula for oblique asymptotes

$$m = \lim_{x \to \pm\infty} \frac{f(x)}{x}, \quad b = \lim_{x \to \pm\infty} (f(x) - mx).$$

Compute $m$:

$$\frac{f(x)}{x} = \frac{2x^2 + 3x + 1}{x(x+1)} = \frac{2x^2 + 3x + 1}{x^2 + x} \to 2 \quad \text{as } x \to \pm\infty \Rightarrow m = 2.$$

Now compute $b$:

$$
\begin{aligned}
f(x) - 2x &= \frac{2x^2 + 3x + 1}{x + 1} - 2x \\
&= \frac{2x^2 + 3x + 1 - 2x(x+1)}{x + 1} \\
&= \frac{2x^2 + 3x + 1 - 2x^2 - 2x}{x + 1} \\
&= \frac{x + 1}{x + 1} \\
&= 1.
\end{aligned}
$$

So $b = 1$, and the oblique asymptote is

$$y = 2x + 1. \qquad \square$$

**Example 12.8** (Finding horizontal asymptote using L'Hospital's rule). Evaluate

$$\lim_{x \to \infty} \frac{\ln x}{x}$$

and determine whether it gives a horizontal asymptote.

*Solution.* This limit is in the indeterminate form $\frac{\infty}{\infty}$, so we apply L'Hospital's rule

$$\lim_{x \to \infty} \frac{\ln x}{x} = \lim_{x \to \infty} \frac{\frac{1}{x}}{1} = \lim_{x \to \infty} \frac{1}{x} = 0.$$

Therefore, the horizontal asymptote is

$$y = 0. \qquad \square$$

**Example 12.9** (Complete asymptote analysis). Analyze all asymptotes of

$$f(x) = \frac{x^3 + 1}{x^2 - 1}.$$

*Solution.*

## 1. Vertical asymptotes

Vertical asymptotes occur where the denominator is zero, except at removable discontinuities.

$$x^2 - 1 = 0 \quad \Longrightarrow \quad x = 1, \quad x = -1.$$

Check the numerator at these points

$$x = 1: \quad (1)^3 + 1 = 2 \neq 0, \quad x = -1: \quad (-1)^3 + 1 = -1 + 1 = 0.$$

So, at $x = -1$ both numerator and denominator are zero; let's check the limit:

$$\lim_{x \to -1} \frac{x^3 + 1}{x^2 - 1} = \lim_{x \to -1} \frac{(x + 1)(x^2 - x + 1)}{(x + 1)(x - 1)}$$

$$= \lim_{x \to -1} \frac{x^2 - x + 1}{x - 1}$$

$$= \frac{(-1)^2 - (-1) + 1}{-1 - 1}$$

$$= \frac{1 + 1 + 1}{-2}$$

$$= \frac{3}{-2}.$$

So, at $x = -1$ we have a removable discontinuity (hole), not a vertical asymptote.
At $x = 1$,

$$\lim_{x \to 1^-} f(x) = -\infty, \quad \lim_{x \to 1^+} f(x) = +\infty.$$

Thus, there is a vertical asymptote at $x = 1$.

## 2. Horizontal and oblique asymptotes

Degree of numerator = 3, degree of denominator = 2.
- If degree(numerator) > degree(denominator) by one: oblique (slant) asymptote.
- If equal: horizontal asymptote.
- If numerator degree > denominator degree by more than 1: no finite asymptote.

Here, numerator degree is greater by 1, so compute the slant (oblique) asymptote.
    Perform polynomial division

$$\frac{x^3 + 1}{x^2 - 1} = x + \frac{x + 1}{x^2 - 1}.$$

So, as $x \to \pm\infty, f(x) \sim x$.
    Thus, the oblique asymptote is $y = x$.

As $x \to \pm\infty$, the remainder $\frac{x+1}{x^2-1} \to 0$, so $f(x) \to x$.
No horizontal asymptote.

### 3. Summary
- Vertical asymptote: $x = 1$ -
- Removable discontinuity: at $x = -1$ -
- Oblique asymptote: $y = x$. □

## 12.3 Python and AI

**AI Request**

Write a Python code that computes all the asymptotes of a given function. Plot all of them.

## 12.4 Exercises

### True or false
Determine whether each statement is **true** or **false**. Justify your answer.
1. Every rational function has a horizontal asymptote.
2. A vertical asymptote occurs whenever the denominator of a rational function is zero.
3. If $f(x)$ has an oblique asymptote, then it cannot have a horizontal asymptote.
4. The line $y = 0$ is a horizontal asymptote of $f(x) = \frac{\sin x}{x}$ as $x \to +\infty$.
5. If $\lim_{x \to a^-} f(x) = +\infty$ and $\lim_{x \to a^+} f(x) = -\infty$, then $x = a$ is a vertical asymptote.

### Multiple choice questions
Choose the correct option (A, B, C, or D) for each question.
1. Which function has a horizontal asymptote at $y = 0$?
   A. $f(x) = \frac{x^2}{x+1}$    B. $f(x) = e^{-x}$
   C. $f(x) = x$      D. $f(x) = \tan x$
   Answer: _____
2. What is the oblique asymptote of $f(x) = \frac{2x^2+1}{x}$?
   A. $y = 2x$       B. $y = x$
   C. $y = 2x + 1$   D. $y = 0$
   Answer: _____
3. For which of the following does $x = 0$ give a vertical asymptote?
   A. $f(x) = \frac{1}{x}$    B. $f(x) = \ln x$
   C. $f(x) = \frac{1}{x^2}$    D. All of the above
   Answer: _____
4. What is the degree of the polynomial asymptote of $f(x) = \frac{x^4+1}{x^2}$ as $x \to \infty$?

A. 1    B. 2
C. 3    D. 4
**Answer:** _____

5.  Which of the following best describes the behavior of $f(x) = \frac{x}{x^2+1}$ as $x \to \infty$?
    A. Increases without bound    B. Approaches zero
    C. Has a vertical asymptote    D. Has an oblique asymptote
    **Answer:** _____

## Matching exercise

Match each function with its correct asymptote.

1. $f(x) = \frac{2x+3}{x-2}$        A. Oblique asymptote $y = 2$
2. $f(x) = \frac{1}{x^2}$           B. Horizontal asymptote $y = 0$
3. $f(x) = x^2 + \frac{1}{x}$       C. Polynomial asymptote $y = x^2$
4. $f(x) = \frac{x^3}{x^2+1}$       D. Oblique asymptote $y = x$
5. $f(x) = \tan x$                  E. Infinite number of vertical asymptotes

**Answers:** 1. _____    2. _____    3. _____    4. _____    5. _____

## Fill in the blanks

Complete each sentence with the most appropriate word or phrase.

1.  A vertical asymptote occurs at $x = a$ if $\lim_{x \to a} f(x) =$ _____.
2.  The line $y = mx + b$ is an _____ asymptote of $f(x)$ if $f(x) - (mx + b) \to 0$ as $x \to \infty$.
3.  If $f(x) = \frac{p(x)}{q(x)}$ and $\deg p < \deg q$, then the horizontal asymptote is $y =$ _____.
4.  A function may have a polynomial asymptote of degree greater than 1 if the degree of the numerator is _____ than that of the denominator.
5.  The function $f(x) = \frac{x^2+1}{x-2}$ has a vertical asymptote at $x =$ _____.

## Theoretical/computational problems

Solve each problem step-by-step.

1.  Find all vertical and horizontal asymptotes of $f(x) = \frac{x-3}{x+2}$.
2.  Find the oblique asymptote of $f(x) = \frac{x^2+x+1}{x}$.
3.  Determine all asymptotes (vertical, horizontal, or oblique) of $f(x) = \frac{2x^3-x}{x^2-4}$.
4.  Find the degree and explicit expression of the polynomial asymptote of $f(x) = \frac{x^4+x^2+1}{x^2}$.
5.  Compute the limit $\lim_{x \to \infty}[f(x) - y_{asymp}(x)]$ where $f(x) = \frac{x^3+2x}{x^2-1}$ and $y_{asymp}(x)$ is its oblique asymptote.

## Python-based exercises

Use `sympy` or `matplotlib` to plot and verify asymptotes numerically.

1. Plot $f(x) = \frac{1}{x-1}$ and show numerically how $f(x)$ behaves near its vertical asymptote.

2. For $f(x) = \frac{x^2}{x^2+1}$, plot and show numerically its approach to its horizontal asymptote as $x \to \infty$.

3. Use symbolic computation to find the oblique asymptote of $f(x) = \frac{2x^2+3x+1}{x+1}$.

4. Numerically plot $f(x) = x^3 + 2x$ and its polynomial asymptote for large $|x|$.

5. Write a Python function that takes a rational function and returns all its vertical and horizontal asymptotes.

# 13 Derivative application on approximations

## Contents

Derivatives provide the foundation for constructing accurate approximations to functions that are otherwise difficult to analyze directly. By leveraging Taylor's theorem, we can represent smooth functions locally by polynomials, allowing for systematic error control through the Lagrange remainder. This chapter explores how derivatives are used in Taylor expansions, polynomial interpolation, and root-finding algorithms such as the Newton–Raphson method. We examine both the theoretical guarantees and practical limitations of these techniques. Worked examples and graphical visualizations are included to highlight the strengths and subtleties of derivative-based approximation methods. Through this, you will gain tools to approach a wide range of applied problems where precise calculations are impractical.

## 13.1 Introduction

The power of calculus lies not only in its ability to compute rates of change and local properties of functions, but also in its capacity to produce accurate approximations of complicated functions using simpler, well-understood expressions. The most important tool for this purpose is the derivative, which, through Taylor's theorem and related constructions, forms the theoretical backbone of approximation techniques in both pure and applied mathematics.

This chapter explores in depth how derivatives are harnessed to approximate functions, analyze errors, and design efficient numerical algorithms. We begin with the fundamental principle that, locally, a smooth function can be closely approximated by its tangent line—an idea formalized as linear (first-order) approximation. From there, we ascend to higher-order Taylor polynomials, which provide increasingly precise representations by incorporating higher derivatives. The key result, Taylor's theorem with the Lagrange remainder, quantifies the error between the function and its polynomial approximation, ensuring control over accuracy and convergence.

The study then extends to interpolation, where the goal is to construct a polynomial that passes exactly through a set of prescribed data points. We present the classical Lagrange interpolation formula and its elegant connection to systems of linear equations

https://doi.org/10.1515/9783112228289-013

via the Vandermonde matrix. The theory is complemented with a rigorous analysis of the interpolation error, revealing how both the smoothness of the underlying function and the distribution of interpolation points affect approximation quality.

A crucial application of derivatives in approximation is the solution of nonlinear equations. We present the Newton–Raphson method, one of the most widely used iterative algorithms for finding roots. Its remarkable quadratic convergence is a direct consequence of Taylor's theorem, and we compare its performance to that of the more robust, but slower, bisection method. Through detailed worked examples and visualizations, we elucidate the geometric and computational intuition behind these methods.

Throughout the chapter, every major result is illustrated with concrete examples, graphical interpretations, and step-by-step computations. We highlight practical issues such as the tradeoff between simplicity and accuracy, the risk of divergence in iterative methods, and the subtleties of error estimation. Numerous exercises, both theoretical and computational (including Python-based activities), provide the opportunity to solidify understanding and develop computational fluency.

By mastering the concepts in this chapter, you will acquire a comprehensive toolkit for tackling real-world problems where exact solutions are unavailable and intelligent approximation is essential. The ideas developed here underpin countless applications in physics, engineering, economics, data science, and beyond. You can find more on [4].

## 13.2 Taylor's approximation

The following theorem can be seen as a generalization of the mean value Theorem 8.4.

---

**Theorem 13.1** (Taylor's theorem with Lagrange remainder). *Let $f \in C^{n+1}[a,b]$, i. e., $f$ is $n + 1$ times continuously differentiable on $[a,b]$. Then for any $x,y \in [a,b]$ there exists some $\xi \in (\min\{x,y\}, \max\{x,y\}) \subset (a,b)$ such that*

$$f(y) = f(x) + \frac{f'(x)}{1!}(y-x) + \frac{f''(x)}{2!}(y-x)^2 + \cdots + \frac{f^{(n)}(x)}{n!}(y-x)^n + \frac{f^{(n+1)}(\xi)}{(n+1)!}(y-x)^{n+1}.$$

---

*Proof.* Let's fix $x$ and $y$ in $[a,b]$. Without loss of generality, assume $x < y$. If $y < x$, the argument follows similarly.

Consider the auxiliary function $F(t)$ defined for $t \in [x,y]$ as

$$F(t) = f(y) - \left[ f(t) + \frac{f'(t)}{1!}(y-t) + \frac{f''(t)}{2!}(y-t)^2 + \cdots + \frac{f^{(n)}(t)}{n!}(y-t)^n \right] - K\frac{(y-t)^{n+1}}{(n+1)!}$$

where $K$ is a constant we will determine.

First, evaluate $F(y)$:

$$F(y) = f(y) - \left[ f(y) + \frac{f'(y)}{1!}(y-y) + \cdots + \frac{f^{(n)}(y)}{n!}(y-y)^n \right] - K\frac{(y-y)^{n+1}}{(n+1)!}.$$

Since all terms involving $(y - y)$ are zero, we get $F(y) = 0$.

Next, evaluate $F(x)$:

$$F(x) = f(y) - \left[ f(x) + \frac{f'(x)}{1!}(y - x) + \frac{f''(x)}{2!}(y - x)^2 + \cdots + \frac{f^{(n)}(x)}{n!}(y - x)^n \right] - K\frac{(y - x)^{n+1}}{(n + 1)!}.$$

We choose the constant $K$ such that $F(x) = F(y)$. Since $F(y) = 0$, we set $F(x) = 0$. This implies

$$K\frac{(y - x)^{n+1}}{(n + 1)!} = f(y) - \left[ f(x) + \frac{f'(x)}{1!}(y - x) + \cdots + \frac{f^{(n)}(x)}{n!}(y - x)^n \right].$$

This choice of $K$ ensures $F(x) = F(y) = 0$.

Since $f \in C^{n+1}[a, b]$, the function $F(t)$ is continuous on $[x, y]$ and differentiable on $(x, y)$. By Rolle's theorem, there exists some $\xi \in (x, y)$ such that $F'(\xi) = 0$.

Let's compute the derivative $F'(t)$. The derivative of the sum part involves repeated cancelations. Let $P_n(t) = f(t) + \frac{f'(t)}{1!}(y - t) + \frac{f''(t)}{2!}(y - t)^2 + \cdots + \frac{f^{(n)}(t)}{n!}(y - t)^n$. Differentiating $P_n(t)$ with respect to $t$ using the product rule for each term $\frac{f^{(k)}(t)}{k!}(y - t)^k$,

$$\frac{d}{dt}\left( \frac{f^{(k)}(t)}{k!}(y - t)^k \right) = \frac{f^{(k+1)}(t)}{k!}(y - t)^k + \frac{f^{(k)}(t)}{k!} \cdot k(y - t)^{k-1}(-1)$$

$$= \frac{f^{(k+1)}(t)}{k!}(y - t)^k - \frac{f^{(k)}(t)}{(k - 1)!}(y - t)^{k-1}.$$

Summing these derivatives from $k = 0$ to $n$ (where $f^{(0)}(t) = f(t)$ and $(y - t)^0 = 1$ for $k = 0$): $P'_n(t) = (f'(t)) + (\frac{f''(t)}{1!}(y - t) - f'(t)) + (\frac{f'''(t)}{2!}(y - t)^2 - \frac{f''(t)}{1!}(y - t)) + \cdots + (\frac{f^{(n+1)}(t)}{n!}(y - t)^n - \frac{f^{(n)}(t)}{(n-1)!}(y - t)^{n-1})$. Notice that all intermediate terms cancel out in a telescoping sum. For example, the term $f'(t)$ from the derivative of $f(t)$ cancels with $-f'(t)$ from the derivative of $\frac{f'(t)}{1!}(y - t)$. Similarly, $\frac{f''(t)}{1!}(y - t)$ cancels with $-\frac{f''(t)}{1!}(y - t)$. This continues until the last term. Thus, $P'_n(t)$ simplifies to

$$P'_n(t) = \frac{f^{(n+1)}(t)}{n!}(y - t)^n.$$

Now, let's differentiate the last term of $F(t)$

$$\frac{d}{dt}\left( -K\frac{(y - t)^{n+1}}{(n + 1)!} \right) = -K\frac{(n + 1)(y - t)^n}{(n + 1)!}(-1) = K\frac{(y - t)^n}{n!}.$$

Combining these, the derivative of $F(t)$ is

$$F'(t) = -P'_n(t) + K\frac{(y - t)^n}{n!},$$

$$F'(t) = -\frac{f^{(n+1)}(t)}{n!}(y - t)^n + K\frac{(y - t)^n}{n!},$$

$$F'(t) = \frac{(y-t)^n}{n!}(K - f^{(n+1)}(t)).$$

Since $F'(\xi) = 0$ for some $\xi \in (x, y)$, and given that $y \neq \xi$ (so $(y-\xi)^n \neq 0$) and $n! \neq 0$,

$$\frac{(y-\xi)^n}{n!}(K - f^{(n+1)}(\xi)) = 0 \quad \Longrightarrow \quad K - f^{(n+1)}(\xi) = 0 \quad \Longrightarrow \quad K = f^{(n+1)}(\xi).$$

Finally, substitute this value of $K$ back into the equation for $F(x) = 0$:

$$f(y) - \left[ f(x) + \frac{f'(x)}{1!}(y-x) + \cdots + \frac{f^{(n)}(x)}{n!}(y-x)^n \right] - \frac{f^{(n+1)}(\xi)}{(n+1)!}(y-x)^{n+1} = 0.$$

Rearranging the terms, we obtain the desired result

$$f(y) = f(x) + \frac{f'(x)}{1!}(y-x) + \frac{f''(x)}{2!}(y-x)^2 + \cdots + \frac{f^{(n)}(x)}{n!}(y-x)^n + \frac{f^{(n+1)}(\xi)}{(n+1)!}(y-x)^{n+1}.$$

The location of $\xi$ is strictly between $x$ and $y$, i. e., $\xi \in (\min\{x, y\}, \max\{x, y\})$, which is a subset of $(a, b)$ since $x, y \in [a, b]$. □

**Example 13.1.** Let
$$f(x) = x e^x - \left(1 + e^x\right) \ln\left(1 + e^x\right), \quad x \in (0, \infty).$$

One computes
$$f'(x) = e^x\left(x - \ln\left(1 + e^x\right)\right) \le 0, \quad f''(x) = e^x\left(x - \ln\left(1 + e^x\right) + \frac{1}{1 + e^x}\right) \le 0 \quad \text{for } x > 0.$$

Apply Taylor's theorem about $x = 1$
$$f(x) = f(1) + f'(1)(x - 1) + \frac{1}{2}f''(\xi)(x - 1)^2, \quad \xi \in (1, x).$$

Since $f''(\xi) \le 0$, it follows
$$f(x) \le f(1) + f'(1)(x - 1).$$

But $f'(1) < 0$, so $\lim_{x \to \infty} f(x) = -\infty$.

**Example 13.2** (Taylor expansion with Lagrange remainder). Find the second-order Taylor polynomial of $f(x) = e^x$ around $x = 0$, and write the Lagrange remainder term.

*Solution.*
**Step 1: Compute derivatives of $f(x) = e^x$**
The function $f(x) = e^x$ and all its derivatives are

$$f(x) = e^x, \quad f'(x) = e^x, \quad f''(x) = e^x, \quad f^{(n)}(x) = e^x \quad \text{for all } n.$$

Evaluating at $x = 0$,

$$f(0) = 1, \quad f'(0) = 1, \quad f''(0) = 1.$$

**Step 2: Taylor polynomial of degree 2 at $x = 0$**

The general form of the Taylor polynomial of degree 2 centered at $x = 0$ is

$$P_2(x) = f(0) + f'(0)x + \frac{f''(0)}{2!}x^2.$$

Substitute values

$$P_2(x) = 1 + x + \frac{1}{2}x^2.$$

**Step 3: Lagrange remainder term**

The Lagrange form of the remainder is

$$R_2(x) = \frac{f^{(3)}(c)}{3!}x^3$$

for some $c$ between 0 and $x$. Since $f^{(3)}(x) = e^x$, we get

$$R_2(x) = \frac{e^c}{6}x^3 \quad \text{for some } c \in (0, x).$$

**Final answer**

$$f(x) = e^x = 1 + x + \frac{1}{2}x^2 + \frac{e^c}{6}x^3, \quad \text{where } c \in (0, x).$$

We compare the function $f(x) = e^x$ with its second-order Taylor polynomial around $x = 0$

$$P_2(x) = 1 + x + \frac{1}{2}x^2.$$

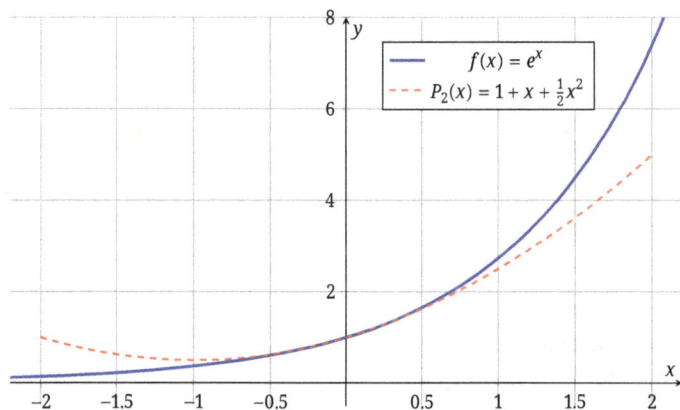

**Observation:**

- Near $x = 0$, the approximation is very accurate.
- As $x$ moves further from 0 (especially beyond $|x| > 1$), the error increases.
- This visualizes the role of the **Lagrange remainder term** and how higher-order terms improve accuracy. □

**Example 13.3** (Error in Taylor approximation). Let $f(x) = \ln(1+x)$. Compute the third-degree Taylor polynomial centered at $x = 0$, and bound the error on the interval $x \in [-0.5, 0.5]$.

*Solution.*

**Step 1: Compute derivatives**

Let $f(x) = \ln(1 + x)$. We compute the first few derivatives:

$$f(x) = \ln(1 + x),$$
$$f'(x) = \frac{1}{1 + x},$$
$$f''(x) = -\frac{1}{(1 + x)^2},$$
$$f^{(3)}(x) = \frac{2}{(1 + x)^3}.$$

Evaluate at $x = 0$:

$$f(0) = 0, \quad f'(0) = 1, \quad f''(0) = -1, \quad f^{(3)}(0) = 2.$$

**Step 2: Third-degree Taylor polynomial at $x = 0$**

The general Taylor expansion is

$$T_3(x) = f(0) + f'(0)x + \frac{f''(0)}{2!}x^2 + \frac{f^{(3)}(0)}{3!}x^3.$$

Substitute values

$$T_3(x) = 0 + x - \frac{1}{2}x^2 + \frac{2}{6}x^3 = x - \frac{1}{2}x^2 + \frac{1}{3}x^3$$

$$\boxed{T_3(x) = x - \frac{1}{2}x^2 + \frac{1}{3}x^3}.$$

**Step 3: Error bound using Lagrange remainder**

The remainder in Lagrange form is

$$R_3(x) = \frac{f^{(4)}(c)}{4!}x^4 \quad \text{for some } c \in (0, x).$$

We compute

$$f^{(4)}(x) = -\frac{6}{(1+x)^4} \quad \Rightarrow \quad |f^{(4)}(x)| = \frac{6}{(1+x)^4}.$$

On $x \in [-0.5, 0.5]$, the maximum value of $|f^{(4)}(c)|$ occurs at $x = -0.5$

$$|f^{(4)}(c)| \le \frac{6}{(1-0.5)^4} = \frac{6}{0.5^4} = \frac{6}{1/16} = 96.$$

Therefore, the error satisfies

$$|R_3(x)| \le \frac{96}{4!}|x|^4 = \frac{96}{24}|x|^4 = 4|x|^4.$$

**Conclusion**

$$\ln(1+x) = x - \frac{1}{2}x^2 + \frac{1}{3}x^3 + R_3(x), \quad |R_3(x)| \le 4|x|^4 \quad \text{for } x \in [-0.5, 0.5].$$

We compare the function $f(x) = \ln(1+x)$ with its third-degree Taylor polynomial centered at $x = 0$:

$$T_3(x) = x - \frac{1}{2}x^2 + \frac{1}{3}x^3.$$

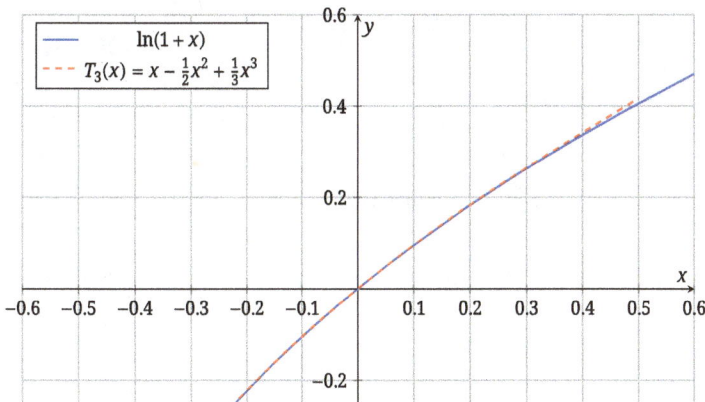

**Observation:**

- Near $x = 0$, the approximation is highly accurate.
- The two graphs begin to diverge slightly as $|x|$ increases toward 0.5.
- This matches the error bound $|R_3(x)| \le 4|x|^4$. $\qquad\qquad$ $\square$

**Example 13.4** (Taylor approximation of $\sin(\frac{1}{2})$). Approximate $\sin(\frac{1}{2})$ using the third-degree Taylor polynomial of $f(x) = \sin x$ centered at $x = 0$. Estimate the error using the Lagrange remainder.

*Solution.*

**Step 1: Compute Taylor polynomial of degree 3 at $x = 0$**

The Taylor expansion of $\sin x$ at $x = 0$ up to degree 3 is

$$\sin x = x - \frac{x^3}{6} + R_3(x).$$

**Step 2: Substitute $x = \frac{1}{2}$**

$$P_3\left(\frac{1}{2}\right) = \frac{1}{2} - \frac{1}{6}\left(\frac{1}{2}\right)^3 = \frac{1}{2} - \frac{1}{6}\cdot\frac{1}{8} = \frac{1}{2} - \frac{1}{48} = \frac{24-1}{48} = \frac{23}{48} \approx 0.4792.$$

**Step 3: Estimate the Lagrange remainder**

The Lagrange remainder term is

$$R_3\left(\frac{1}{2}\right) = \frac{f^{(4)}(c)}{4!}\left(\frac{1}{2}\right)^4$$

for some $c$ between 0 and $\frac{1}{2}$.

Since $f^{(4)}(x) = \sin x$, and $|\sin x| \leq 1$, we have

$$\left|R_3\left(\frac{1}{2}\right)\right| \leq \frac{1}{24}\cdot\left(\frac{1}{2}\right)^4 = \frac{1}{24}\cdot\frac{1}{16} = \frac{1}{384} \approx 0.0026.$$

**Step 4: Comparison with actual value**

The actual value is

$$\sin\left(\frac{1}{2}\right) \approx 0.4794.$$

The Taylor polynomial gives 0.4792, with an error of about 0.0002, which is within the estimated bound.

**Conclusion**

$$\boxed{\sin\left(\frac{1}{2}\right) \approx \frac{23}{48} \approx 0.4792 \quad \text{with error } |R_3| \leq 0.0026}$$

**Example 13.5** (Application to differential equations). We will derive the centered finite difference approximation for the second derivative. Expand $u(x + h)$ and $u(x - h)$ in Taylor series about $x$:

$$u(x + h) = u(x) + u'(x)h + \frac{u''(x)}{2}h^2 + \frac{u^{(3)}(x)}{6}h^3 + \frac{u^{(4)}(\xi_1)}{24}h^4,$$

$$u(x - h) = u(x) - u'(x)h + \frac{u''(x)}{2}h^2 - \frac{u^{(3)}(x)}{6}h^3 + \frac{u^{(4)}(\xi_2)}{24}h^4,$$

for some $\xi_1 \in (x, x + h)$ and $\xi_2 \in (x - h, x)$. Adding these two expansions causes the odd-order terms to cancel, yielding

$$u(x + h) + u(x - h) = 2u(x) + u''(x)h^2 + \frac{u^{(4)}(\xi_1) + u^{(4)}(\xi_2)}{24}h^4.$$

Solving for $u''(x)$ gives

$$u''(x) = \frac{u(x + h) - 2u(x) + u(x - h)}{h^2} - \frac{u^{(4)}(\xi)}{12}h^2,$$

where we used the fact that the average $\frac{u^{(4)}(\xi_1) + u^{(4)}(\xi_2)}{2} = u^{(4)}(\xi)$ for some $\xi \in (x - h, x + h)$. Thus the central difference formula

$$u''(x) \approx \frac{u(x + h) - 2u(x) + u(x - h)}{h^2}$$

has a truncation error of order $O(h^2)$.

**Application to partial differential equations**
Consider the 1D Poisson equation

$$-u''(x) = f(x), \quad x \in [0, 1],$$

with boundary conditions $u(0) = u(1) = 0$.

Discretize the domain with step size $h = \frac{1}{N}$, where $x_i = ih, i = 0, 1, \ldots, N$. Using the finite difference approximation

$$-u''(x_i) \approx -\frac{u(x_{i+1}) - 2u(x_i) + u(x_{i-1})}{h^2}.$$

Substituting into the Poisson equation

$$-\frac{u(x_{i+1}) - 2u(x_i) + u(x_{i-1})}{h^2} = f(x_i).$$

Rearranging

$$-u(x_{i+1}) + 2u(x_i) - u(x_{i-1}) = h^2 f(x_i).$$

This forms a tridiagonal system

$$2u(x_1) - u(x_2) = h^2 f(x_1),$$
$$-u(x_{i-1}) + 2u(x_i) - u(x_{i+1}) = h^2 f(x_i), \quad i = 2, \ldots, N - 1,$$
$$-u(x_{N-1}) + 2u(x_N) = h^2 f(x_N),$$

with boundary conditions $u(x_0) = u(x_N) = 0$. Solving the above system, we have the values $u(x_i)$ which approximate the solution of the above differential equation.

## 13.3 Lagrange interpolation

Lagrange interpolation aims to find a unique polynomial $P(x)$ that passes through a given set of $n + 1$ distinct data points $(x_0, y_0), (x_1, y_1), \ldots, (x_n, y_n)$. Instead of constructing the polynomial using Lagrange basis functions, we can directly determine its coefficients by forming and solving a system of linear equations.

Let the interpolating polynomial be expressed in its standard form

$$P(x) = a_0 + a_1 x + a_2 x^2 + \cdots + a_n x^n.$$

Since the polynomial must pass through each data point $(x_i, y_i)$, substituting each point into the polynomial equation gives us

$$a_0 + a_1 x_i + a_2 x_i^2 + \cdots + a_n x_i^n = y_i.$$

Applying this condition for all $n + 1$ points results in a system of $n + 1$ linear equations with $n + 1$ unknown coefficients $(a_0, a_1, \ldots, a_n)$.

$$a_0 + a_1 x_0 + a_2 x_0^2 + \cdots + a_n x_0^n = y_0,$$
$$a_0 + a_1 x_1 + a_2 x_1^2 + \cdots + a_n x_1^n = y_1,$$
$$\vdots$$
$$a_0 + a_1 x_n + a_2 x_n^2 + \cdots + a_n x_n^n = y_n.$$

This system can be concisely represented in matrix form as $V\mathbf{a} = \mathbf{y}$, where:

-   $V$ is the **Vandermonde matrix**

$$V = \begin{pmatrix} 1 & x_0 & x_0^2 & \cdots & x_0^n \\ 1 & x_1 & x_1^2 & \cdots & x_1^n \\ \vdots & \vdots & \vdots & \ddots & \vdots \\ 1 & x_n & x_n^2 & \cdots & x_n^n \end{pmatrix}.$$

-   $\mathbf{a}$ is the column vector of the unknown coefficients

$$\mathbf{a} = \begin{pmatrix} a_0 \\ a_1 \\ \vdots \\ a_n \end{pmatrix}.$$

-   $\mathbf{y}$ is the column vector of the given $y$-values

$$\mathbf{y} = \begin{pmatrix} y_0 \\ y_1 \\ \vdots \\ y_n \end{pmatrix}.$$

Solving this linear system $\mathbf{a} = V^{-1}\mathbf{y}$ (or using other methods like Gaussian elimination) yields the desired coefficients of the interpolating polynomial.

## Approximating $f(x) = e^x$ using a system of equations

Let's approximate the function $f(x) = e^x$ using a quadratic polynomial ($n = 2$) derived from three data points. We will use the same points as before:

- $P_0 : (x_0, y_0) = (0, e^0) = (0, 1)$
- $P_1 : (x_1, y_1) = (1.5, e^{1.5}) \approx (1.5, 4.482)$
- $P_2 : (x_2, y_2) = (3, e^3) \approx (3, 20.086)$

The polynomial we seek is $P(x) = a_0 + a_1 x + a_2 x^2$.

Substituting each point into the polynomial equation:

1. For $(x_0, y_0) = (0, 1)$,

$$a_0 + a_1(0) + a_2(0)^2 = 1 \quad \Longrightarrow \quad \mathbf{a_0 = 1}.$$

2. For $(x_1, y_1) = (1.5, 4.482)$,

$$a_0 + a_1(1.5) + a_2(1.5)^2 = 4.482.$$

Substituting $a_0 = 1$,

$$1 + 1.5a_1 + 2.25a_2 = 4.482 \quad \Longrightarrow \quad 1.5a_1 + 2.25a_2 = 3.482. \quad \text{(equation 1)}$$

3. For $(x_2, y_2) = (3, 20.086)$,

$$a_0 + a_1(3) + a_2(3)^2 = 20.086.$$

Substituting $a_0 = 1$,

$$1 + 3a_1 + 9a_2 = 20.086 \quad \Longrightarrow \quad 3a_1 + 9a_2 = 19.086. \quad \text{(equation 2)}$$

Now we have a system of two linear equations with two unknowns ($a_1, a_2$):

$$1.5a_1 + 2.25a_2 = 3.482,$$
$$3a_1 + 9a_2 = 19.086.$$

To solve this system using elimination, multiply the first equation by 2

$$3a_1 + 4.5a_2 = 6.964. \quad \text{(modified equation 1)}$$

Subtract modified equation 1 from equation 2:

$$(3a_1 + 9a_2) - (3a_1 + 4.5a_2) = 19.086 - 6.964,$$
$$4.5a_2 = 12.122,$$
$$a_2 = \frac{12.122}{4.5} \approx \mathbf{2.6937}.$$

Substitute the value of $a_2$ back into equation 1:

$$1.5a_1 + 2.25(2.6937) = 3.482,$$
$$1.5a_1 + 6.0608 \approx 3.482,$$
$$1.5a_1 \approx 3.482 - 6.0608,$$
$$1.5a_1 \approx -2.5788,$$
$$a_1 \approx \frac{-2.5788}{1.5} \approx -1.7192.$$

Thus, the coefficients of the interpolating polynomial are

$$a_0 = 1,$$
$$a_1 \approx -1.7192,$$
$$a_2 \approx 2.6937.$$

The interpolating polynomial is $P(x) \approx 1 - 1.7192x + 2.6937x^2$.

### Visualization of the approximation

The following figure illustrates the original function $f(x) = e^x$ and the Lagrange interpolating polynomial $P(x)$ with the coefficients derived above.

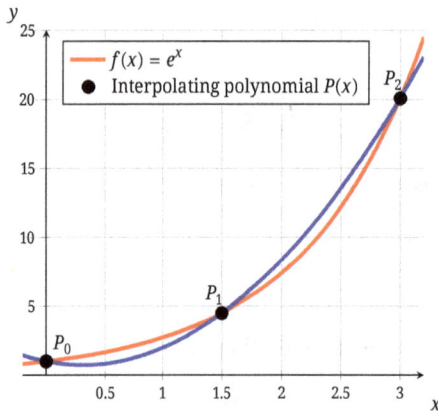

Figure 13.1: Lagrange interpolation of $f(x) = e^x$ using three points with coefficients derived from a linear system. The blue curve ($P(x)$) accurately passes through the black data points, approximating the red curve ($f(x)$).

The resulting polynomial, as seen in Figure 13.1, successfully interpolates the chosen points from $f(x) = e^x$.

> **Theorem 13.2** (Lagrange interpolation theorem). *Let $x_0, x_1, \ldots, x_n \in \mathbb{R}$ be distinct points and $y_0, y_1, \ldots, y_n \in \mathbb{R}$. Then there exists exactly one polynomial $p(x)$ of degree at most $n$ satisfying*
>
> $$p(x_i) = y_i, \quad i = 0, 1, \ldots, n.$$

*Proof.* This follows directly from the uniqueness of the solution to the corresponding linear system. The determinant of the coefficient matrix (Vandermonde determinant) is nonzero since the points $x_i$ are distinct, ensuring a unique solution. □

We now study the interpolation error given by the following theorem. The proof uses Rolle's theorem.

> **Theorem 13.3** (Lagrange interpolation error). *Let $n \in \mathbb{N}$ and $f \in C^{n+1}[a, b]$, that is, $f$ is $n + 1$ times contin-uously differentiable. Let $x_0, \ldots, x_n \in [a, b]$ be distinct points, and let $p(x)$ be the interpolation polynomial of $f$ at the points $x_0, \ldots, x_n$. Then, for each $x \in [a, b]$, there exists $\xi \in [a, b]$ such that*
>
> $$f(x) - p(x) = \frac{f^{(n+1)}(\xi)}{(n+1)!} \prod_{i=0}^{n}(x - x_i).$$
>
> *Moreover*
>
> $$\|f - p\|_\infty \le \max_{x \in [a,b]} \left| \prod_{i=0}^{n}(x - x_i) \right| \frac{\|f^{(n+1)}\|_\infty}{(n+1)!},$$
>
> *where $\|f\|_\infty = \max_{x \in [a,b]} |f(x)|$.*

*Proof.* Clearly, equation above holds at points $x_0, \ldots, x_n$.

Let $x \in [a, b] \setminus \{x_0, \ldots, x_n\}$. Define

$$\Phi(t) = \prod_{i=0}^{n}(t - x_i), \quad t \in [a, b]$$

and

$$\phi(t) = f(t) - p(t) - \frac{f(x) - p(x)}{\Phi(x)} \Phi(t), \quad t \in [a, b].$$

Then $\phi(t)$ has at least $n+2$ distinct roots in the interval $[a, b]$. By repeated application of Rolle's theorem, we find that $\phi^{(n+1)}(\xi) = 0$ for some $\xi \in (a, b)$.

But

$$\phi^{(n+1)}(t) = f^{(n+1)}(t) - \frac{f(x) - p(x)}{\Phi(x)}(n + 1)!.$$

Evaluating at $\xi$, we obtain

$$0 = f^{(n+1)}(\xi) - \frac{f(x) - p(x)}{\Phi(x)}(n + 1)!,$$

from which the required formula easily follows. □

## 13.4 Newton–Raphson method

**Theorem 13.4** (Fixed point approximation). *Let the sequence $x_{n+1} = \phi(x_n)$ be defined for $n \geq 0$ and a given initial value $x_0$. Assume that*

(i)  $\phi : [a, b] \to I \subseteq [a, b]$.

(ii)  $\phi \in C^1([a, b])$, i. e., $\phi$ is differentiable with continuous derivative on $[a, b]$.

(iii)  *There exists $K < 1$ such that $|\phi'(x)| \leq K$ for all $x \in [a, b]$.*

*Then the function $\phi$ has a unique fixed point $\xi \in [a, b]$ (i. e., $\xi = \phi(\xi)$), and the sequence $(x_n)$ converges to $\xi$ for any initial value $x_0 \in [a, b]$. Furthermore,*

$$\lim_{k \to \infty} \frac{x_{k+1} - \xi}{x_k - \xi} = \phi'(\xi).$$

*Proof.* By Theorem 5.7, there exists $\xi \in [a, b]$ such that $\xi = \phi(\xi)$. Under the given assumptions, we will show uniqueness. Indeed, suppose $\xi_1, \xi_2 \in [a, b]$ with $\xi_i = \phi(\xi_i)$ for $i = 1, 2$. By the mean value theorem, there is $x \in (\xi_1, \xi_2)$ such that

$$\xi_1 - \xi_2 = \phi(\xi_1) - \phi(\xi_2) = \phi'(x)(\xi_1 - \xi_2).$$

Thus,

$$|\xi_1 - \xi_2| = |\phi'(x)| \cdot |\xi_1 - \xi_2|.$$

If $\xi_1 \neq \xi_2$, then $|\phi'(x)| < 1$ by hypothesis, which leads to a contradiction unless $\xi_1 = \xi_2$.

Next, consider the sequence $x_{n+1} = \phi(x_n)$. By the mean value theorem again, for each $n$ there exists $\eta_n \in [x_n, \xi]$ such that

$$x_{n+1} - \xi = \phi(x_n) - \phi(\xi) = \phi'(\eta_n)(x_n - \xi),$$

so

$$|x_{n+1} - \xi| \leq K|x_n - \xi| \leq K^{n+1}|x_0 - \xi|.$$

Since $K < 1$, we conclude that $|x_{n+1} - \xi| \to 0$ as $n \to \infty$; thus, the sequence converges to the unique fixed point.

Moreover, from the previous relation,

$$\lim_{n \to \infty} \frac{x_{n+1} - \xi}{x_n - \xi} = \lim_{n \to \infty} \phi'(\eta_n) = \phi'(\xi)$$

because $\eta_n \to \xi$. $\qquad\square$

**Remark 16.** The number $|\phi'(\xi)|$ describes the **asymptotic rate of convergence** of the sequence $(x_n)$. If $|\phi'(\xi)|$ is small, the convergence is fast; if it is close to 1, convergence is slow. $\qquad\square$

**Theorem 13.5** (Newton's method). *Let f be twice continuously differentiable on an interval containing a simple root $x^*$, i.e.,*

$$f(x^*) = 0, \quad f'(x^*) \neq 0.$$

*Then there exists a closed interval I containing $x^*$ such that for any initial guess $x_0 \in I$ the Newton iterates*

$$x_{n+1} = x_n - \frac{f(x_n)}{f'(x_n)}, \quad n = 0, 1, 2, \ldots$$

*remain in I and converge to $x^*$. Moreover, the convergence is quadratic, in fact*

$$\lim_{n \to \infty} \frac{x_{n+1} - x^*}{(x_n - x^*)^2} = \frac{f''(x^*)}{2f'(x^*)}.$$

*Proof.* Define the Newton map

$$\phi(x) = x - \frac{f(x)}{f'(x)}.$$

Since $f'(x^*) \neq 0$, by continuity there is a small closed interval

$$I = [x^* - \varepsilon, \, x^* + \varepsilon]$$

on which $f'(x) \neq 0$ and $\phi$ is $C^1$. Moreover $\phi'(x^*) = 0$, so by shrinking $I$ if necessary we may arrange

$$L = \max_{x \in I} |\phi'(x)| < 1.$$

Then, for every $x \in I$,

$$|\phi(x) - x^*| = |\phi(x) - \phi(x^*)| \leq L |x - x^*| < |x - x^*|,$$

which shows $\phi(I) \subset I$ and that $\phi$ is a contraction on $I$. Hence the sequence $x_{n+1} = \phi(x_n)$ converges to the unique fixed point of $\phi$ in $I$, namely $x^*$, and so $x_n \to x^*$.

To see quadratic convergence, expand $f$ and $f'$ in Taylor series about $x^*$. For each $n$ there exist $\xi_1, \xi_2 \in (x_n, x^*)$ such that

$$f(x_n) = f'(x^*)(x_n - x^*) + \frac{1}{2}f''(\xi_1)(x_n - x^*)^2, \quad f'(x_n) = f'(x^*) + f''(\xi_2)(x_n - x^*).$$

Substituting into $x_{n+1} - x^* = x_n - x^* - \frac{f(x_n)}{f'(x_n)}$, one finds

$$x_{n+1} - x^* = (x_n - x^*)^2 \frac{f''(\xi_2) - \frac{1}{2}f''(\xi_1)}{f'(x^*) + f''(\xi_2)(x_n - x^*)}.$$

Since $x_n \to x^*$ and $\xi_1, \xi_2 \to x^*$, taking the limit gives

$$\lim_{n \to \infty} \frac{x_{n+1} - x^*}{(x_n - x^*)^2} = \frac{f''(x^*)}{2f'(x^*)},$$

as claimed. $\square$

**Remark 17.** As one can see, in order to use Newton's method we should choose $x_0$ close to the root $x^*$. To compute such a $x_0$, we can use the bisection method. Note that we can not use the bisection method to compute the root of $f(x) = x^2$, but we can using Newton's method. See Table 13.1 for a comparison of the two methods.

Table 13.1: Comparison of the Newton–Raphson and bisection methods.

| Method | Advantages | Disadvantages |
|---|---|---|
| Newton–Raphson | – Quadratic convergence near a simple root (very fast)<br>– Often reaches high accuracy in few iterations<br>– Requires only one function evaluation and one derivative evaluation per step | – Convergence not guaranteed; may diverge if the initial guess is poor<br>– Needs an analytic expression (or approximation) for $f'(x)$<br>– Fails or stalls if $f'(x_n) = 0$ at any iterate<br>– Can converge to an unintended root if multiple roots exist |
| Bisection | – Guaranteed convergence if $f(a)$ and $f(b)$ have opposite signs<br>– No derivative required, only function values<br>– Error halves each iteration: $|x_n - x^*| \leq (b - a)/2^n$ | – Only linear convergence (relatively slow)<br>– Requires a valid bracketing interval $[a, b]$ with a sign change<br>– Does not exploit curvature; always bisects regardless of local behavior |

## Application to the function $f(x) = x^2$

Let us apply the Newton–Raphson method to a simple function, $f(x) = x^2$, to find its root. We can easily see that the only root of this function is at $x = 0$.

First, we need to find the first derivative of the function

$$f(x) = x^2 \quad \Longrightarrow \quad f'(x) = 2x.$$

Now, we substitute $f(x)$ and $f'(x)$ into the Newton–Raphson formula

$$x_{n+1} = x_n - \frac{x_n^2}{2x_n}.$$

This simplifies to

$$x_{n+1} = x_n - \frac{x_n}{2} = \frac{x_n}{2}.$$

This simplified formula shows that for this specific function, each iteration simply halves the current value, leading to extremely fast convergence towards the root at $x = 0$.

## Iterative steps and visualization

Let's start with an initial guess of $x_0 = 2$ and perform the first two iterations.

### First iteration

- Start at the point $(x_0, f(x_0)) = (2, 2^2) = (2, 4)$.
- The derivative at this point is $f'(x_0) = 2(2) = 4$.
- The tangent line at $(2, 4)$ has the equation $y-4 = 4(x-2)$, which simplifies to $y = 4x-4$.
- This tangent line intersects the x-axis when $y = 0$, which gives $0 = 4x - 4 \implies x = 1$.
- Our first approximation is therefore $x_1 = 1$.

### Second iteration

- Our new point is $(x_1, f(x_1)) = (1, 1^2) = (1, 1)$.
- The derivative at this point is $f'(x_1) = 2(1) = 2$.
- The tangent line at $(1, 1)$ has the equation $y-1 = 2(x-1)$, which simplifies to $y = 2x-1$.
- This tangent line intersects the x-axis when $y = 0$, which gives $0 = 2x-1 \implies x = 0.5$.
- Our second approximation is therefore $x_2 = 0.5$.

The process would continue, with $x_3 = 0.25$, $x_4 = 0.125$, and so on, rapidly approaching the root at $x = 0$.

### Conclusion

As demonstrated both computationally and visually in Figure 13.2, the Newton–Raphson method provides a highly efficient way to approximate the root of the function $f(x) = x^2$. The sequence of approximations, $2, 1, 0.5, \ldots$, quickly approaches the true root at $x = 0$. The geometric interpretation of using tangent lines helps to intuitively understand why this method is so effective, especially for well-behaved functions.

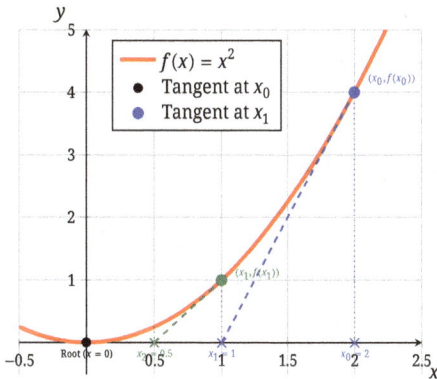

Figure 13.2: Two iterations of the Newton–Raphson method for $f(x) = x^2$ starting with $x_0 = 2$. The method converges rapidly to the root at $x = 0$.

**Example 13.6** (Newton–Raphson method for $f(x) = \cos x - x$ with visualization (Figure 13.3)). Use the Newton–Raphson method to approximate the root of $f(x) = \cos x - x = 0$, starting from $x_0 = 0.5$. Perform two iterations and illustrate them on the graph.

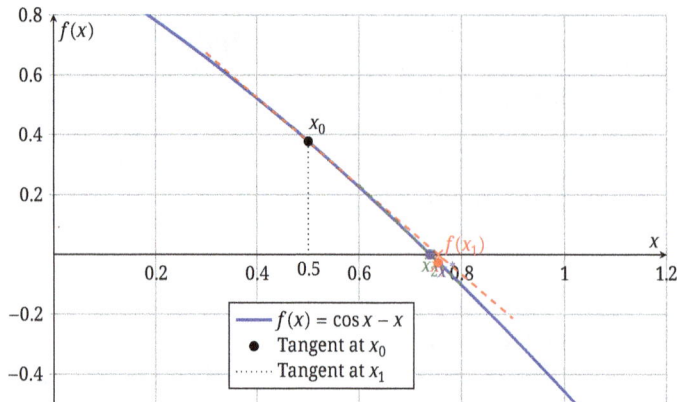

Figure 13.3: Two iterations of the Newton–Raphson method for $f(x) = \cos x - x$ starting at $x_0 = 0.5$. The method converges rapidly to the root $x^* \approx 0.7391$.

*Solution.* We use the Newton–Raphson formula

$$x_{n+1} = x_n - \frac{f(x_n)}{f'(x_n)}$$

for $f(x) = \cos x - x$ and $f'(x) = -\sin x - 1$.

**Step 1: First iteration ($x_0 = 0.5$):**

$$f(0.5) \approx 0.3776, \quad f'(0.5) \approx -1.4794,$$

$$x_1 = 0.5 - \frac{0.3776}{-1.4794} \approx 0.7553.$$

**Step 2: Second iteration ($x_1 \approx 0.7553$):**

$$f(0.7553) \approx -0.0286, \quad f'(0.7553) \approx -1.6853,$$

$$x_2 = 0.7553 - \frac{-0.0286}{-1.6853} \approx 0.7383.$$

**Conclusion:** The Newton–Raphson method rapidly converges to the root $x^* \approx 0.7391$, as shown by the intersection points of the tangent lines with the $x$-axis. $\square$

## 13.5 Python and AI

**AI Request**

Given a function $f$ I want a Python code that gives the $n$ first terms of the Taylor expansion around $x_0$. Plot the function and the approximation. Ah, provide me also with any historical information about Taylor.

## 13.6 Exercises

### True or false

Determine whether each statement is **true** or **false**. Justify your answer.

1.  The Taylor polynomial of degree $n$ for a function $f(x)$ always equals $f(x)$ for all $x$ if $f$ is a polynomial of degree less than or equal to $n$.
2.  The error of the Newton–Raphson method always decreases with each iteration, regardless of the function or initial guess.
3.  The Lagrange remainder in Taylor's theorem depends on the $(n+1)$-th derivative of $f$ at some intermediate point.
4.  The Lagrange interpolation polynomial of degree $n$ passes exactly through $n + 1$ given data points.
5.  If $f(x)$ is twice differentiable, Newton's method for $f(x) = 0$ always converges quadratically for any initial guess $x_0$.

### Multiple choice questions

Choose the correct answer.

1.  The second-order Taylor polynomial of $f(x) = \sin x$ at $x = 0$ is
    A. $x$          B. $x - \frac{x^3}{6}$
    C. $x - \frac{x^2}{2}$    D. $x + \frac{x^2}{2}$
    **Answer:** _____

2.  If the Newton–Raphson iteration $x_{n+1} = x_n - \frac{f(x_n)}{f'(x_n)}$ is applied to $f(x) = x^2 - 4$ with $x_0 = 3$, what is $x_1$?
    A. 2    B. 1
    C. 2.5    D. 4
    **Answer:** _____

3.  Which of the following best describes the interpolation error for a degree $n$ polynomial?
    A. It is always zero for all $x$.
    B. It depends on $f^{(n+1)}(x)$.
    C. It is bounded by the difference between $f(x)$ and $p(x)$ only at the data points.
    D. It increases with the number of data points.
    **Answer:** _____

4.  For which $f(x)$ does Newton's method fail to converge if $x_0$ is chosen so that $f'(x_0) = 0$?
    A. $f(x) = x^3$    B. $f(x) = x^2$
    C. $f(x) = e^x$    D. $f(x) = \sin x$
    **Answer:** _____

5.  Which is the correct error bound for the Taylor approximation of $f(x)$ about $x = a$?
    A. $|R_n(x)| \leq \frac{|x-a|^{n+1}}{(n+1)!} \cdot \max_c |f^{(n+1)}(c)|$    B. $|R_n(x)| = 0$
    C. $|R_n(x)| \leq n! \cdot |x - a|$          D. $|R_n(x)| \leq |x - a|$
    **Answer:** _____

## Matching

Match the concept to its description.

1. Taylor polynomial          A. Iterative root-finding using derivatives
2. Lagrange interpolation       B. Passes through $n + 1$ given points exactly
3. Newton–Raphson method    C. Approximates $f(x)$ near a point using derivatives
4. Bisection method            D. Guaranteed convergence for sign-changing intervals

**Answers: 1.** _____    **2.** _____    3. _____    4. _____

## Fill in the blanks

1. The error of a Taylor polynomial is given by the _____ form of the remainder.
2. The Newton–Raphson method uses the _____ of the function at each step.
3. The Lagrange interpolation polynomial of degree $n$ is unique because the associated Vandermonde matrix is _____.
4. If $f''(c) > 0$, the second-order Taylor polynomial at $x = c$ approximates $f(x)$ from _____.
5. For the Newton–Raphson method to converge quadratically, $f'(x^*)$ must be _____ at the root.

## Theoretical/computational exercises

1. Compute the third-order Taylor polynomial for $f(x) = \ln(1 + x)$ about $x = 0$.
2. Estimate the error $|f(x) - P_2(x)|$ for $f(x) = \sin x$, $P_2(x)$ the Taylor polynomial of degree 2 about $x = 0$, and $x = 0.5$. Use the Lagrange form of the remainder.
3. Find the interpolating polynomial of degree 2 passing through $(0, 1)$, $(1, 2)$, $(2, 5)$.
4. Apply two iterations of the Newton–Raphson method to $f(x) = x^3 - 2x - 5$ with $x_0 = 2$.
5. For $f(x) = e^x$, use Lagrange interpolation with points $x_0 = 0$, $x_1 = 1$ to write the interpolating polynomial and estimate $f(0.5)$.

## Python/computational questions

1. Write a Python code (using sympy or numpy) to compute and plot the Taylor approximation of $f(x) = e^{-x^2}$ at $x = 0$ up to degree 4.
2. Use numpy to solve a Vandermonde system for points $(0, 1)$, $(1, 3)$, $(2, 8)$ and give the resulting quadratic polynomial.
3. Implement Newton's method for $f(x) = \cos x - x$ and $x_0 = 1$, and show the result after four iterations.
4. Plot $f(x) = x^2$ and the sequence of Newton approximations starting from $x_0 = 2$ for three iterations.
5. Write a function that returns the Lagrange interpolation polynomial through given data points and evaluates it at a given $x$.

# 14 Range of a function: global extrema and bounds

## Contents

Understanding the range of a real function is fundamental for analyzing its possible values, global extrema, and overall behavior. This chapter presents systematic methods for determining the range and global bounds of a function, including the identification of all critical points, analysis of endpoint behavior, and computation of local and global maxima and minima. We show how to construct tables of extrema and use limits to handle open or infinite domains. The approach applies to a wide variety of functions, including rational, piecewise, and transcendental cases. Through detailed examples and exercises, you will develop reliable techniques for describing the range and extremal values of real functions in diverse contexts.

## 14.1 Introduction

The **range** (image) of a function is a fundamental concept in mathematical analysis and calculus: it describes all the possible values that a function can attain as its argument varies over the domain. Understanding the range is crucial not only for a complete description of the function itself, but also for applications in optimization, mathematical modeling, and real-world problem solving, where it is essential to know the possible outcomes or constraints of a system.

In this chapter, we develop a systematic approach (see Table 14.1) for determining the range of real functions, with particular attention to global extrema (maximum and minimum values) and global bounds. The methods presented combine classical calculus tools—such as derivatives, monotonicity, and limits—with a structured tabular approach that clarifies how local extrema, endpoint behavior, and the continuity of the function contribute to its global behavior.

Table 14.1: Summary of steps for finding the range of a real function.

| Step | Description |
|------|-------------|
| 1. | Find the domain of $f(x)$ |
| 2. | Compute limits at open endpoints and infinity if included in the domain |
| 3. | Find local extrema |
| 4. | Evaluate $f(x)$ at closed endpoints |
| 5. | Use all the above to describe the range of $f$ |

https://doi.org/10.1515/9783112228289-014

You will learn how to:
- Identify the domain and decompose it into intervals where the function is continuous and well-defined.
- Find all critical points by solving $f'(x) = 0$ and analyzing points of non-differentiability or discontinuity.
- Evaluate the function at closed endpoints and compute the limits at open endpoints or infinity.
- Use this information to fill in a table of extrema and bounds, which organizes all key values (including limits at infinity) in a single, clear view.
- Assemble the range for each subinterval and deduce the overall range of the function as a union of these sets.

Special attention is paid to piecewise-defined functions, rational and transcendental cases, and situations where the range consists of multiple disjoint intervals. Several worked examples demonstrate the method in detail, including the use of limits to capture asymptotic values, and careful analysis of local and global extrema.

A thorough collection of exercises, both theoretical and computational (including Python-based numerical investigations), will reinforce your understanding and allow you to master the procedure for a wide variety of functions. By the end of the chapter, you will be equipped with a reliable, repeatable framework for determining the range, global extrema, and bounds of real-valued functions—a foundational skill in analysis, modeling, and applied mathematics.

The **range** (or *image*) of a real function $f : D \to \mathbb{R}$ is the set of all real numbers $y$ for which there exists $x \in D$ such that $f(x) = y$

$$\text{Range}(f) = \{y \in \mathbb{R} : \exists x \in D, \ f(x) = y\}.$$

## General procedure

1. **Determine the domain** $D$ of $f$ and write it as the union of intervals in which the function is continuous. By the intermediate value theorem the range $f(I)$ will be also an interval.
2. **Compute limits** of $f(x)$ as $x$ approaches any open endpoints, including $\pm\infty$ if these belong to the domain

$$\lim_{x \to a^+} f(x), \quad \lim_{x \to b^-} f(x), \quad \lim_{x \to \pm\infty} f(x).$$

3. **Find critical points:** Find all interior points where the function attains local extrema.
4. **Compute the function** $f$ at closed endpoints.
5. **Assemble the range** using all the above information: the minimum and maximum values attained, possible infinite limits, and the continuity of $f$.

## Constructing the table of extrema and bounds

Suppose the domain $D$ of the function is a union of intervals in which $f$ is continuous. Work on each interval $I_i$ separately and compute the range $f(I_i)$. The table of extrema and bounds (see Table 14.2) is constructed as follows:

- Include all points where the first derivative vanishes and the function attains a local extremum (i. e., the monotonicity changes around that point). In the first column, write the point $x_0$; in the second, $f(x_0)$; and in the third, the characterization (local maximum or minimum).
- Also include all points where the function is not differentiable and the monotonicity changes. Again, record $x_0, f(x_0)$, and the characterization.
- For closed endpoints, write the endpoint $x_0$ and its value $f(x_0)$; for open endpoints (including $\pm\infty$), write the endpoint $a$ and $\lim_{x\to a} f(x)$.
- In the last column, indicate whether each value is a global extremum or a bound (i. e., whether the function is bounded above or below at that point).

Table 14.2: Table of critical points, non-differentiable points, and endpoint behavior for global extrema and bounds.

| $x_0$ | $f(x_0)$ | Local extrema | Global extrema/bounds |
|---|---|---|---|
| points where $f'$ vanishes | $f(x_0)$ | characterization | characterization |
| points where $f$ is not differentiable | $f(x_0)$ | characterization | characterization |
| closed endpoint | $f(x_0)$ | – | characterization |
| open endpoint $a$ | $\lim_{x\to a} f(x)$ | – | characterization |

The range $f(I_i)$ of the function on each interval can now be found easily: examine the second column to find the smallest and largest values (for example, $a$ and $b$), and the range is $[a, b]$. If one of these is a limit, the corresponding endpoint is open; otherwise, it is closed.

The overall range of $f$ is the union of the ranges over all intervals: $f(D) = \bigcup_i f(I_i)$.

**Remark 18.** Note that to compute the range of the function we do not need the 3rd and 4th columns, we need only the first and the second.

**Example 14.1.** Find the range of the function $f(x) = \frac{x}{x^2+1}, x \in \mathbb{R}$ (Figure 14.1).

*Solution.* We wish to find all real values $y$ for which there exists $x \in \mathbb{R}$ such that $f(x) = y$.

**Step-by-step solution:**
- **Domain:**
  - The denominator $x^2 + 1 > 0$ for all $x \in \mathbb{R}$, so $f$ is defined everywhere.
  - Domain: $\mathbb{R}$.

**Graph of** $f(x) = \frac{x}{x^2+1}$

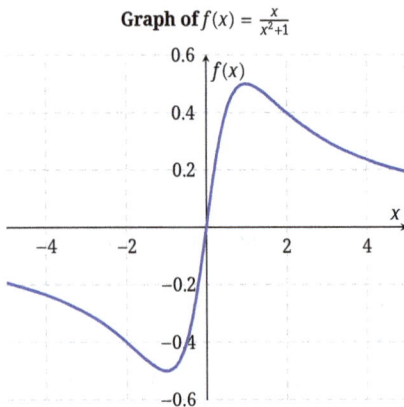

Figure 14.1: This figure displays the graph of the function $f(x) = \frac{x}{x^2+1}$, where $x \in \mathbb{R}$. As shown above, the function exhibits key characteristics: it passes through the origin $(0,0)$, has a local maximum at $(1, 0.5)$, and a local minimum at $(-1, -0.5)$. Furthermore, the function approaches zero as $x$ tends towards positive or negative infinity, indicating horizontal asymptotes along the x-axis.

The function $f$ is continuous in its domain.

- **Limits at infinity:**
  - $\lim_{x \to +\infty} \frac{x}{x^2+1} = 0.$
  - $\lim_{x \to -\infty} \frac{x}{x^2+1} = 0.$
- **Critical points (local extrema):**
  - Compute the derivative

$$f'(x) = \frac{(x^2+1) \cdot 1 - x \cdot 2x}{(x^2+1)^2} = \frac{x^2+1-2x^2}{(x^2+1)^2} = \frac{1-x^2}{(x^2+1)^2}.$$

  - Set $f'(x) = 0$

$$1 - x^2 = 0 \quad \Longrightarrow \quad x = \pm 1.$$

  - Evaluate $f$ at these points

$$f(1) = \frac{1}{2}, \quad f(-1) = -\frac{1}{2}.$$

  - Examine the sign of $f'(x)$
    * For $|x| < 1, 1 - x^2 > 0 \implies f'(x) > 0$ (increasing).
    * For $|x| > 1, 1 - x^2 < 0 \implies f'(x) < 0$ (decreasing).
  - So, $x = 1$ is a local maximum ($f(1) = \frac{1}{2}$), $x = -1$ is a local minimum ($f(-1) = -\frac{1}{2}$).
- **Closed or open endpoints:**
  - The function is defined for all $x$; there are no finite endpoints. At infinity, we already computed the limits.

## Table of extrema and bounds:

Table 14.3: Extrema and bounds for $f(x) = \frac{x}{x^2+1}$.

| $x_0$ | $f(x_0)$ | Local extrema | Global extrema/bounds |
|---|---|---|---|
| $-1$ | $-\frac{1}{2}$ | local minimum | global minimum |
| $1$ | $\frac{1}{2}$ | local maximum | global maximum |
| $+\infty$ | $0$ | – | horizontal bound |
| $-\infty$ | $0$ | – | horizontal bound |

## Conclusion (Table 14.3):

- The maximum value is $f(1) = \frac{1}{2}$, the minimum value is $f(-1) = -\frac{1}{2}$.
- As $x \to \pm\infty$, $f(x) \to 0$, but never attains 0 except at $x = 0$ ($f(0) = 0$).
- The function is continuous and achieves all intermediate values by the intermediate value theorem.

$$\text{Range}(f) = \left[-\frac{1}{2}, \frac{1}{2}\right]$$

□

**Example 14.2.** Compute the range of $f(x) = \frac{2x+1}{x-1}$ (Figure 14.2).

**Graph of $f(x) = \frac{2x+1}{x-1}$**

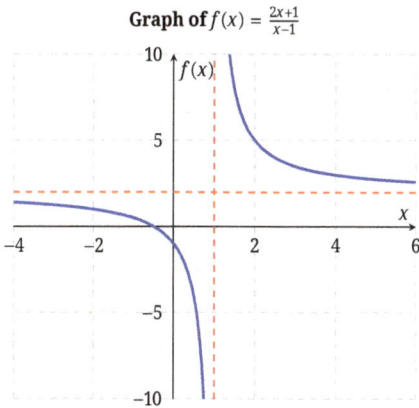

Figure 14.2: This figure displays the graph of the rational function $f(x) = \frac{2x+1}{x-1}$, defined for $x \in \mathbb{R}$, $x \neq 1$. As shown above, the function has a vertical asymptote at $x = 1$ and a horizontal asymptote at $y = 2$. The graph approaches these asymptotes but never crosses them, illustrating the characteristic behavior of rational functions with first-degree polynomials in both numerator and denominator.

*Solution.* We seek the range of

$$f(x) = \frac{2x + 1}{x - 1}$$

using only limits and local extrema.

**Step 1: Domain**

The function is defined everywhere except where the denominator vanishes

$$x - 1 = 0 \quad \Longrightarrow \quad x = 1.$$

Thus,

$$D = \mathbb{R} \setminus \{1\}.$$

We split into two intervals: $(-\infty, 1)$ and $(1, +\infty)$ in which the function is continuous.

**A. For $x < 1$**

- Endpoints:
  - As $x \to -\infty$,

$$f(x) \sim \frac{2x}{x} = 2.$$

  More precisely,

$$f(x) = 2 + \frac{3}{x - 1} \to 2 \quad \text{from below.}$$

  - As $x \to 1^-$,

$$\lim_{x \to 1^-} f(x) = -\infty.$$

- Derivative:

$$f'(x) = \frac{(2)(x - 1) - (2x + 1)(1)}{(x - 1)^2} = \frac{2x - 2 - 2x - 1}{(x - 1)^2} = \frac{-3}{(x - 1)^2} < 0.$$

  So $f$ is strictly decreasing on $(-\infty, 1)$.
- **No critical points** in $(-\infty, 1)$.

  **Table for $x < 1$:**

  Table 14.4: Extrema and bounds for $f(x)$ on $(-\infty, 1)$.

  | $x_0$ | $f(x_0)$ | Local extrema | Global extrema/bounds |
  |---|---|---|---|
  | $-\infty$ | 2 | - | upper bound (not attained) |
  | $1^-$ | $-\infty$ | - | lower bound (not attained) |

  **Subrange:** $\boxed{(-\infty, 2)}$

**B. For** $x > 1$

- **Endpoints:**
  - As $x \to 1^+$,

$$\lim_{x \to 1^+} f(x) = +\infty.$$

  - As $x \to +\infty$,

$$f(x) = 2 + \frac{3}{x-1} \to 2 \quad \text{from above.}$$

- **Derivative:**

$$f'(x) = \frac{-3}{(x-1)^2} < 0.$$

  So $f$ is strictly decreasing on $(1, +\infty)$.
- **No critical points** in $(1, +\infty)$.

  **Table for** $x > 1$:

  Table 14.5: Extrema and bounds for $f(x)$ on $(1, +\infty)$.

  | $x_0$ | $f(x_0)$ | Local extrema | Global extrema/bounds |
  |---|---|---|---|
  | $1^+$ | $+\infty$ | - | upper bound (not attained) |
  | $+\infty$ | 2 | - | lower bound (not attained) |

  **Subrange:** $(2, +\infty)$

## C. Total range

The range of $f$ is the union of the two subranges (Tables 14.4, 14.5):

$$\text{Range}(f) = (-\infty, 2) \cup (2, +\infty).$$

(*Note:* $y = 2$ is never attained.) ☐

---

**Example 14.3** (Finding the range of a piecewise function). Find the range of the function (Figure 14.3)

$$f(x) = \begin{cases} x^2, & -1 \le x \le 1, \\ \frac{1}{x}, & 1 < x \le 3. \end{cases}$$

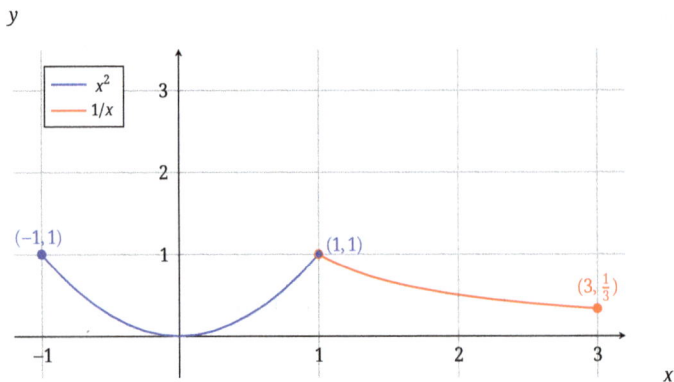

**Figure 14.3:** Scheme of the piecewise function $f(x)$ defined as $f(x) = x^2$ for $-1 \leq x \leq 1$ (blue curve), and $f(x) = \frac{1}{x}$ for $1 < x \leq 3$ (red curve). As shown, the function is continuous at $x = 1$, with both branches meeting at the point $(1, 1)$. The closed circles indicate inclusion of endpoints, while the open circle at $(1, 1)$ for the red curve indicates that this point is approached but not included in its domain, yet visually it aligns with the first branch.

*Solution.*

### Step 1: Domain and intervals

The domain is $[-1, 3]$ in which the function is continuous.

### Local extrema

Finding the first derivative we see that $f'(0) = 0$ and therefore $0$ is a critical point.
It is easy to see that $f$ is not differentiable at $x = 1$, so this is another critical point.

| $x_0$ | $f(x_0)$ | Local extrema | Global extrema—bounds |
|---|---|---|---|
| $x = 0$ | $0$ | Local minimum | Global minimum |
| $x = 1$ | $1$ | Local maximum | – |
| $x = -1$ | $1$ | – | |
| $x = 3$ | $\frac{1}{3}$ | – | |

**Final answer:**

$$\boxed{\text{Range}(f) = [0, 1]}$$

**Example 14.4** (Finding the range of a piecewise function). Find the range of the function (Figure 14.4)

$$f(x) = \begin{cases} \frac{x^2}{1+x^2} + \frac{1}{2}, & x \leq 1, \\ \frac{1}{x}, & 1 < x \leq 2, \\ -\frac{x^2}{1+x} + \frac{11}{6}, & x > 2. \end{cases}$$

y

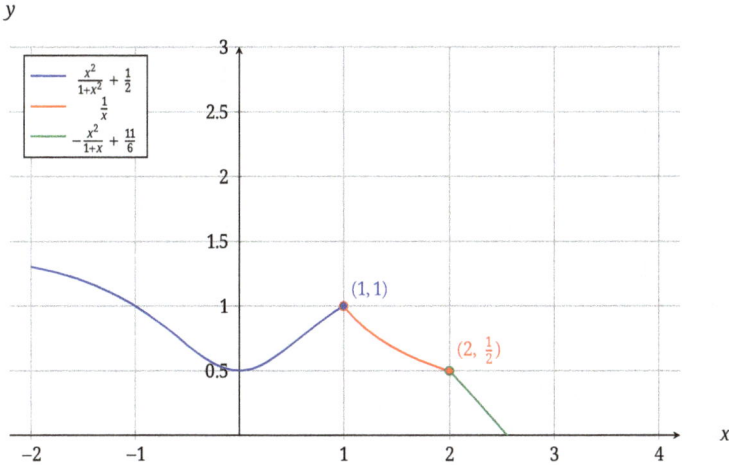

**Figure 14.4:** Scheme of the piecewise function $f(x)$, composed of three branches: $\frac{x^2}{1+x^2} + \frac{1}{2}$ for $x \leq 1$ (blue curve), $\frac{1}{x}$ for $1 < x \leq 2$ (red curve), and $-\frac{x^2}{1+x} + \frac{11}{6}$ for $x > 2$ (dark green curve). As shown above, the function is continuous at both transition points. At $x = 1$, both the blue and red branches meet at $(1, 1)$. At $x = 2$, the red and dark green branches meet at $(2, \frac{1}{2})$. The closed circles indicate inclusion of endpoints for the respective domains, while open circles signify points excluded from a domain but approached by the function.

*Solution.* The domain of this function is $(-\infty, \infty)$ in which is continuous. We can easily check that it is not differentiable at $x = 1$ and $x = 2$, so these are critical points. Moreover, $f'(0) = 0$ therefore 0 is another critical point. The points 0, 1 are local extrema.

| $x_0$ | $f(x_0)$ | Local extrema | Global extrema—bounds |
|---|---|---|---|
| $x = 0$ | $\frac{1}{2}$ | Local minimum | – |
| $x = 1$ | $1$ | Local maximum | – |
| $-\infty$ | $\frac{3}{2}$ | – | Upper bounded by $\frac{3}{2}$ |
| $+\infty$ | $-\infty$ | – | Not lower bounded |

That is, the range is

$$\text{Range}(f) = (-\infty, 3/2). \qquad \square$$

**Example 14.5** (Finding the range of a piecewise function). Find the range (set of values) of the function (Figure 14.5)

$$f(x) = \begin{cases} \frac{x^2+2}{2x^2+1}, & x \leq 3, \\ x^2, & 3 < x \leq 4. \end{cases}$$

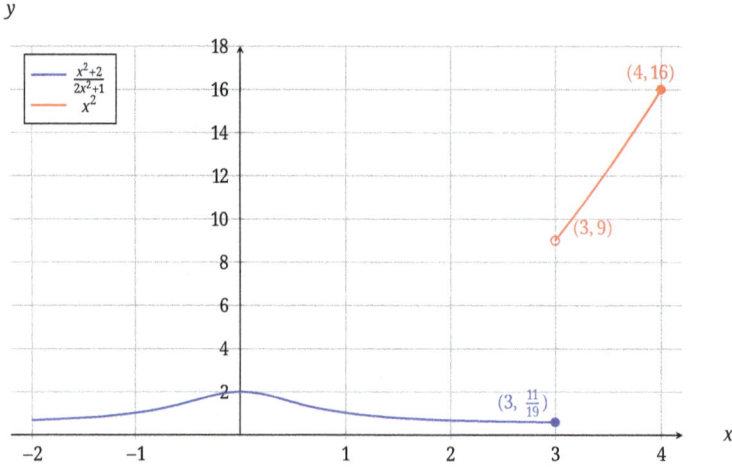

**Figure 14.5:** Scheme of the piecewise function $f(x)$, defined as $f(x) = \frac{x^2+2}{2x^2+1}$ for $x \leq 3$ (blue curve), and $f(x) = x^2$ for $3 < x \leq 4$ (red curve). As shown above, the function exhibits a **jump discontinuity** at $x = 3$. The blue curve ends at a closed circle at $(3, \frac{11}{19})$, while the red curve begins with an open circle at $(3, 9)$, illustrating the jump. The red curve then continues to a closed circle at $(4, 16)$, marking the end of its defined domain.

*Solution.*

**Domain**

The domain of this function is $(-\infty, 3] \cup (3, 4]$ and is continuous in each interval. We compute the range by analyzing each interval separately.

**1. Interval $x \leq 3$**

Let $f_1(x) = \frac{x^2+2}{2x^2+1}$ for $x \leq 3$.

**Critical points:**

Compute the derivative

$$f_1'(x) = \frac{2x(2x^2 + 1) - 2x(x^2 + 2)}{(2x^2 + 1)^2} = \frac{2x(2x^2 + 1 - x^2 - 2)}{(2x^2 + 1)^2} = \frac{2x(x^2 - 1)}{(2x^2 + 1)^2}.$$

Set $f_1'(x) = 0$

$$2x(x^2 - 1) = 0 \implies x = 0, \ x = 1, \ x = -1.$$

**Compute values at these points:**

$$f_1(0) = \frac{0^2 + 2}{2 \cdot 0^2 + 1} = \frac{2}{1} = 2,$$

$$f_1(1) = \frac{1 + 2}{2 + 1} = \frac{3}{3} = 1,$$

$$f_1(-1) = \frac{1 + 2}{2 + 1} = \frac{3}{3} = 1.$$

**Boundary values:**

$$f_1(3) = \frac{9+2}{18+1} = \frac{11}{19},$$

$$\lim_{x \to -\infty} f_1(x) = \lim_{x \to -\infty} \frac{x^2 + 2}{2x^2 + 1} = \lim_{x \to -\infty} \frac{1 + 2/x^2}{2 + 1/x^2} = \frac{1}{2}.$$

**Table of critical points, endpoints, and limits (for $x \le 3$):**

Table 14.6: Extrema and bounds for $f(x)$ on $x \le 3$.

| $x_0$ | $f(x_0)$ | Local extrema | Global extrema/bounds |
|---|---|---|---|
| −1 | 1 | local | – |
| 0 | 2 | local | maximum |
| 1 | 1 | local | – |
| 3 | $\frac{11}{19}$ | right endpoint | – |
| $-\infty$ | $\frac{1}{2}$ | – | lower bound (not attained) |

**Range for $x \le 3$:**
- The maximum value is 2 at $x = 0$.
- The minimum value is $\frac{1}{2}$ (as $x \to -\infty$, not attained).
- $f(x)$ decreases from 2 at $x = 0$ toward $\frac{1}{2}$ as $x \to -\infty$, and at $x = 3, f(3) = \frac{11}{19}$.
- For $x \le 3$, the range is the open interval $(\frac{1}{2}, 2]$ (since $f(x)$ never actually reaches $\frac{1}{2}$).

**2. Interval $3 < x \le 4$**
Let $f_2(x) = x^2$ for $3 < x \le 4$.

**Endpoints:**

$$f_2(3^+) = 9 \text{ (left endpoint, not included)}, \quad f_2(4) = 16 \text{ (right endpoint, included)}$$

$f_2(x)$ is strictly increasing on $(3, 4]$.

Table 14.7: Extrema and bounds for $f(x)$ on $3 < x \le 4$.

| $x_0$ | $f(x_0)$ | Local extrema | Global extrema/bounds |
|---|---|---|---|
| 4 | 16 | maximum | right endpoint |
| $3^+$ | 9 | – | left bound (not attained) |

**Range for $3 < x \le 4$:**

$$(9, 16].$$

**3. Final answer: union of subranges (Tables 14.6, 14.7)**

$$\text{Range}(f) = \left(\frac{1}{2}, 2\right] \cup (9, 16].$$

– All values from just above $\frac{1}{2}$ up to 2 (including 2).
– All values strictly greater than 9 and up to 16 (including 16).    □

**Example 14.6** (Range of $f(x) = \frac{x + \sqrt{x^2 + x}}{x}$ via limits and extrema). Compute the range of $f(x) = \frac{x + \sqrt{x^2 + x}}{x}$ (Figure 14.6).

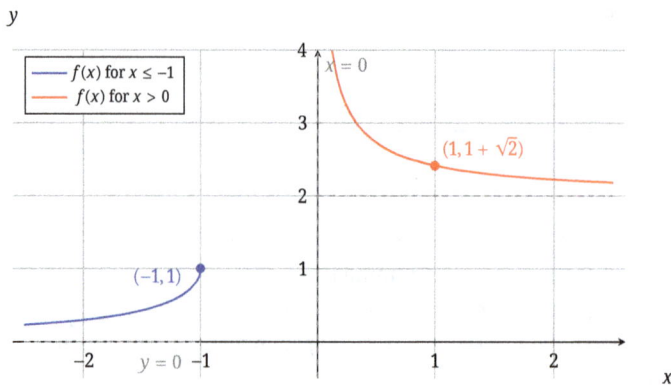

Figure 14.6: Scheme of the function $f(x) = \frac{x + \sqrt{x^2 + x}}{x}$. The function's domain is $x \in (-\infty, -1] \cup (0, \infty)$. As shown, the graph consists of two distinct branches. The blue branch, for $x \leq -1$, approaches the horizontal asymptote $y = 0$ as $x \to -\infty$ and ends at a closed circle at $(-1, 1)$. The red branch, for $x > 0$, exhibits a vertical asymptote at $x = 0$ (approaching $+\infty$ as $x \to 0^+$) and approaches the horizontal asymptote $y = 2$ as $x \to +\infty$. A specific point $(1, 1 + \sqrt{2})$ is marked on the red branch.

*Solution.*
**Step 1: Domain.**
Require $x^2 + x \geq 0$ and $x \neq 0$.

$$x(x + 1) \geq 0 \quad \Longrightarrow \quad x \leq -1 \text{ or } x \geq 0.$$

But $x = 0$ is not allowed.

**Domain:** $D = (-\infty, -1] \cup (0, +\infty)$.

**Step 2: Study each interval separately.**

**A. For $x \leq -1$**

- The function simplifies to

$$f(x) = 1 + \frac{\sqrt{x^2 + x}}{x}.$$

Here, since $x < 0$, $\sqrt{x^2 + x} > 0$, and thus $\frac{\sqrt{x^2+x}}{x} < 0$. So $f(x) < 1$ for $x < -1$.

- **Behavior at endpoints:**
  - As $x \to -1^-$, $x^2 + x \to 0^+$, so $f(x) \to 1$.
  - As $x \to -\infty$,

$$f(x) = 1 + \frac{\sqrt{x^2 + x}}{x} = 1 + \frac{|x|\sqrt{1 + 1/x}}{x} = 1 - \sqrt{1 + \frac{1}{x}} \to 1 - 1 = 0$$

  as $x \to -\infty$.
- **Monotonicity and extrema:**
  - Compute derivative for $x < -1$

$$f(x) = 1 + \frac{\sqrt{x^2 + x}}{x}.$$

Set $u = \sqrt{x^2 + x}$.

$$f'(x) = \frac{d}{dx}\left(1 + \frac{u}{x}\right) = \frac{u'x - u}{x^2}$$

$u' = \frac{2x+1}{2\sqrt{x^2+x}}$. Setting numerator zero:

$$u'x - u = 0 \quad \Longrightarrow \quad xu' = u,$$

$$\Longrightarrow \quad x \cdot \frac{2x + 1}{2u} = u \quad \Longrightarrow \quad (2x + 1)x = 2u^2 = 2(x^2 + x),$$

$$\Longrightarrow \quad 2x^2 + x = 2x^2 + 2x \quad \Longrightarrow \quad x = 2x \quad \Longrightarrow \quad x = 0.$$

But $x = 0$ is **not** in $(-\infty, -1]$, so **no critical points in $x \leq -1$.**
- **Endpoints:**
  - At $x = -1, f(-1) = \frac{-1+0}{-1} = 1$.
  - As $x \to -\infty, f(x) \to 0$.

**Table for $x \leq -1$:**

Table 14.8: Extrema and bounds for $f(x)$ on $(-\infty, -1]$.

| $x_0$ | $f(x_0)$ | Local extrema | Global extrema/bounds |
|---|---|---|---|
| $-1$ | 1 | endpoint | maximum |
| $-\infty$ | 0 | – | lower bound (not attained) |

**Subrange:** $\boxed{(0,1]}$

**B. For $x > 0$**

Let $x > 0$.

$$f(x) = 1 + \sqrt{1 + \frac{1}{x}}.$$

As $x \to 0^+$, $1/x \to +\infty$, so $f(x) \to +\infty$.

As $x \to +\infty$, $1/x \to 0$, so $f(x) \to 1 + 1 = 2$.

- **Derivative:**

$$f'(x) = \frac{1}{2} \cdot \frac{-1}{x^2} \cdot \left(1 + \frac{1}{x}\right)^{-1/2} = -\frac{1}{2x^2 \sqrt{1 + 1/x}} < 0.$$

So $f$ is **strictly decreasing** on $(0, +\infty)$.

- **No critical points** in $(0, +\infty)$.
- **Endpoints:**
  - $x \to 0^+$: $f(x) \to +\infty$ (not attained).
  - $x \to +\infty$: $f(x) \to 2$.

**Table for $x > 0$:**

Table 14.9: Extrema and bounds for $f(x)$ on $(0, +\infty)$.

| $x_0$ | $f(x_0)$ | Local extrema | Global extrema/bounds |
|---|---|---|---|
| $0^+$ | $+\infty$ | – | upper bound (not attained) |
| $+\infty$ | 2 | – | lower bound (not attained) |

**Subrange:** $\boxed{(2, +\infty)}$

## C. Final range (Tables 14.8, 14.9)

The range of $f$ is the union of subranges:

$$\boxed{\text{Range}(f) = (0, 1] \cup (2, +\infty)}$$ □

**Example 14.7** (Computing the range of $f(x) = \frac{(0.4)^x}{\ln 0.4} - \frac{x^2}{2}$ via bisection and Newton–Raphson). Let

$$f(x) = \frac{(0.4)^x}{\ln 0.4} - \frac{x^2}{2}, \quad x \in \mathbb{R}.$$

Find the **range** of $f$ (Figure 14.7), i. e., the set of all possible values of $f(x)$ as $x$ varies over $\mathbb{R}$.

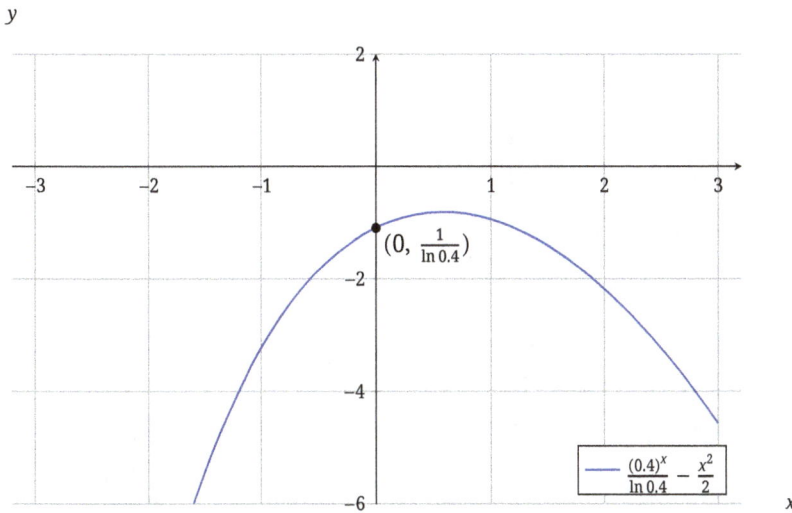

Figure 14.7: Scheme of the function $f(x) = \frac{(0.4)^x}{\ln 0.4} - \frac{x^2}{2}$. This function exhibits a combination of a decreasing exponential term and a downward-opening parabolic term, resulting in the displayed overall shape. A key point marked on the graph is the y-intercept at $(0, \frac{1}{\ln 0.4})$, which is approximately $(0, -1.09)$.

*Solution.*
### Step 1: Behavior at the endpoints
- As $x \to -\infty$,

$$(0.4)^x \to +\infty, \quad \ln 0.4 < 0 \quad \Longrightarrow \quad \frac{(0.4)^x}{\ln 0.4} \to -\infty, \quad -\frac{x^2}{2} \to -\infty.$$

The $-\frac{x^2}{2}$ dominates, so $f(x) \to -\infty$.
- As $x \to +\infty$,

$$(0.4)^x \to 0^+, \quad \frac{(0.4)^x}{\ln 0.4} \to 0, \quad -\frac{x^2}{2} \to -\infty.$$

So again, $f(x) \to -\infty$.

**Step 2: Find the global maximum**

Since $f(x) \to -\infty$ as $x \to \pm\infty$, the function attains its maximum at a unique interior point.

Compute the critical point(s) by solving $f'(x) = 0$:

$$f'(x) = (0.4)^x - x.$$

Set $f'(x) = 0 \implies (0.4)^x = x$.

Define $g(x) = (0.4)^x - x$. Let's find the root numerically.

*Bisection method:*

- $g(0) = 1 - 0 = 1 > 0$
- $g(1) = 0.4 - 1 = -0.6 < 0$
- $g(0.5) \approx 0.632 - 0.5 = 0.132 > 0$
- $g(0.75) \approx 0.514 - 0.75 = -0.236 < 0$
- $g(0.625) \approx 0.569 - 0.625 = -0.056 < 0$
- $g(0.5625) \approx 0.599 - 0.5625 = 0.0365 > 0$
- $g(0.59375) \approx 0.584 - 0.59375 = -0.00975 < 0$

So the root is in $(0.5625, 0.59375)$.

*Newton–Raphson refinement:*

Set $x_0 = 0.58$.

$$g(x_0) = (0.4)^{0.58} - 0.58 \approx 0.589 - 0.58 = 0.009,$$

$$g'(x_0) = (0.4)^{0.58} \ln 0.4 - 1 \approx 0.589 \times (-0.9163) - 1 \approx -0.539 - 1 = -1.539,$$

$$x_1 = x_0 - \frac{g(x_0)}{g'(x_0)} \approx 0.58 - \frac{0.009}{-1.539} \approx 0.58 + 0.00585 = 0.58585.$$

Iterating once more gives $x^* \approx 0.5878$.

**Step 3: Maximum value**

At $x^* \approx 0.5878$,

$$f(x^*) = \frac{(0.4)^{x^*}}{\ln 0.4} - \frac{(x^*)^2}{2},$$

$$(0.4)^{0.5878} \approx 0.5878, \quad \ln 0.4 \approx -0.9163,$$

$$f(x^*) \approx \frac{0.5878}{-0.9163} - \frac{(0.5878)^2}{2} \approx -0.6417 - 0.1727 = -0.8144.$$

**Step 4: Range of $f$**

- $f(x)$ has a unique global maximum at $x^* \approx 0.588$ with $f(x^*) \approx -0.814$.
- As $x \to \pm\infty, f(x) \to -\infty$.

$$\boxed{\text{Range}(f) = (-\infty, \ -0.814]}$$ ☐

**Example 14.8** (Young's inequality). Suppose $p > 1$, $q > 1$ with $\frac{1}{p} + \frac{1}{q} = 1$ and $y \geq 0$. We will prove that

$$xy \leq \frac{x^p}{p} + \frac{y^q}{q}, \quad \text{when } \frac{1}{p} + \frac{1}{q} = 1, \ x \geq 0, \ y \geq 0.$$

*Solution.* Let

$$f(x) = \frac{x^p}{p} + \frac{y^q}{q} - xy$$

on the interval $[0, +\infty)$ for $y \geq 0$.

**Step 1: Compute the derivative.**

$$f'(x) = x^{p-1} - y.$$

Setting $f'(x) = 0$ gives the critical point

$$x = y^{1/(p-1)}.$$

**Step 2: Second derivative and characterization.**

$$f''(x) = (p-1)x^{p-2} > 0 \quad \text{for } x > 0.$$

Thus, $f$ is convex and the critical point is a *local minimum*. Since the function is convex and defined on $[0, +\infty)$, this minimum is also the global minimum.

**Step 3: Limit as $x \to +\infty$ (open right endpoint).**

$$\lim_{x \to +\infty} f(x) = +\infty.$$

**Step 4: Constructing the table of extrema and bounds.**

| $x_0$ | $f(x_0)$ | Local extrema | Global extrema/bounds |
|---|---|---|---|
| $x = y^{1/(p-1)}$ | $0$ | local minimum | global minimum |
| $x = 0$ | $\frac{y^q}{q}$ | – | – |
| $+\infty$ | $+\infty$ | – | unbounded above |

**Conclusion:** Therefore, the function $f$ attains its global minimum value $0$ at $x = y^{1/(p-1)}$. That is,

$$f(x) \geq 0 \quad \text{for all } x \geq 0$$

or, equivalently,

$$xy \leq \frac{x^p}{p} + \frac{y^q}{q}, \quad \text{when } \frac{1}{p} + \frac{1}{q} = 1, \ x \geq 0, \ y \geq 0. \qquad \square$$

## 14.2 Python and AI

---

**AI Request**

Prove that $e^x \geq x^e$ for $x \geq 0$ and give me the Latex code of the solution step-by-step. Write a Python code that computes the range of the function $f(x) = e^x - x^e$ for $x \geq 0$.

---

## 14.3 Exercises

### True or false

Decide whether each statement is **true** or **false**. Justify your answer.

1. If a continuous function $f$ on $[a, b]$ attains its global maximum at an interior point, then $f'(x) = 0$ at that point.
   **Answer:** _____

2. A function with an unbounded domain can still be bounded above and below.
   **Answer:** _____

3. If $f$ is differentiable on $(a, b)$ and $\lim_{x \to a^+} f(x) = +\infty$, then $f$ has no global minimum on $(a, b)$.
   **Answer:** _____

4. Whenever $f'(x) = 0$ and $f''(x) > 0$, the point $x$ yields a global minimum of $f$.
   **Answer:** _____

5. If a continuous function is strictly increasing on its domain, its range is always an open interval.
   **Answer:** _____

### Multiple choice questions

Choose the correct option (A, B, C, or D) for each question.

1. Which of the following is *sufficient* to guarantee that $f$ has both a global maximum and a global minimum on $D$?
   A. $D$ is open and $f$ is continuous
   B. $D$ is closed and $f$ is differentiable
   C. $D$ is closed and bounded, and $f$ is continuous
   D. $f$ is monotone on $D$
   **Answer:** _____

2. If $\lim_{x \to \infty} f(x) = L$ and $\lim_{x \to -\infty} f(x) = -\infty$, the range of $f$ is
   A. $(-\infty, L]$    B. $(-\infty, L)$
   C. $[-\infty, L]$    D. cannot be determined without more data
   **Answer:** _____

3. For $g(x) = x^3 - 3x$ on $(-\infty, \infty)$, which statement is true?
   A. $g$ is bounded below
   B. $g$ is bounded above
   C. $g$ has neither global max nor global min
   D. $g$ reaches its global min at $x = -1$
   **Answer:** _____

4. Let $h(x) = \frac{1}{x}$ on $(0, \infty)$. Its global extrema are

A. max $h = 1$, min $h = 0$          B. max $h = \infty$, min $h = 0$
C. max $h = \infty$, min $h$ does not exist    D. $h$ has no global extrema
**Answer:** _____

5.   If $f$ is differentiable and has exactly one critical point $c$ on $(a, b)$ with $f''(c) < 0$, then
    A. $f(c)$ is a global maximum          B. $f(c)$ is a global minimum
    C. $f(c)$ is neither global max nor min    D. $f$ must be constant
    **Answer:** _____

## Matching exercise

Match each term with its correct description.

1. Global maximum        A. Point where $f$ switches from increasing to decreasing
2. Critical point            B. Highest value $f$ attains on its domain
3. Horizontal asymptote    C. Value approached by $f(x)$ as $x \to \pm\infty$
4. Unbounded below       D. $\inf_{x \in D} f(x) = -\infty$
5. Closed interval         E. Domain type guaranteeing extrema for continuous $f$
**Answers:** 1._____    2._____    3._____    4._____    5._____

## Fill in the blanks

Fill in each blank with the most appropriate word or phrase.

1.   A continuous function on a _____ and bounded interval must attain
    both a maximum and a minimum.
    **Answer:** _____

2.   If $\lim_{x \to a^-} f(x) = +\infty$, then $x = a$ is a _____ vertical bound for $f$.
    **Answer:** _____

3.   The set $\{f(x) : x \in D\}$ is called the _____ of $f$.
    **Answer:** _____

4.   When $f'(x) = 0$ has no real solution but $f$ is continuous, extrema can still occur at

    _____.

    **Answer:** _____

5.   A function whose range is all of $\mathbb{R}$ is said to be _____.
    **Answer:** _____

## Theoretical and computational problems

Solve each problem step-by-step.

1.   Determine the global extrema and range of $f(x) = \frac{x^2 - 4x + 3}{x - 5}$ on $(-\infty, 5) \cup (5, \infty)$.
2.   Prove that the function $g(x) = \arctan x$ is bounded and find its exact range.
3.   Let $h(x) = xe^{-x}$ on $[0, \infty)$. Find the range of $h$ and the $x$-coordinate where the global
    maximum occurs.
4.   For $p(x) = x^4 - 8x^2 + 16$, locate all global minima and determine whether $p$ is bounded
    above.

5. Show that if $f$ is continuous on $(a, \infty)$ and $\lim_{x \to \infty} f(x) = L$, then $f$ is bounded on $(a, \infty)$.
6. Prove that $e^x > x + 1$ for all $x \in \mathbb{R} \setminus \{0\}$.
7. Prove that $(1 + e^x) \ln(1 + e^x) - x(1 + e^x) - 1 > 0$ for $x \in \mathbb{R}$.
8. Prove that $e^x \geq x^e$ for $x \geq 0$.
9. Find the range of the function

$$f(x) = \begin{cases} x^4 + 1, & \text{when } x \in (-\infty, 0), \\ 1, & \text{when } x \in [0, e), \\ \ln x, & \text{when } x \in [e, +\infty). \end{cases}$$

## Python-based exercises

Use Python (`sympy`, `numpy`, `matplotlib`) to tackle the following:
1. Plot $f(x) = \frac{1}{1+x^2}$ for $x \in [-10, 10]$ and confirm numerically that its range is $(0, 1]$.
2. Write a function `global_extrema(f, interval)` that samples a symbolic $f$ on the given interval and returns approximate global maxima and minima.
3. For $g(x) = x^3 - 6x^2 + 9x + 1$, use `sympy` to find critical points, classify them, and plot $g$ marking global extrema.
4. Numerically explore $h(x) = x \sin x$ on $[0, 20]$: estimate its highest peak and lowest trough in that interval.
5. Investigate the function $p(x) = \frac{\ln x}{x}$ on $(1, \infty)$: plot it, locate its global maximum, and approximate its range on that domain.

# 15 Trigonometric functions

## Contents

Trigonometric functions are fundamental mathematical tools that connect geometry, algebra, and real-world applications. Originating from the study of triangles and circles, the sine, cosine, and tangent functions are now central to calculus, physics, engineering, and many sciences. In this chapter, we will define the trigonometric functions using both the right triangle and the unit circle. We will explore their analytic properties, such as continuity and differentiability, and derive essential identities including the Pythagorean, angle addition, and double-angle formulas. Through examples and exercises, you will learn to manipulate trigonometric expressions, solve equations, and understand the graphical behavior and periodicity of these functions.

## 15.1 Introduction

Trigonometric functions are among the oldest and most important tools in mathematics, connecting geometry, algebra, and analysis. Their origins lie in the study of triangles and circles, yet their applications extend far beyond classical geometry, appearing in fields as diverse as physics, engineering, signal processing, and even biology. Understanding trigonometric functions is a foundational milestone for every student of mathematics, providing a gateway to more advanced topics in calculus, differential equations, and mathematical modeling.

In this chapter, we begin by exploring the geometric and algebraic definitions of the basic trigonometric functions—sine, cosine, and tangent—via the unit circle and right triangles. The Pythagorean theorem provides the crucial geometric underpinning, while the unit circle construction naturally introduces the periodicity and symmetry that are hallmarks of trigonometric functions.

We then develop the analytic properties of these functions, establishing their continuity, differentiability, and key limits. Notably, the famous limit $\lim_{x\to 0} \frac{\sin x}{x} = 1$ serves as a cornerstone in the theory of derivatives and series expansions. The chapter includes rigorous proofs, both geometric and analytic, and presents important derivative formulas for both the direct and inverse trigonometric functions.

Central to this chapter are the trigonometric identities (Table 15.1): the Pythagorean, even-odd, angle sum and difference, double- and half-angle formulas, and more. These

https://doi.org/10.1515/9783112228289-015

Table 15.1: Condensed summary of trigonometric identities.

| Category | Definition/identity |
|---|---|
| Basic definitions | $\sin\theta = y$, $\cos\theta = x$ on unit circle; $\tan\theta = \frac{\sin\theta}{\cos\theta}$, $\sec x = \frac{1}{\cos x}$, $\csc x = \frac{1}{\sin x}$, $\cot x = \frac{1}{\tan x}$ |
| Pythagorean identities | $\sin^2 x + \cos^2 x = 1$, $1 + \tan^2 x = \sec^2 x$, $1 + \cot^2 x = \csc^2 x$ |
| Even-odd | $\sin(-x) = -\sin x$, $\cos(-x) = \cos x$, $\tan(-x) = -\tan x$ |
| Angle sum/difference | $\sin(x \pm y) = \sin x \cos y \pm \cos x \sin y$ |
| | $\cos(x \pm y) = \cos x \cos y \mp \sin x \sin y$ |
| | $\tan(x \pm y) = \frac{\tan x \pm \tan y}{1 \mp \tan x \tan y}$ |
| Double-angle | $\sin(2x) = 2\sin x \cos x$ |
| | $\cos(2x) = \cos^2 x - \sin^2 x = 2\cos^2 x - 1 = 1 - 2\sin^2 x$ |
| | $\tan(2x) = \frac{2\tan x}{1-\tan^2 x}$ |
| Half-angle | $\sin(\frac{x}{2}) = \pm\sqrt{\frac{1-\cos x}{2}}$, $\cos(\frac{x}{2}) = \pm\sqrt{\frac{1+\cos x}{2}}$, $\tan(\frac{x}{2}) = \frac{\sin x}{1+\cos x} = \frac{1-\cos x}{\sin x}$ |

identities form an essential toolkit for simplifying expressions, solving equations, and tackling integrals in later courses. Each identity is derived step-by-step, often accompanied by geometric illustrations to aid intuition.

Beyond the basics, we emphasize the power and versatility of trigonometric functions through a wide selection of examples and exercises. Piecewise definitions, the structure of the graphs, periodicity, and symmetry are highlighted, and students are encouraged to explore the rich interplay between algebraic manipulation and geometric interpretation.

The chapter concludes with a comprehensive set of exercises: conceptual (true/false and multiple choice), theoretical (proofs and derivations), and computational (including Python-based explorations). These problems are designed to reinforce understanding and foster fluency in both analytic and computational techniques.

By mastering the content of this chapter, you will gain the ability to:

- Define and interpret the six trigonometric functions in terms of both right triangles and the unit circle.
- Derive and apply the fundamental trigonometric identities.
- Understand the analytic properties—continuity, differentiability, and limits—of trigonometric and inverse trigonometric functions.
- Graph trigonometric functions, recognizing their periodicity, amplitude, and phase shift.
- Solve a wide variety of equations and applied problems using trigonometric methods.
- Approach advanced topics (such as Fourier analysis and complex exponentials) with a strong conceptual foundation.

Trigonometric functions open the door to a vibrant world of mathematical ideas and applications. The tools and intuition you build here will serve as indispensable assets throughout your studies and beyond.

## 15.2  Definitions

The following theorem is one of the most celebrated results in mathematics, relating the lengths of the sides in a right-angled triangle and forming the foundation for much of geometry and trigonometry.

---

**Theorem 15.1** (Pythagorean theorem). *Let ABC be a right-angled triangle with the right angle at C. Then*

$$(AC)^2 + (CB)^2 = (AB)^2.$$

---

*Proof.* We shall use the concept of *similar triangles* (see Figure 15.1).

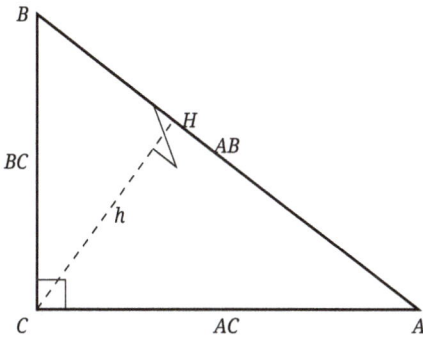

Figure 15.1: Right triangle *ABC* with altitude *CH*.

It is easy to verify that $\triangle ACH$ is similar to $\triangle ABC$, and $\triangle ABC$ is similar to $\triangle BHC$. Hence

$$\frac{AC}{AB} = \frac{AH}{AC}, \quad \frac{BC}{AB} = \frac{BH}{BC}.$$

Consequently,

$$(AC)^2 = AH \cdot AB, \quad (BC)^2 = BH \cdot AB.$$

Adding these equalities gives

$$(AC)^2 + (BC)^2 = AH \cdot AB + BH \cdot AB = AB\,(AH + BH) = (AB)^2. \qquad \square$$

---

**Definition 15.1** (Definition of sin *x*, cos *x*, and tan *x* functions).  As shown in Figure 15.2, the function sin $\theta$ is defined as the distance of the intersection point between the radius that makes an angle $\theta$ with the *x*-axis and the unit circle, measured from the *x*-axis. Equivalently, it is the length of the corresponding circular arc. The definition of cos $\theta$ is analogous. Finally, we define

$$\tan \theta = \frac{\sin \theta}{\cos \theta}.$$

---

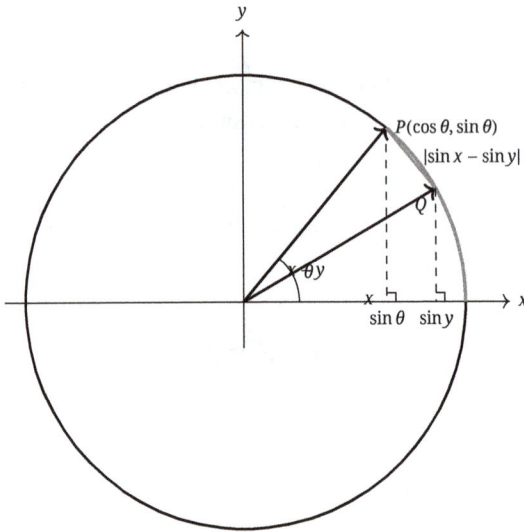

Figure 15.2: Unit circle, the definitions of sin $\theta$, cos $\theta$, and the geometric proof that $|\sin x - \sin y| \leq |x - y|$.

**Remark 19.** In Figure 15.4 one can see what exactly tan $x$ equals to. This is true because similar right-triangles share the same ratio of opposite to adjacent side.

We now prove—using the geometry of the unit circle—that sin $x$ (where $x$ denotes the length of the corresponding arc) is a *continuous* function of $x$.

**Theorem 15.2** (Continuity of trigonometric functions). *The trigonometric functions* sin $x$ *and* cos $x$ *are continuous in the interval* $[0, \frac{\pi}{2}]$.

*Proof.* Figure 15.2 shows that sin $x \leq x$ for every arc-length $x > 0$: the vertical segment from the point on the circle to the $x$-axis is shorter than the arc itself, therefore sin $x \to 0$ as $x \to 0$. Consider two arc lengths with $x > y > 0$. The same reasoning applied to the smaller arc $x - y$ gives

$$|\sin x - \sin y| \leq |x - y|, \quad x, y \in \left(0, \frac{\pi}{2}\right),$$

so sin is Lipschitz, and hence continuous, on that interval.

Continuity at the endpoints 0 and $\pi/2$ follows from the above inequality and the elementary limits $\lim_{h \to 0} \sin h = 0$ and $\lim_{h \to 0} \sin(\frac{\pi}{2} - h) = 1$. Using $\sin^2 x + \cos^2 x = 1$, the continuity of cos $x$ follows on the same interval, and one handles the endpoints with the symmetry $\cos(\pi - y) = -\cos y$. $\square$

**Remark 20.** Using symmetry and periodicity of the trigonometric functions, we arrive at the conclusion that they are continuous everywhere.

**Theorem 15.3.** *It holds that*

$$\lim_{x \to 0} \frac{\sin x}{x} = 1.$$

*Proof.* Refer to Figure 15.4. Let the unit circle be centered at the origin $O = (0, 0)$. Fix a (small) positive arc-length $x$. Denote

- $A = (1, 0)$.
- $B = (\cos x, \sin x)$, the point where the radius making an angle $x$ meets the circle.
- $G$, the intersection of the ray $OB$ with the tangent line to the circle at $A$.

Three regions of interest are formed:

$$\text{Area}(\triangle OAB) = \frac{1}{2} \sin x, \quad \text{Area(\ sector } OAB) = \frac{x}{2}, \quad \text{Area}(\triangle OAG) = \frac{1}{2} \tan x.$$

Because the triangle $OAB$ is contained in the circular sector, which in turn is contained in $OAG$,

$$\frac{1}{2} \sin x \le \frac{x}{2} \le \frac{1}{2} \tan x \quad \Longrightarrow \quad \sin x \le x \le \tan x \quad (x > 0).$$

Dividing by $\sin x$ and letting $x \to 0^{+}$,

$$1 \le \frac{x}{\sin x} \le \frac{\tan x}{\sin x} = \frac{1}{\cos x} \longrightarrow 1,$$

so by the squeeze theorem

$$\lim_{x \to 0} \frac{\sin x}{x} = 1. \qquad \square$$

**Theorem 15.4** (Trigonometric addition formulas (Figure 15.3)). *It holds that*

$$\cos(u + v) = \cos u \cos v - \sin u \sin v \quad and \quad \sin(u + v) = \sin u \cos v + \cos u \sin v.$$

*Proof.* We will use a geometric approach involving the unit circle and the distance formula.

Consider a unit circle centered at the origin $O(0, 0)$. Let's define four points on this circle based on angles measured counterclockwise from the positive $x$-axis:

- Point $A$: Corresponds to angle $u$. Its coordinates are $A(\cos u, \sin u)$.
- Point $B$: Corresponds to angle $-v$. Its coordinates are $B(\cos(-v), \sin(-v)) = B(\cos v, -\sin v)$.
- Point $C$: Corresponds to angle $u + v$. Its coordinates are $C(\cos(u + v), \sin(u + v))$.
- Point $D$: This is the reference point $(1, 0)$, corresponding to an angle of $0$.

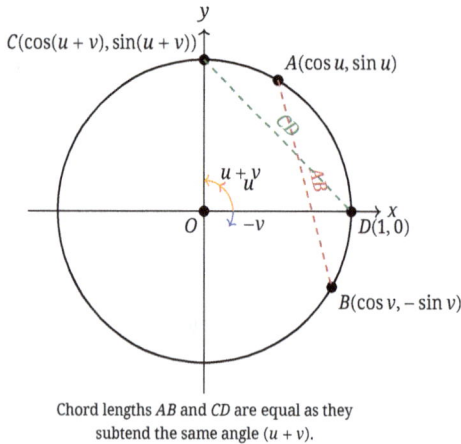

Figure 15.3: Geometric setup for proving angle addition formulas. The lengths of chords $AB$ and $CD$ are equal.

Chord lengths $AB$ and $CD$ are equal as they subtend the same angle $(u + v)$.

The arc length from point $B$ to point $A$ is $u - (-v) = u + v$. The arc length from point $D$ to point $C$ is $(u + v) - 0 = u + v$.

Since these arc lengths are equal, the chords connecting these points must also be equal in length. Therefore, the distance between $A$ and $B$ must be equal to the distance between $C$ and $D$. We use the distance formula $d = \sqrt{(x_2 - x_1)^2 + (y_2 - y_1)^2}$, or more conveniently, $d^2 = (x_2 - x_1)^2 + (y_2 - y_1)^2$.

**Calculate $AB^2$**

Points are $A(\cos u, \sin u)$ and $B(\cos v, - \sin v)$.

$$
\begin{aligned}
AB^2 &= (\cos u - \cos v)^2 + (\sin u - (-\sin v))^2 \\
&= (\cos u - \cos v)^2 + (\sin u + \sin v)^2 \\
&= (\cos^2 u - 2 \cos u \cos v + \cos^2 v) + (\sin^2 u + 2 \sin u \sin v + \sin^2 v) \\
&= (\cos^2 u + \sin^2 u) + (\cos^2 v + \sin^2 v) - 2 \cos u \cos v + 2 \sin u \sin v.
\end{aligned}
$$

Using the Pythagorean identity $\cos^2 \theta + \sin^2 \theta = 1$,

$$
AB^2 = 1 + 1 - 2 \cos u \cos v + 2 \sin u \sin v = 2 - 2 \cos u \cos v + 2 \sin u \sin v. \qquad (*)
$$

**Calculate $DC^2$**

Points are $D(1, 0)$ and $C(\cos(u + v), \sin(u + v))$.

$$
\begin{aligned}
DC^2 &= (\cos(u + v) - 1)^2 + (\sin(u + v) - 0)^2 \\
&= \cos^2(u + v) - 2 \cos(u + v) + 1 + \sin^2(u + v).
\end{aligned}
$$

Using the Pythagorean identity $\cos^2 \theta + \sin^2 \theta = 1$,

$$
DC^2 = (\cos^2(u+v) + \sin^2(u+v)) - 2 \cos(u+v) + 1 = 1 - 2 \cos(u+v) + 1 = 2 - 2 \cos(u+v). \qquad (**)
$$

**Equate** $AB^2$ **and** $DC^2$
Since $AB^2 = DC^2$,

$$2 - 2 \cos u \cos v + 2 \sin u \sin v = 2 - 2 \cos(u + v).$$

Subtract 2 from both sides

$$-2 \cos u \cos v + 2 \sin u \sin v = -2 \cos(u + v).$$

Divide by $-2$

$$\boxed{\cos(u + v) = \cos u \cos v - \sin u \sin v}.$$

**Deriving** $\sin(u + v)$ **from** $\cos(u + v)$
We can use the previously proven cosine addition formula and the co-function identity
$\sin \theta = \cos(\frac{\pi}{2} - \theta)$.
    Let $\theta = u + v$. Then

$$\sin(u + v) = \cos\left(\frac{\pi}{2} - (u + v)\right).$$

Rearrange the terms inside the cosine

$$\sin(u + v) = \cos\left(\left(\frac{\pi}{2} - u\right) - v\right).$$

Now, apply the cosine addition formula $\cos(A - B) = \cos A \cos B + \sin A \sin B$, where
$A = \frac{\pi}{2} - u$ and $B = v$,

$$\sin(u + v) = \cos\left(\frac{\pi}{2} - u\right) \cos v + \sin\left(\frac{\pi}{2} - u\right) \sin v.$$

Using the co-function identities $\cos(\frac{\pi}{2} - u) = \sin u$ and $\sin(\frac{\pi}{2} - u) = \cos u$,

$$\boxed{\sin(u + v) = \sin u \cos v + \cos u \sin v}.$$

These identities are fundamental in trigonometry and have wide applications in various
fields of mathematics and physics. □

**Theorem 15.5** (Derivatives of trigonometric functions). *It holds that*

$$\cos' x = -\sin x \qquad \sin' x = \cos x.$$

*Proof.* Using Theorem 15.3 and Theorem 15.4, we obtain for $x \in (0, \frac{\pi}{2})$

$$\cos' x = \lim_{h \to 0} \frac{\cos(x + h) - \cos x}{h} = \lim_{h \to 0} \frac{-2 \sin \frac{h}{2} \sin(x + \frac{h}{2})}{h} = -\sin x,$$

$$\sin' x = \lim_{h \to 0} \frac{\sin(x + h) - \sin x}{h} = \lim_{h \to 0} \frac{2 \sin \frac{h}{2} \cos(x + \frac{h}{2})}{h} = \cos x.$$

By symmetry and $2\pi$-periodicity these formulas hold for all $x \in \mathbb{R}$; consequently, $\sin x$ and $\cos x$ are differentiable (hence continuous) everywhere. □

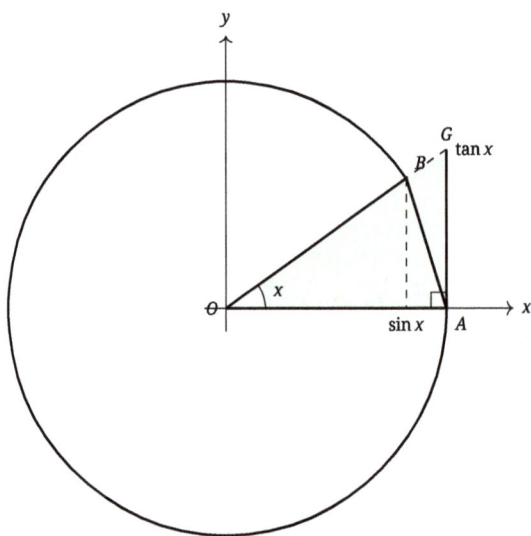

Figure 15.4: For small positive $x$, one has $\sin x < x < \tan x$; dividing by $\sin x$ and letting $x \to 0$ yields $\lim_{x \to 0} \frac{\sin x}{x} = 1$.

**Theorem 15.6** (The derivatives of inverses). *If $-1 < x < 1$ then*
- $(\arcsin x)' = \frac{1}{\sqrt{1-x^2}}$.
- $(\arccos x)' = -\frac{1}{\sqrt{1-x^2}}$.
- $(\arctan x)' = \frac{1}{1+x^2}$.

*Proof.* We start with arcsin $x$:

$$(\arcsin x)' = (\sin^{-1} x)' = \frac{1}{\cos(\arcsin x)}.$$

Because

$$(\sin(\arcsin x))^2 + (\cos(\arcsin x))^2 = 1 \quad \Longrightarrow \quad x^2 + (\cos(\arcsin x))^2 = 1,$$

it follows that $\cos(\arcsin x) = \sqrt{1-x^2}$, proving the first formula. The second one is analogous.

Set $\sec x := 1/\cos x$. For arctan we have

$$(\arctan x)' = (\tan^{-1} x)' = \frac{1}{\tan'(\arctan x)} = \frac{1}{\sec^2(\arctan x)}.$$

Since $\tan^2 u + 1 = \sec^2 u$, $\sec^2(\arctan x) = x^2 + 1$, giving $(\arctan x)' = 1/(1+x^2)$. ☐

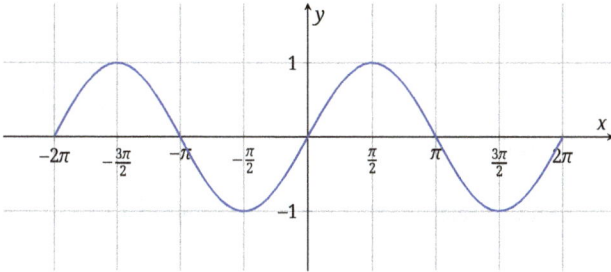

$$\text{—— } y = \sin x$$

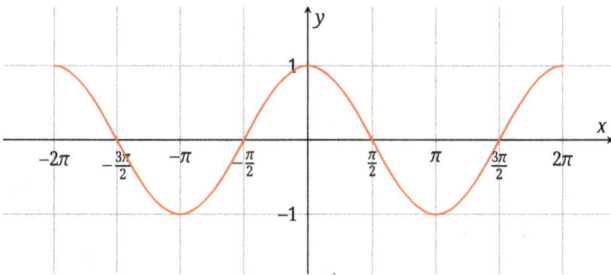

$$\text{—— } y = \cos x$$

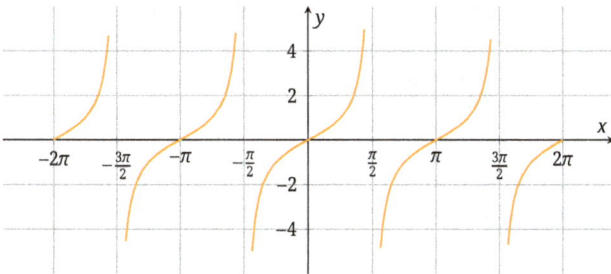

$$\text{—— } y = \tan x$$

## 15.3 Trigonometric identities

**Definition 15.2** (Reciprocal trigonometric identities). We define the following functions:

$$\sec x = \frac{1}{\cos x}, \quad \csc x = \frac{1}{\sin x}, \quad \cot x = \frac{1}{\tan x}.$$

**Theorem 15.7** (Pythagorean trigonometric identities). *It holds that*

$$\sin^2 x + \cos^2 x = 1, \quad 1 + \tan^2 x = \sec^2 x, \quad 1 + \cot^2 x = \csc^2 x.$$

*Proof.* The identity

$$\sin^2 x + \cos^2 x = 1$$

follows from the definition of the unit circle: every point on it satisfies $x^2 + y^2 = 1$, and the coordinates are $(\cos x, \sin x)$.

Dividing both sides by $\cos^2 x$ (when $\cos x \neq 0$) gives

$$\frac{\sin^2 x}{\cos^2 x} + \frac{\cos^2 x}{\cos^2 x} = \frac{1}{\cos^2 x} \quad \Rightarrow \quad \tan^2 x + 1 = \sec^2 x.$$

Similarly, dividing $\sin^2 x + \cos^2 x = 1$ by $\sin^2 x$ (when $\sin x \neq 0$)

$$1 + \frac{\cos^2 x}{\sin^2 x} = \frac{1}{\sin^2 x} \quad \Rightarrow \quad 1 + \cot^2 x = \csc^2 x. \qquad \square$$

**Theorem 15.8** (Even-odd trigonometric identities). *It holds that*

$$\sin(-x) = -\sin x, \quad \cos(-x) = \cos x, \quad \tan(-x) = -\tan x,$$
$$\csc(-x) = -\csc x, \quad \sec(-x) = \sec x, \quad \cot(-x) = -\cot x.$$

*Proof.* Using the unit circle, the point at angle $-x$ is the reflection of the point at angle $x$ across the $x$-axis. Thus:

– The $x$-coordinate (cosine) remains unchanged

$$\cos(-x) = \cos x.$$

– The $y$-coordinate (sine) changes sign

$$\sin(-x) = -\sin x.$$

Hence,

$$\tan(-x) = \frac{\sin(-x)}{\cos(-x)} = \frac{-\sin x}{\cos x} = -\tan x.$$

The reciprocal identities follow immediately:

$$\csc(-x) = \frac{1}{\sin(-x)} = -\csc x, \quad \sec(-x) = \frac{1}{\cos(-x)} = \sec x,$$

$$\cot(-x) = \frac{1}{\tan(-x)} = -\cot x. \qquad \square$$

**Theorem 15.9** (Angle sum and difference identities). *It holds that*

$$\tan(x \pm y) = \frac{\tan x \pm \tan y}{1 \mp \tan x \tan y}, \quad when \ \cos x \cos y \neq 0.$$

*Proof.* To derive the tangent identity, divide both sides of the sine and cosine addition formulas

$$\tan(x + y) = \frac{\sin(x + y)}{\cos(x + y)} = \frac{\sin x \cos y + \cos x \sin y}{\cos x \cos y - \sin x \sin y}.$$

Divide numerator and denominator by $\cos x \cos y$:

$$\tan(x + y) = \frac{\frac{\sin x}{\cos x} + \frac{\sin y}{\cos y}}{1 - \frac{\sin x}{\cos x} \cdot \frac{\sin y}{\cos y}} = \frac{\tan x + \tan y}{1 - \tan x \tan y}.$$

The negative case follows similarly. $\qquad \square$

**Theorem 15.10** (Double-angle identities). *It holds that*

$$\sin(2x) = 2 \sin x \cos x,$$
$$\cos(2x) = \cos^2 x - \sin^2 x = 2 \cos^2 x - 1 = 1 - 2 \sin^2 x,$$
$$\tan(2x) = \frac{2 \tan x}{1 - \tan^2 x}, \quad when \ \tan x \neq \pm 1.$$

*Proof.* Use the addition identities with $x = y$:
-   Sine:

$$\sin(2x) = \sin(x + x) = \sin x \cos x + \cos x \sin x = 2 \sin x \cos x.$$

-   Cosine:

$$\cos(2x) = \cos(x + x) = \cos^2 x - \sin^2 x.$$

Using the Pythagorean identity $\sin^2 x + \cos^2 x = 1$, we substitute

$$\cos(2x) = 1 - 2 \sin^2 x = 2 \cos^2 x - 1.$$

–   Tangent:

$$\tan(2x) = \frac{2\tan x}{1 - \tan^2 x},$$

derived from the tangent sum formula with $x = y$. ☐

**Theorem 15.11** (Half-angle identities). *Let $x \in \mathbb{R}$. Then*

$$\sin\left(\frac{x}{2}\right) = \pm\sqrt{\frac{1 - \cos x}{2}},$$

$$\cos\left(\frac{x}{2}\right) = \pm\sqrt{\frac{1 + \cos x}{2}},$$

$$\tan\left(\frac{x}{2}\right) = \frac{\sin x}{1 + \cos x} = \frac{1 - \cos x}{\sin x}.$$

*Proof.* Start with the double-angle identity

$$\cos(2\theta) = 1 - 2\sin^2\theta.$$

Solving for $\sin^2\theta$ gives

$$\sin^2\theta = \frac{1 - \cos(2\theta)}{2}.$$

Let $x = 2\theta$, so $\theta = \frac{x}{2}$. Then

$$\sin^2\left(\frac{x}{2}\right) = \frac{1 - \cos x}{2} \quad \Rightarrow \quad \sin\left(\frac{x}{2}\right) = \pm\sqrt{\frac{1 - \cos x}{2}}.$$

Similarly,

$$\cos^2\left(\frac{x}{2}\right) = \frac{1 + \cos x}{2} \quad \Rightarrow \quad \cos\left(\frac{x}{2}\right) = \pm\sqrt{\frac{1 + \cos x}{2}}.$$

For tangent,

$$\tan\left(\frac{x}{2}\right) = \frac{\sin x}{1 + \cos x},$$

derived by dividing numerator and denominator of the identity $\sin x = 2\sin(x/2)\cos(x/2)$, $\cos x = \cos^2(x/2) - \sin^2(x/2)$, then simplifying. ☐

## 15.4 Python and AI

**AI Request**

Find the exact value of $\sin(\frac{\pi}{12})$ using the angle subtraction formula and give me the Latex code of the solution step-by-step. Next write a Python code that computes the above so as I can compare the results.

---

**AI Request**
Provide me with an in-depth historical overview of Pythagoras of Samos.

---

## 15.5 Exercises

### True or false

Decide whether each statement is **true** or **false**. Justify your answer.

1.  If $\sin x = 0$, then $x = n\pi$ for all integers $n$.

    **Answer:** _____

2.  The tangent function is continuous everywhere.

    **Answer:** _____

3.  The range of $\cos x$ is $[-1, 1]$.

    **Answer:** _____

4.  The identity $\tan^2 x + 1 = \sec^2 x$ holds only when $\cos x \neq 0$.

    **Answer:** _____

5.  The sine function is an odd function, while the cosine function is even.

    **Answer:** _____

### Multiple choice questions

Choose the correct option (A, B, C, or D) for each question.

1.  Which of the following is equivalent to $\sin(2x)$?

    A. $2 \sin x \cos x$    B. $\sin^2 x - \cos^2 x$

    C. $1 - 2\sin^2 x$    D. $2\cos^2 x - 1$

    **Answer:** _____

2.  The value of $\lim_{x \to 0} \frac{\sin x}{x}$ is

    A. 0              B. 1

    C. undefined    D. $\infty$

    **Answer:** _____

3.  Which of the following is a valid expression for $\tan(x + y)$?

    A. $\frac{\tan x + \tan y}{1 - \tan x \tan y}$              B. $\frac{\sin x + \sin y}{\cos x + \cos y}$

    C. $\sin x \cos y + \cos x \sin y$    D. $\cos x \cos y - \sin x \sin y$

    **Answer:** _____

4.  The derivative of $\arcsin x$ is

    A. $\frac{1}{\sqrt{1-x^2}}$    B. $-\frac{1}{\sqrt{1-x^2}}$

    C. $\frac{1}{1+x^2}$    D. $-\frac{1}{1+x^2}$

    **Answer:** _____

5.  The half-angle identity for $\cos\left(\frac{x}{2}\right)$ is

A. $\pm\sqrt{\frac{1+\cos x}{2}}$    B. $\pm\sqrt{\frac{1-\cos x}{2}}$

C. $\frac{\sin x}{1+\cos x}$    D. $\frac{1-\cos x}{\sin x}$

**Answer:** _____

## Matching exercise
Match each term with its correct description.

1. Pythagorean identity            A. Function whose value repeats after every $2\pi$
2. Periodic function                B. $\sin^2 x + \cos^2 x = 1$
3. Even function                    C. Function symmetric about the $y$-axis
4. Odd function                      D. Function antisymmetric about the origin
5. Trigonometric addition formula    E. $\sin(x + y) = \sin x \cos y + \cos x \sin y$

**Answers:** 1._____    2._____    3._____    4._____    5._____

## Fill in the blanks
Fill in each blank with the most appropriate word or phrase.

1. The reciprocal of $\cos x$ is called the _____.

   **Answer:** _____

2. The angle addition formula for $\sin(x + y)$ involves both _____ and

   _____ terms.

   **Answer:** _____

3. The limit $\lim_{x \to 0} \frac{\sin x}{x}$ demonstrates that $\sin x$ is approximately equal to _____

   near $x = 0$.

   **Answer:** _____

4. The domain of $\tan x$ excludes points where $\cos x =$ _____.

   **Answer:** _____

5. The inverse of $\sin x$ is denoted as _____, and its derivative is given

   by _____.

   **Answer:** _____

## Theoretical and computational problems
Solve each problem step-by-step.

1. Prove that $\sin^2 x + \cos^2 x = 1$ using the unit circle definition of trigonometric functions.

2. Derive the double-angle identity for $\sin(2x)$ starting from the angle addition formula.

3. Show that $\lim_{x \to 0} \frac{\sin x}{x} = 1$ geometrically using the unit circle.

4. Using the Pythagorean identity, derive the formula for $\tan^2 x + 1 = \sec^2 x$.

5. Find the exact value of $\sin(\frac{\pi}{12})$ using the angle subtraction formula.

6. Prove that $(\arcsin x)' = \frac{1}{\sqrt{1-x^2}}$.

7. Solve the equation $\sin x = \cos x$ for $x \in [0, 2\pi]$.

8. Determine the range of the function $f(x) = \sin(2x) + \cos(2x)$.
9. Verify the half-angle identity for $\sin(\frac{x}{2})$ algebraically.
10. Using the chain rule, find the derivative of $g(x) = \sin(\cos x)$.

### Python-based exercises

Use Python (`sympy`, `numpy`, `matplotlib`) to tackle the following:

1. Plot the functions $\sin x$, $\cos x$, and $\tan x$ on the interval $[-2\pi, 2\pi]$. Mark their zeros and asymptotes.
2. Write a Python function to compute the value of $\sin(x)$ using the Taylor series expansion up to the $n$-th term.
3. Use `sympy` to verify the angle addition formula for $\sin(x + y)$ symbolically.
4. Numerically approximate the roots of the equation $\sin x = \cos x$ on the interval $[0, 2\pi]$.
5. Investigate the behavior of the function $h(x) = \sin(1/x)$ on the interval $(0, 1]$. Plot it and describe its oscillatory nature.

# 16 Hyperbolic functions

## Contents

Hyperbolic functions, much like their trigonometric counterparts, arise naturally in geometry, calculus, and many applications across mathematics and physics. While trigonometric functions relate to circles, hyperbolic functions are connected to hyperbolas and exponential growth. The primary hyperbolic functions are the hyperbolic sine and cosine, denoted $\sinh x$ and $\cosh x$, and are defined in terms of exponentials. These functions possess elegant identities, derivatives, and addition formulas analogous to those in trigonometry. Hyperbolic functions describe the shape of hanging cables (catenaries), model rapid growth, and appear in solutions of differential equations. In this chapter, we will define the basic hyperbolic functions, explore their algebraic and analytic properties, and study their applications through identities, graphs, and exercises.

## 16.1 Introduction

Hyperbolic functions form a fascinating family of functions closely related to the exponential function and play a central role in both pure and applied mathematics. Unlike the circular (trigonometric) functions, which are defined via the geometry of the unit circle, hyperbolic functions are intimately connected to the geometry of the hyperbola. The basic hyperbolic functions, $\sinh x$ and $\cosh x$, are defined by

$$\sinh x = \frac{e^x - e^{-x}}{2}, \quad \cosh x = \frac{e^x + e^{-x}}{2},$$

and generate, through their algebraic properties, the remaining hyperbolic functions: $\tanh x$, $\coth x$, $\operatorname{sech} x$, and $\operatorname{csch} x$. Many identities satisfied by trigonometric functions have striking analogues among hyperbolic functions, including addition and double-angle formulas, but with key differences—most notably, the signs in the fundamental identity

$$\cosh^2 x - \sinh^2 x = 1$$

in contrast to the circular identity $\cos^2 x + \sin^2 x = 1$.

Hyperbolic functions are not just algebraic curiosities; they model real-world phenomena. For example, the shape of a hanging cable or chain, known as a catenary,

https://doi.org/10.1515/9783112228289-016

is described by the hyperbolic cosine. Solutions to many differential equations, especially those involving exponential growth and decay, often involve hyperbolic sines and cosines. In special relativity, hyperbolic functions appear in the description of Lorentz transformations, illustrating their deep connections to geometry and physics.

From a calculus perspective, hyperbolic functions are simple to differentiate and integrate. Their graphs reveal smooth, exponential growth and decay, with $\sinh x$ being an odd function and $\cosh x$ being even. Unlike trigonometric functions, hyperbolic functions are unbounded and nonperiodic, but they share many structural similarities.

In this chapter, we introduce the hyperbolic functions, establish their key identities and properties (Table 16.1), explore their inverses and their graphical behavior, and solve problems to build a strong conceptual and computational foundation. We also draw parallels and contrasts with trigonometric functions, emphasizing both the analogies and the distinctive features that make hyperbolic functions a powerful mathematical tool.

**Table 16.1:** Summary of hyperbolic functions and identities.

| Function | Definition | Key properties |
|---|---|---|
| $\sinh x$ | $\frac{e^x - e^{-x}}{2}$ | odd func., $(\sinh x)' = \cosh x$ |
| | | $\sinh(-x) = -\sinh x$ |
| | | $\sinh(x + y) = \sinh x \cosh y + \cosh x \sinh y$ |
| $\cosh x$ | $\frac{e^x + e^{-x}}{2}$ | even func., $(\cosh x)' = \sinh x$ |
| | | $\cosh(-x) = \cosh x$ |
| | | $\cosh(x + y) = \cosh x \cosh y + \sinh x \sinh y$ |
| $\tanh x$ | $\frac{\sinh x}{\cosh x}$ | odd func., $(\tanh x)' = \operatorname{sech}^2 x$ |
| | | $\tanh(-x) = -\tanh x$ |
| | | $1 - \tanh^2 x = \operatorname{sech}^2 x$ |
| $\coth x$ | $\frac{\cosh x}{\sinh x}$ | odd func., $(\coth x)' = -\operatorname{csch}^2 x$ |
| | | $\coth(-x) = -\coth x$ |
| | | $\coth^2 x - 1 = \operatorname{csch}^2 x$ |
| $\operatorname{sech} x$ | $\frac{1}{\cosh x}$ | even func., $(\operatorname{sech} x)' = -\operatorname{sech} x \tanh x$ |
| | | $\operatorname{sech}(-x) = \operatorname{sech} x$ |
| $\operatorname{csch} x$ | $\frac{1}{\sinh x}$ | odd func., $(\operatorname{csch} x)' = -\operatorname{csch} x \coth x$ |
| | | $\operatorname{csch}(-x) = -\operatorname{csch} x$ |

*Fundamental identities*
$\cosh^2 x - \sinh^2 x = 1$
$1 - \tanh^2 x = \operatorname{sech}^2 x$
$\coth^2 x - 1 = \operatorname{csch}^2 x$

*Double angle formulas*
$\sinh 2x = 2 \sinh x \cosh x$
$\cosh 2x = \cosh^2 x + \sinh^2 x = 2 \cosh^2 x - 1 = 1 + 2 \sinh^2 x$
$\tanh 2x = \frac{2 \tanh x}{1 + \tanh^2 x}$

*Relationship with exponential function*
$e^x = \cosh x + \sinh x$
$e^{-x} = \cosh x - \sinh x$

## 16.2 Definitions

**Definition 16.1** (Hyperbolic functions). We introduce two new functions:

$$\cosh x = \frac{e^x + e^{-x}}{2}, \quad \sinh x = \frac{e^x - e^{-x}}{2}.$$

We also define

$$\tanh x = \frac{\sinh x}{\cosh x}, \quad \coth x = \frac{\cosh x}{\sinh x}, \quad \operatorname{sech} x = \frac{1}{\cosh x}, \quad \operatorname{csch} x = \frac{1}{\sinh x}.$$

**Theorem 16.1** (Derivative of hyperbolic cosine). *It holds that*

$$(\cosh x)' = \sinh x.$$

*Proof.* To find the derivative of $\cosh x$, we differentiate its definition with respect to $x$:

$$\begin{aligned}
(\cosh x)' &= \left(\frac{e^x + e^{-x}}{2}\right)' \\
&= \frac{1}{2}(e^x + e^{-x})' \\
&= \frac{1}{2}(e^x + (-1)e^{-x}) \\
&= \frac{e^x - e^{-x}}{2} \\
&= \sinh x.
\end{aligned}$$

Thus, we have proven that $(\cosh \mathbf{x})' = \sinh \mathbf{x}$. $\qquad\square$

**Theorem 16.2** (Derivative of hyperbolic sine). *It holds that*

$$(\sinh x)' = \cosh x.$$

*Proof.* Similarly, to find the derivative of $\sinh x$, we differentiate its definition with respect to $x$:

$$\begin{aligned}
(\sinh x)' &= \left(\frac{e^x - e^{-x}}{2}\right)' \\
&= \frac{1}{2}(e^x - e^{-x})' \\
&= \frac{1}{2}(e^x - (-1)e^{-x}) \\
&= \frac{e^x + e^{-x}}{2} \\
&= \cosh x.
\end{aligned}$$

Thus, we have proven that $(\sinh \mathbf{x})' = \cosh \mathbf{x}$. $\qquad\square$

**Theorem 16.3** (Fundamental hyperbolic identity). *It holds that*

$$\cosh^2 x - \sinh^2 x = 1.$$

*Proof.* We will substitute the definitions of $\cosh x$ and $\sinh x$ into the left-hand side of the identity:

$$\cosh^2 x - \sinh^2 x = \left(\frac{e^x + e^{-x}}{2}\right)^2 - \left(\frac{e^x - e^{-x}}{2}\right)^2$$

$$= \frac{(e^x)^2 + 2e^x e^{-x} + (e^{-x})^2}{4} - \frac{(e^x)^2 - 2e^x e^{-x} + (e^{-x})^2}{4}$$

$$= \frac{e^{2x} + 2 + e^{-2x}}{4} - \frac{e^{2x} - 2 + e^{-2x}}{4}$$

$$= \frac{(e^{2x} + 2 + e^{-2x}) - (e^{2x} - 2 + e^{-2x})}{4}$$

$$= \frac{e^{2x} + 2 + e^{-2x} - e^{2x} + 2 - e^{-2x}}{4}$$

$$= \frac{4}{4}$$

$$= 1.$$

Thus, we have proven that $\cosh^2 \mathbf{x} - \sinh^2 \mathbf{x} = \mathbf{1}$. This identity is analogous to the Pythagorean identity for trigonometric functions, $\cos^2 x + \sin^2 x = 1$. □

**Theorem 16.4** (Double angle identity for hyperbolic cosine (sum of squares form)). *It holds that*

$$\cosh 2x = \cosh^2 x + \sinh^2 x.$$

*Proof.* We will substitute the definitions of $\cosh x$ and $\sinh x$ into the right-hand side of the identity:

$$\cosh^2 x + \sinh^2 x = \left(\frac{e^x + e^{-x}}{2}\right)^2 + \left(\frac{e^x - e^{-x}}{2}\right)^2$$

$$= \frac{e^{2x} + 2e^x e^{-x} + e^{-2x}}{4} + \frac{e^{2x} - 2e^x e^{-x} + e^{-2x}}{4}$$

$$= \frac{e^{2x} + 2 + e^{-2x}}{4} + \frac{e^{2x} - 2 + e^{-2x}}{4}$$

$$= \frac{(e^{2x} + 2 + e^{-2x}) + (e^{2x} - 2 + e^{-2x})}{4}$$

$$= \frac{2e^{2x} + 2e^{-2x}}{4}$$

$$= \frac{2(e^{2x} + e^{-2x})}{4}$$

$$= \frac{e^{2x} + e^{-2x}}{2}$$

$$= \cosh 2x.$$

Thus, we have proven that $\cosh \mathbf{2x} = \cosh^2 \mathbf{x} + \sinh^2 \mathbf{x}$. □

**Theorem 16.5** (Double angle identity for hyperbolic sine (product form)). *It holds that*

$$\sinh 2x = 2\sinh x \cosh x.$$

*Proof.* We will substitute the definitions of $\sinh x$ and $\cosh x$ into the right-hand side of the identity:

$$2\sinh x \cosh x = 2\left(\frac{e^x - e^{-x}}{2}\right)\left(\frac{e^x + e^{-x}}{2}\right)$$

$$= 2\frac{(e^x - e^{-x})(e^x + e^{-x})}{4}$$

$$= \frac{1}{2}\left((e^x)^2 - (e^{-x})^2\right)$$

$$= \frac{1}{2}(e^{2x} - e^{-2x})$$

$$= \sinh 2x.$$

Thus, we have proven that $\sinh \mathbf{2x} = \mathbf{2}\sinh \mathbf{x} \cosh \mathbf{x}$. □

**Theorem 16.6** (Hyperbolic identity: $1 - \tanh^2 x = \operatorname{sech}^2 x$). *It holds that*

$$1 - \tanh^2 x = \operatorname{sech}^2 x.$$

*Proof.* We start with the fundamental identity $\cosh^2 x - \sinh^2 x = 1$. Divide all terms by $\cosh^2 x$:

$$\frac{\cosh^2 x}{\cosh^2 x} - \frac{\sinh^2 x}{\cosh^2 x} = \frac{1}{\cosh^2 x},$$

$$1 - \left(\frac{\sinh x}{\cosh x}\right)^2 = \left(\frac{1}{\cosh x}\right)^2,$$

$$1 - \tanh^2 x = \operatorname{sech}^2 x.$$

Thus, we have proven that $\mathbf{1} - \tanh^2 \mathbf{x} = \operatorname{sech}^2 \mathbf{x}$. □

**Theorem 16.7** (Hyperbolic identity: $\coth^2 x - 1 = \operatorname{csch}^2 x$). *It holds that*

$$\coth^2 x - 1 = \operatorname{csch}^2 x.$$

*Proof.* Again, starting with $\cosh^2 x - \sinh^2 x = 1$. Divide all terms by $\sinh^2 x$:

$$\frac{\cosh^2 x}{\sinh^2 x} - \frac{\sinh^2 x}{\sinh^2 x} = \frac{1}{\sinh^2 x},$$

$$\left(\frac{\cosh x}{\sinh x}\right)^2 - 1 = \left(\frac{1}{\sinh x}\right)^2,$$

$$\coth^2 x - 1 = \operatorname{csch}^2 x.$$

Thus, we have proven that $\coth^2 \mathbf{x} - \mathbf{1} = \operatorname{csch}^2 \mathbf{x}$. $\square$

**Theorem 16.8** (Hyperbolic parity: $\sinh(-x) = -\sinh x$). *The hyperbolic sine function is an odd function, meaning*

$$\sinh(-x) = -\sinh x.$$

*Proof.* Using the definition of $\sinh x$,

$$\sinh(-x) = \frac{e^{-x} - e^{-(-x)}}{2}$$

$$= \frac{e^{-x} - e^{x}}{2}$$

$$= -\frac{e^{x} - e^{-x}}{2}$$

$$= -\sinh x.$$

Thus, we have proven that $\sinh(-\mathbf{x}) = -\sinh \mathbf{x}$. $\square$

**Theorem 16.9** (Hyperbolic parity: $\cosh(-x) = \cosh x$). *The hyperbolic cosine function is an even function, meaning*

$$\cosh(-x) = \cosh x.$$

*Proof.* Using the definition of $\cosh x$,

$$\cosh(-x) = \frac{e^{-x} + e^{-(-x)}}{2}$$

$$= \frac{e^{-x} + e^{x}}{2}$$

$$= \frac{e^{x} + e^{-x}}{2}$$

$$= \cosh x.$$

Thus, we have proven that $\cosh(-\mathbf{x}) = \cosh \mathbf{x}$. $\square$

**Theorem 16.10** (Hyperbolic parity: $\tanh(-x) = -\tanh x$). *The hyperbolic tangent function is an odd function, meaning*

$$\tanh(-x) = -\tanh x.$$

*Proof.* Using the definition of $\tanh x$ and the parity properties of $\sinh x$ and $\cosh x$,

$$\tanh(-x) = \frac{\sinh(-x)}{\cosh(-x)}$$
$$= \frac{-\sinh x}{\cosh x}$$
$$= -\frac{\sinh x}{\cosh x}$$
$$= -\tanh x.$$

Thus, we have proven that $\tanh(-\mathbf{x}) = -\tanh \mathbf{x}$. $\qquad\square$

**Theorem 16.11** (Hyperbolic addition formula: $\sinh(x+y) = \sinh x \cosh y + \cosh x \sinh y$). *It holds that*

$$\sinh(x+y) = \sinh x \cosh y + \cosh x \sinh y.$$

*Proof.* We expand the left-hand side using the definitions

$$\sinh(x+y) = \frac{e^{x+y} - e^{-(x+y)}}{2}$$
$$= \frac{e^x e^y - e^{-x} e^{-y}}{2}.$$

Now, expand the right-hand side

$$\sinh x \cosh y + \cosh x \sinh y$$
$$= \left(\frac{e^x - e^{-x}}{2}\right)\left(\frac{e^y + e^{-y}}{2}\right) + \left(\frac{e^x + e^{-x}}{2}\right)\left(\frac{e^y - e^{-y}}{2}\right)$$
$$= \frac{1}{4}(e^x e^y + e^x e^{-y} - e^{-x} e^y - e^{-x} e^{-y}) + \frac{1}{4}(e^x e^y - e^x e^{-y} + e^{-x} e^y - e^{-x} e^{-y})$$
$$= \frac{1}{4}(2e^{x+y} - 2e^{-(x+y)})$$
$$= \frac{e^{x+y} - e^{-(x+y)}}{2}.$$

Since both sides are equal, we have proven that

$$\sinh(\mathbf{x}+\mathbf{y}) = \sinh \mathbf{x} \cosh \mathbf{y} + \cosh \mathbf{x} \sinh \mathbf{y}. \qquad\square$$

**Theorem 16.12** (Hyperbolic addition formula: $\cosh(x + y) = \cosh x \cosh y + \sinh x \sinh y$). *It holds that*

$$\cosh(x + y) = \cosh x \cosh y + \sinh x \sinh y.$$

*Proof.* We expand the left-hand side using the definition of cosh

$$\cosh(x + y) = \frac{e^{x+y} + e^{-(x+y)}}{2}$$
$$= \frac{e^x e^y + e^{-x} e^{-y}}{2}.$$

Next, we expand the right-hand side

$$\cosh x \cosh y + \sinh x \sinh y$$
$$= \left(\frac{e^x + e^{-x}}{2}\right)\left(\frac{e^y + e^{-y}}{2}\right) + \left(\frac{e^x - e^{-x}}{2}\right)\left(\frac{e^y - e^{-y}}{2}\right)$$
$$= \frac{1}{4}(e^x e^y + e^x e^{-y} + e^{-x} e^y + e^{-x} e^{-y}) + \frac{1}{4}(e^x e^y - e^x e^{-y} - e^{-x} e^y + e^{-x} e^{-y})$$
$$= \frac{1}{4}(2e^{x+y} + 2e^{-(x+y)})$$
$$= \frac{e^{x+y} + e^{-(x+y)}}{2}.$$

Since both sides are equal, we have proven that

$$\cosh(\mathbf{x} + \mathbf{y}) = \cosh \mathbf{x} \cosh \mathbf{y} + \sinh \mathbf{x} \sinh \mathbf{y}. \qquad \square$$

**Theorem 16.13** (Hyperbolic double angle: $\cosh 2x = 2\cosh^2 x - 1$). *It holds that*

$$\cosh 2x = 2\cosh^2 x - 1.$$

*Proof.* Start with the identity $\cosh 2x = \cosh^2 x + \sinh^2 x$. From the fundamental identity $\cosh^2 x - \sinh^2 x = 1$, we can write $\sinh^2 x = \cosh^2 x - 1$. Substitute this into the double angle formula for $\cosh 2x$:

$$\cosh 2x = \cosh^2 x + (\cosh^2 x - 1)$$
$$= 2\cosh^2 x - 1.$$

Thus, we have proven that $\cosh \mathbf{2x} = 2\cosh^2 \mathbf{x} - 1$. $\qquad \square$

**Theorem 16.14** (Hyperbolic double angle: $\cosh 2x = 1 + 2\sinh^2 x$). *It holds that*

$$\cosh 2x = 1 + 2\sinh^2 x.$$

*Proof.* Start with the identity $\cosh 2x = \cosh^2 x + \sinh^2 x$. From the fundamental identity $\cosh^2 x - \sinh^2 x = 1$, we can write $\cosh^2 x = 1 + \sinh^2 x$. Substitute this into the double angle formula for $\cosh 2x$:

$$\cosh 2x = (1 + \sinh^2 x) + \sinh^2 x$$
$$= 1 + 2\sinh^2 x.$$

Thus, we have proven that $\cosh \mathbf{2x} = \mathbf{1} + \mathbf{2}\sinh^2 \mathbf{x}$. □

**Theorem 16.15** (Exponential relation: $e^x = \cosh x + \sinh x$). *It holds that*

$$e^x = \cosh x + \sinh x.$$

*Proof.* Substitute the definitions of $\cosh x$ and $\sinh x$ into the right-hand side

$$\cosh x + \sinh x = \frac{e^x + e^{-x}}{2} + \frac{e^x - e^{-x}}{2}$$
$$= \frac{e^x + e^{-x} + e^x - e^{-x}}{2}$$
$$= \frac{2e^x}{2}$$
$$= e^x.$$

Thus, we have proven that $\mathbf{e^x} = \cosh \mathbf{x} + \sinh \mathbf{x}$. □

**Theorem 16.16** (Exponential relation: $e^{-x} = \cosh x - \sinh x$). *It holds that*

$$e^{-x} = \cosh x - \sinh x.$$

*Proof.* Substitute the definitions of $\cosh x$ and $\sinh x$ into the right-hand side

$$\cosh x - \sinh x = \frac{e^x + e^{-x}}{2} - \frac{e^x - e^{-x}}{2}$$
$$= \frac{e^x + e^{-x} - (e^x - e^{-x})}{2}$$
$$= \frac{e^x + e^{-x} - e^x + e^{-x}}{2}$$
$$= \frac{2e^{-x}}{2}$$
$$= e^{-x}.$$

Thus, we have proven that $\mathbf{e^{-x}} = \cosh \mathbf{x} - \sinh \mathbf{x}$. □

## 16.3 Inverses of hyperbolic functions

The function $\sinh: \mathbb{R} \to \mathbb{R}$ (the *hyperbolic sine*, Figure 16.1) is strictly increasing, hence bijective and possesses an inverse $\sinh^{-1}: \mathbb{R} \to \mathbb{R}$. Solving $y = \frac{1}{2}(e^x - e^{-x})$ for $x$ gives

$$\sinh^{-1} x = \ln(x + \sqrt{x^2 + 1}).$$

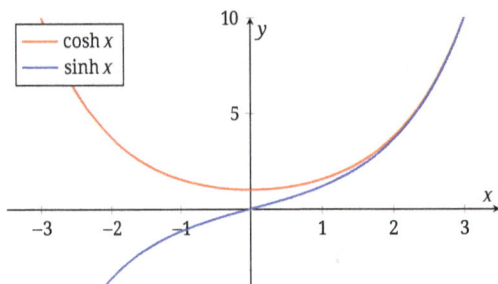

Figure 16.1: Graphs of the hyperbolic cosine (red) and hyperbolic sine (blue). Both grow rapidly for large $|x|$; on this scale the two curves appear to coincide for $x \gg 1$ because $\cosh x \sim \sinh x \sim \frac{1}{2}e^x$.

The hyperbolic cosine $\cosh x$ is strictly increasing on $[0, \infty)$ and strictly decreasing on $(-\infty, 0]$. Consequently, it is invertible on each of those intervals. Writing $y = \frac{1}{2}(e^x + e^{-x})$ and solving for $x$ yields

$$\cosh^{-1} x = \begin{cases} \ln(x - \sqrt{x^2 - 1}), & x \le 0, \\ \ln(x + \sqrt{x^2 - 1}), & x \ge 0. \end{cases}$$

In other words,

$$\cosh^{-1} x = \ln(x \pm \sqrt{x^2 - 1}),$$

where the minus sign corresponds to the branch $(-\infty, 0] \to [1, \infty)$ and the plus sign to $[0, \infty) \to [1, \infty)$.

## 16.4 Python and AI

**AI Request**

Evaluate cosh(ln 3) and sinh(ln 2) and give me the Latex code step-by-step. Write a Python code for the above calculation.

## 16.5 Exercises

### True or false

Decide whether each statement is **true** or **false**. Justify your answer.

1. The range of $\cosh x$ is $[-1, 1]$.

   **Answer:** _____

2. $\sinh x$ is an odd function.

   **Answer:** _____

3. The identity $\cosh^2 x - \sinh^2 x = 1$ is analogous to the Pythagorean identity in trigonometry.

   **Answer:** _____

4. $(\tanh x)' = \operatorname{sech}^2 x$.

   **Answer:** _____

5. For all $x \in \mathbb{R}$, $e^x = \cosh x - \sinh x$.

   **Answer:** _____

### Multiple choice questions

Choose the correct option (A, B, C, or D) for each question.

1. Which of the following is equivalent to $\cosh 2x$?

   A. $2 \sinh x \cosh x$    B. $\cosh^2 x - \sinh^2 x$

   C. $1 - 2 \sinh^2 x$      D. $2 \cosh^2 x - 1$

   **Answer:** _____

2. The derivative of $\operatorname{sech} x$ is

   A. $\operatorname{sech} x \tanh x$    B. $-\operatorname{sech} x \tanh x$

   C. $\operatorname{csch} x \coth x$    D. $-\operatorname{csch} x \coth x$

   **Answer:** _____

3. Which of the following is the correct expression for $\sinh(x - y)$?

   A. $\sinh x \cosh y + \cosh x \sinh y$    B. $\cosh x \cosh y - \sinh x \sinh y$

   C. $\sinh x \cosh y - \cosh x \sinh y$    D. $\cosh x \sinh y - \sinh x \cosh y$

   **Answer:** _____

4. The value of $\sinh^{-1}(x)$ is

   A. $\ln(x + \sqrt{x^2 - 1})$    B. $\ln(x - \sqrt{x^2 - 1})$

   C. $\ln(x + \sqrt{x^2 + 1})$    D. $\ln(x + \sqrt{1 - x^2})$

   **Answer:** _____

5. The value of $\coth^2 x - 1$ is

   A. $\tanh^2 x$    B. $\operatorname{sech}^2 x$

   C. $\operatorname{csch}^2 x$    D. $-\operatorname{csch}^2 x$

   **Answer:** _____

### Matching exercise

Match each term with its correct description or identity.

1. Hyperbolic sine      A. $\cosh x = \frac{e^x + e^{-x}}{2}$

2. Hyperbolic cosine      B. $\sinh(-x) = -\sinh x$

3. Reciprocal of $\sinh x$      C. $\cosh 2x = \cosh^2 x + \sinh^2 x$

4. Odd function property      D. $\operatorname{csch} x$

5. Double angle identity      E. $\sinh x = \frac{e^x - e^{-x}}{2}$

**Answers: 1.**_____    **2.**_____    **3.**_____    **4.**_____    **5.**_____

## Fill in the blanks

Fill in each blank with the most appropriate word or phrase.

1. The reciprocal of $\cosh x$ is called the _____.

   **Answer:** _____

2. The identity $\cosh^2 x - \sinh^2 x =$ _____ is a fundamental property of hyperbolic functions.

   **Answer:** _____

3. The hyperbolic function $\cosh x$ is an _____ function, meaning $\cosh(-x) =$ _____.

   **Answer:** _____

4. The domain of $\tanh x$ is _____, while its range is _____.

   **Answer:** _____

5. The inverse of $\cosh x$ is denoted as $\cosh^{-1} x$, and for $x \geq 1$, it is given by $\ln(x +$ _____$)$.

   **Answer:** _____

## Theoretical and computational problems

Solve each problem step-by-step.

1. Prove that $(\tanh x)' = \operatorname{sech}^2 x$.

2. Show that $\cosh(x-y) = \cosh x \cosh y - \sinh x \sinh y$ using the exponential definitions of $\cosh x$ and $\sinh x$.

3. If $\sinh x = 4$, find the exact value of $\cosh x$.

4. Solve the equation $\cosh x = 2$ for $x$.

5. Derive the identity $\cosh 2x = 2 \sinh^2 x + 1$ from the definition of $\cosh 2x$ and the fundamental identity $\cosh^2 x - \sinh^2 x = 1$.

6. Prove that $(\operatorname{sech} x)' = -\operatorname{sech} x \tanh x$.

7. Evaluate $\cosh(\ln 3)$ and $\sinh(\ln 2)$.

8. Determine the domain and range of $\tanh x$.

9. Using the identity $\cosh 2x = 2\cosh^2 x - 1$, express $\cosh^2 x$ in terms of $\cosh 2x$. This is analogous to the power-reducing formula for $\cos^2 x$.

10. Verify the identity $e^{-x} = \cosh x - \sinh x$ using the definitions of $\cosh x$ and $\sinh x$.

## Python-based exercises

Use Python (`sympy`, `numpy`, `matplotlib`) to tackle the following:

1. Plot the functions $\cosh x$, $\sinh x$, and $\tanh x$ on the interval $[-3,3]$. Comment on their behavior at large positive and negative $x$.

    *Hint: You will need to import numpy for numerical functions and* `matplotlib.pyplot` *for plotting.*

2. Write a Python function that approximates $\cosh x$ using the first $n$ terms of its Taylor series expansion (centered at $x = 0$). Compare the approximation to 'math.cosh(x)' for various $x$ and $n$.

    *Hint: The Taylor series for* $\cosh x$ *is* $1 + \frac{x^2}{2!} + \frac{x^4}{4!} + \cdots = \sum_{k=0}^{\infty} \frac{x^{2k}}{(2k)!}$. *You will need* `math.factorial`.

3. Use `sympy` to verify the identity $\coth^2 x - 1 = \mathrm{csch}^2 x$ symbolically.

4. Numerically solve the equation $\sinh x = x^2$ for $x \in [-5,5]$ using 'scipy.optimize.fsolve' or 'scipy.optimize.root'.

    *Hint: Define a function* $f(x) = \sinh x - x^2$ *and find its roots.*

5. Plot $\sinh^{-1} x$ and $\cosh^{-1} x$ (the principal branch) along with their corresponding exponential forms $\ln(x + \sqrt{x^2 + 1})$ and $\ln(x + \sqrt{x^2 - 1})$ to visually confirm their equivalence. Specify appropriate domains for plotting.

    *Hint: For* $\cosh^{-1} x$, *the domain starts at* $x = 1$.

# 17 Sequences of numbers

## Contents

Sequences play a central role in mathematical analysis, serving as the foundation for the study of limits, continuity, and infinite processes. A sequence is an ordered list of numbers, typically indexed by the natural numbers, whose behavior as the index increases reveals deep properties about functions and the real number system. Through sequences, we develop the notions of convergence, divergence, boundedness, monotonicity, and accumulation points. The study of subsequences, limits superior and inferior, and recursive constructions provides powerful tools for understanding more advanced topics such as series, continuity, and compactness. In this chapter, we present the key definitions, main theorems, and illustrative examples related to sequences. A variety of exercise—ranging from true/false and multiple choice to theoretical and Python-based problems—will help build intuition and computational skill.

## 17.1 Introduction

The concept of a *sequence* is one of the most fundamental and versatile notions in mathematics. At its core, a sequence is simply an ordered list of numbers indexed by the natural numbers. However, this simple idea provides the backbone for much of mathematical analysis, connecting the finite with the infinite, the discrete with the continuous, and the intuitive with the rigorously defined.

   The study of sequences is crucial for understanding limits, one of the cornerstones of calculus and analysis. Through the rigorous language of sequences, we can make precise what it means for numbers or functions to "approach" a certain value, and lay the groundwork for concepts such as continuity, differentiability, and the convergence of series. Sequences allow us to explore the behavior of functions at infinity, to formalize the notion of approximation, and to investigate the deep structure of the real number system.

https://doi.org/10.1515/9783112228289-017

In this chapter, we begin by introducing the basic definitions related to sequences, such as monotonicity and boundedness. We then present the notion of a subsequence, a powerful tool that enables us to analyze the limiting behavior of sequences even when they do not converge in the usual sense. The pivotal concepts of limit superior (lim sup) and limit inferior (lim inf) provide a framework for understanding oscillating or non-convergent sequences, and offer criteria for convergence.

We will also explore recursive sequences—sequences defined in terms of previous terms—which appear naturally in mathematical models, algorithms, and countless applications. Recursive definitions often lead to elegant mathematical properties and beautiful limiting behaviors.

A significant portion of this chapter is dedicated to fundamental results such as the uniqueness of limits, the relationship between convergence and boundedness, the characterization of limits via subsequences, and the celebrated monotone convergence theorem. We will learn various techniques for proving convergence, such as the squeeze theorem and the algebraic properties of limits, and discuss important tests like the ratio and root tests.

To deepen your understanding, the chapter is enriched with numerous examples and a wide array of exercises, from true/false and multiple choice to matching, fill-in-the-blank, theoretical problems, and computational tasks involving Python. These exercises are designed not only to solidify theoretical knowledge, but also to cultivate the analytical and computational skills essential for further study in mathematics and its applications.

By mastering the material in this chapter, you will be equipped with essential tools and perspectives that will serve as a foundation for all subsequent topics in real analysis and beyond.

## 17.2 Definitions

**Definition 17.1** (Monotone sequence). A sequence of real numbers $(a_n)$ is called *monotonically increasing* if for every $n \in \mathbb{N}$ we have

$$a_n \leq a_{n+1}.$$

Similarly, it is called *monotonically decreasing* if the inequality is reversed.

**Definition 17.2** (Bounded sequence). A sequence $(a_n)$ is called *bounded below* if there exists a real number $c \in \mathbb{R}$ such that

$$c \leq a_n \quad \text{for all } n \in \mathbb{N}.$$

The number $c$ is called a *lower bound* of the sequence. Similarly, the sequence is called *bounded above* if there exists $d \in \mathbb{R}$ such that

$$a_n \leq d \quad \text{for all } n \in \mathbb{N}.$$

Consider a sequence $(a_n)$. The index $n$ runs through the set of natural numbers $\mathbb{N}$, i.e., takes values starting from 1 and increasing without bound. If we consider a subset of $\mathbb{N}$, for example the even numbers $\{2, 4, 6, \ldots\}$, then we define a subsequence of $a_n$, i.e. $a_{x_n} = a_{2n}$. Clearly, any sequence is a subsequence of itself by choosing $x_n = n$. The subsequence can be viewed as a selection of terms of the original sequence indexed by an increasing sequence of natural numbers.

---

**Definition 17.3** (Subsequence). Let $(x_n)_{n\in\mathbb{N}}$ be a sequence of real numbers. A **subsequence** of $(x_n)$ is a sequence of the form $(x_{n_k})_{k\in\mathbb{N}}$, where $(n_k)_{k\in\mathbb{N}}$ is a strictly increasing sequence of natural numbers, i.e., $n_1 < n_2 < n_3 < \cdots$ and $n_k \in \mathbb{N}$ for all $k$.

In other words, a subsequence is obtained by selecting a subset of the terms of the original sequence, preserving their order but possibly skipping some elements.

---

**Definition 17.4** (Limit of a sequence). Let $(a_n)$ be a sequence of real numbers.
- The sequence $(a_n)$ is said to *converge* to the real number $a \in \mathbb{R}$ if for every $\varepsilon > 0$ there exists $n_0 \in \mathbb{N}$ such that for all $n > n_0$ we have

$$|a_n - a| < \varepsilon.$$

In this case, we write

$$\lim_{n\to\infty} a_n = a.$$

- The sequence $(a_n)$ is said to *diverge to* $+\infty$ (respectively to $-\infty$) if for every $M > 0$ there exist some $N > 0$ such that for all $n > N$

$$a_n > M \quad (\text{respectively } a_n < -M).$$

In this case, we write

$$\lim_{n\to\infty} a_n = +\infty \quad \text{or} \quad \lim_{n\to\infty} a_n = -\infty.$$

If a sequence has a limit, it is called *convergent*. Otherwise, it is called *divergent*.

---

## 17.3 Basic results

The following results establish foundational properties of sequences in real analysis, including uniqueness of limits, convergence behavior via subsequences, and basic examples. Together, they provide essential tools for understanding how sequences behave, when they converge, and how their limits are characterized.

---

**Proposition 17.1** (Uniqueness of the limit). *If a sequence converges, then its limit is unique.*

---

*Proof.* Suppose that a sequence $a_n$ converges to two distinct limits $a \neq b$. By the definition of the limit, we have

$$\begin{cases} \forall \varepsilon_1 > 0, & \exists n_1 : |a_n - a| \le \varepsilon_1 \quad \forall n \ge n_1, \\ \forall \varepsilon_2 > 0, & \exists n_2 : |a_n - b| \le \varepsilon_2 \quad \forall n \ge n_2. \end{cases}$$

Taking $n_0 = \max\{n_1, n_2\}$, both inequalities hold simultaneously for all $n \ge n_0$, so

$$|a_n - a| \le \varepsilon_1 \quad \text{and} \quad |a_n - b| \le \varepsilon_2, \quad \forall n \ge n_0.$$

By the triangle inequality,

$$|b - a| = |(b - a_n) + (a_n - a)| \le |b - a_n| + |a_n - a| \le \varepsilon_1 + \varepsilon_2, \quad \forall n \ge n_0.$$

Since $\varepsilon_1, \varepsilon_2$ can be chosen arbitrarily small, this implies $b = a$, a contradiction. $\square$

**Theorem 17.1** (Characterization of convergence via subsequences). *A sequence $a_n$ converges to a real number $a \in \mathbb{R}$ **if and only if** every subsequence of $a_n$ converges to the same number $a$. Similarly, the sequence $a_n$ diverges to $\pm\infty$ **if and only if** every subsequence diverges in the same way.*

*Proof.* Assume that $a_n \to a \in \mathbb{R}$ as $n \to \infty$, and let $a_{k_n}$ be any subsequence of $a_n$. We will prove that $a_{k_n} \to a$ as $n \to \infty$.

Since $a_n \to a$, for every $\varepsilon > 0$ there exists $N > 0$ such that

$$|a_n - a| < \varepsilon \quad \text{for all } n > N.$$

Because $k_n \to \infty$ as $n \to \infty$, there exists $N' > 0$ such that

$$k_n > N \quad \text{for all } n > N'.$$

Hence,

$$|a_{k_n} - a| < \varepsilon \quad \text{for all } n > N'.$$

Conversely, if every subsequence of $a_n$ converges to $a$, then the sequence $a_n$ itself converges to $a$, since $a_n$ is a subsequence of itself.

If $a_n \to +\infty$ (respectively $a_n \to -\infty$), then every subsequence of $a_n$ diverges to $+\infty$ (respectively to $-\infty$). Indeed, for every $M > 0$, there exists $N$ such that

$$a_n > M \quad \text{(respectively } a_n < -M) \quad \text{for all } n > N.$$

If $a_{k_n}$ is a subsequence of $a_n$, then there exists $N' > 0$ such that

$$k_n > N \quad \text{for all } n > N',$$

implying

$$a_{k_n} > M \quad \text{(respectively } a_{k_n} < -M) \quad \text{for all } n > N'.$$

The converse direction follows similarly. $\square$

**Remark 21.** If a sequence has two subsequences converging to different limits, then obviously the sequence itself does not converge, since by the previous theorem every subsequence of a convergent sequence must converge to the same limit. For example, the sequence $a_n = (-1)^n$ does not converge because the subsequence of even terms converges to 1, while the subsequence of odd terms converges to $-1$. □

**Proposition 17.2** (Every convergent sequence is bounded). *Every convergent sequence is bounded. The converse is not necessarily true.*

*Proof.* Let $a_n \to a$. Then

$$\forall \varepsilon > 0, \quad \exists n_0 : |a_n - a| \le \varepsilon \quad \forall n \ge n_0.$$

Therefore, for all $n \ge n_0$,

$$|a_n| = |(a_n - a) + a| \le |a_n - a| + |a| \le \varepsilon + |a|.$$

This shows that the sequence $(a_n)$ is bounded eventually (i. e., from some index onward). If we define

$$m = \max\{|a_1|, |a_2|, \ldots, |a_{n_0}|, |a| + 1\},$$

then $(a_n)$ is absolutely bounded by $m$.

To see that the converse is false, consider the sequence $a_n = (-1)^n$, which is bounded but does not converge since it has two different convergent subsequences. □

**Proposition 17.3** (Examining the convergence of some simple sequences). *The sequence $a_n = \frac{1}{n}$ converges to 0, the sequence $a_n = n$ diverges to $+\infty$, while the sequence $a_n = (-1)^n$ does not converge to any real number or diverge to $\pm\infty$.*

*Proof.* We will use the definition of convergence of a sequence. Let $\varepsilon > 0$. By the Archimedean property of natural numbers, there exists $n_0 \in \mathbb{N}$ such that $\frac{1}{n_0} < \varepsilon$. For each $\varepsilon > 0$, choosing such $n_0(\varepsilon)$, it follows that for all $n > n_0(\varepsilon)$,

$$|a_n - 0| < \varepsilon.$$

The sequence $a_n = n$ diverges to $+\infty$ because for any arbitrarily large $M > 0$, we can find $N > 0$ (in fact by choosing $N = \lfloor M \rfloor + 1$) such that

$$a_n = n > M \quad \text{for all } n > N.$$

The sequence $a_n = (-1)^n$ is bounded, hence it clearly does not diverge to $\pm\infty$. However, it does not converge to any real number $a \in \mathbb{R}$. Indeed, suppose there exists $a \in \mathbb{R}$

such that $\lim a_n = a$. Then, for every $\varepsilon > 0$, there exists some $N > 0$ such that for all $n > N$

$$|a_n - a| < \varepsilon.$$

But this is a contradiction because if we choose $n = 2N$ then

$$|1 - a| < \varepsilon \quad \Longrightarrow \quad 1 - \varepsilon < a < 1 + \varepsilon,$$

while if we choose $n = 2N + 1$ then

$$|-1 - a| < \varepsilon \quad \Longrightarrow \quad -1 - \varepsilon < a < -1 + \varepsilon,$$

which is impossible. $\qquad\qquad\square$

The following theorem provides a powerful characterization of discontinuity, offering several equivalent formulations, from the classical $\varepsilon - \delta$ perspective to sequence-based criteria, that reveal how failure of continuity can manifest through persistent deviations in function values near a point.

---

**Theorem 17.2** (Discontinuity criterion for functions). *Let $f : I \subseteq \mathbb{R} \to \mathbb{R}$ and $x_0 \in I$. The following statements are equivalent:*

(i)   *$f$ is discontinuous at $x_0$.*

(ii)  *There exists $\varepsilon > 0$ such that for every $\delta > 0$ there exists $x_\delta$ with*

$$|x_\delta - x_0| < \delta \quad \text{and} \quad \left|f(x_\delta) - f(x_0)\right| \geq \varepsilon.$$

(iii) *There exists $\varepsilon > 0$ and a sequence $(a_n) \subseteq I$ such that $a_n \to x_0$ as $n \to \infty$ and*

$$\left|f(a_n) - f(x_0)\right| \geq \varepsilon.$$

(iv) *There exists a sequence $(a_n) \subseteq I$ such that $a_n \to x_0$ as $n \to \infty$ and*

$$f(a_n) \not\to f(x_0).$$

---

*Proof.* Recalling the definition of continuity, statement (ii) is the negation of continuity, so (i) and (ii) are equivalent. Similarly, (iii) and (iv) are equivalent. It remains to prove that (ii) and (iii) are equivalent.

Assume (ii) holds. Set $\delta = \frac{1}{n}$. Then there exists some $a_n$ with

$$|a_n - x_0| < \frac{1}{n} \quad \text{and} \quad \left|f(a_n) - f(x_0)\right| \geq \varepsilon,$$

which is exactly (iii).

Assume (iii) holds. For any $\delta > 0$, there exists $n_\delta \in \mathbb{N}$ such that

$$|a_{n_\delta} - x_0| < \delta \quad \text{and} \quad \left|f(a_{n_\delta}) - f(x_0)\right| \geq \varepsilon.$$

Set $x_\delta := a_{n_\delta}$ to obtain (ii). $\qquad\qquad\square$

**Corollary 17.1** (Continuity via sequences). *A function $f$ is continuous at the point $x_0$ if and only if for every sequence $a_n \to x_0$ we have*

$$f(a_n) \to f(x_0).$$

**Proposition 17.4** (Properties of limits of sequences). *Let $(a_n)$, $(b_n)$ be convergent sequences with limits $a_n \to a$ and $b_n \to b$, and let $k \in \mathbb{R}$. Then:*

(i)   $a_n + b_n \to a + b$,
(ii)  *for every continuous function $f$ at $a$, $f(a_n) \to f(a)$,*
(iii) $a_n b_n \to ab$,
(iv)  $k a_n \to ka$,
(v)   *if $b \neq 0$, then there exists $n_0 \in \mathbb{N}$ such that $b_n \neq 0$ for all $n \geq n_0$, and*

$$\frac{a_n}{b_n} \to \frac{a}{b}.$$

*Proof.* (i)   To prove $a_n + b_n \to a + b$, let $\varepsilon > 0$. There exist integers $n_1, n_2$ such that

$$|a_n - a| < \frac{\varepsilon}{2} \quad \text{for all } n > n_1, \quad \text{and} \quad |b_n - b| < \frac{\varepsilon}{2} \quad \text{for all } n > n_2.$$

Set $n_0 = \max\{n_1, n_2\}$. Then, for all $n > n_0$,

$$|(a_n + b_n) - (a + b)| \leq |a_n - a| + |b_n - b| < \varepsilon.$$

(ii)  This follows from Corollary 17.1.

(iii) From (ii), if $a_n \to a$ then $a_n^2 \to a^2$. Since

$$a_n b_n = \frac{(a_n + b_n)^2 - a_n^2 - b_n^2}{2},$$

using (i) and (ii) it follows that $a_n b_n \to ab$.

(iv)  Taking $b_n = k$ constant and applying (iii) yields the result.

(v)   If $f(x) = \frac{1}{x}$, then by applying (ii) and (iii) to $a_n f(b_n)$, the conclusion follows.   $\square$

## 17.3.1 Exercises

### True or false

Decide whether each statement is **true** or **false**. Justify your answer.

1.   If a sequence $(a_n)$ is bounded and monotone, then it must converge.
     **Answer:** _____

2.   Every convergent sequence is bounded.
     **Answer:** _____

3.   If every subsequence of $(a_n)$ converges, then they all must converge to the same limit.
     **Answer:** _____

4.  The sequence $a_n = (-1)^n$ has a subsequence that converges to 1.
    **Answer:** _____
5.  If $\lim_{n\to\infty} a_n = +\infty$, then no subsequence of $(a_n)$ can converge to a finite limit.
    **Answer:** _____

## Multiple choice questions

Choose the correct option (A, B, C, or D) for each question.
1.  Which of the following sequences is monotonically increasing?
    A. $a_n = \frac{1}{n}$      B. $a_n = (-1)^n$
    C. $a_n = n^2 - n$    D. $a_n = \sin(n)$
    **Answer:** _____
2.  A sequence $(a_n)$ converges to $a$ if and only if
    A. there exists $\varepsilon > 0$ such that $|a_n - a| < \varepsilon$ for all $n$
    B. for every $\varepsilon > 0$, there exists $N \in \mathbb{N}$ such that $|a_n - a| < \varepsilon$ for all $n > N$
    C. $|a_n - a|$ decreases as $n$ increases
    D. $a_n = a$ for all sufficiently large $n$
    **Answer:** _____
3.  Consider the sequence $a_n = \frac{n}{n+1}$. What is $\lim_{n\to\infty} a_n$?
    A. 0    B. $\frac{1}{2}$
    C. 1    D. Does not exist
    **Answer:** _____
4.  Which of the following is a valid subsequence of $(a_n) = (\frac{1}{n})$?
    A. $(\frac{1}{2n-1})$    B. $(\frac{1}{n^2})$
    C. $(\frac{1}{\sqrt{n}})$    D. $(\frac{1}{(-n)})$
    **Answer:** _____
5.  Let $(a_n)$ be a sequence such that $a_n \to a$. Which of the following must also converge to $a$?
    A. Any bounded sequence      B. Any rearrangement of $(a_n)$
    C. Any subsequence of $(a_n)$    D. The sequence $(a_n^2)$
    **Answer:** _____

## Matching exercise

Match each sequence with its correct behavior or property.

| | |
|---|---|
| 1. $a_n = \frac{1}{n}$ | A. Diverges to $+\infty$ |
| 2. $a_n = (-1)^n$ | B. Converges to 0 |
| 3. $a_n = n$ | C. Bounded but divergent |
| 4. $a_n = \frac{n+1}{n}$ | D. Converges to 1 |
| 5. $a_n = \sqrt{n+1} - \sqrt{n}$ | E. Converges to 0 (after rationalization) |

**Answers:** 1._____    2._____    3._____    4._____    5._____

## Fill in the blanks

Fill in each blank with the most appropriate word or phrase.

1. A sequence $(a_n)$ is _____ if $a_n \leq a_{n+1}$ for all $n$.

   **Answer:** _____

2. A sequence is bounded if it is both bounded above and _____.

   **Answer:** _____

3. The sequence $a_n = (-1)^n$ does not converge because it has two subsequences converging to _____.

   **Answer:** _____

4. If $a_n \to a$, then for every $\varepsilon > 0$, there exists $N$ such that all terms $a_n$ for $n > N$ lie within the interval _____.

   **Answer:** _____

5. A strictly increasing sequence of indices $(n_k)$ defines a _____ of the original sequence.

   **Answer:** _____

## Theoretical and computational problems

Solve each problem step-by-step.

1. Prove that the sequence $a_n = \frac{1}{n}$ converges to 0 using the $\varepsilon$-definition of a limit.
2. Show that the sequence $a_n = \frac{n}{n+1}$ is monotonically increasing and bounded. Deduce that it converges.
3. Let $(a_n)$ be a sequence such that $a_n \to a$. Prove that every subsequence of $(a_n)$ also converges to $a$.
4. Consider the sequence $a_n = (-1)^n + \frac{1}{n}$. Determine whether it converges. If not, find two subsequences that converge to different limits.
5. Prove that if a sequence $(a_n)$ diverges to $+\infty$, then no subsequence can converge to a finite limit.
6. Use the definition to show that $\lim_{n \to \infty} n^2 = +\infty$.
7. Let $f : \mathbb{R} \to \mathbb{R}$ be defined by $f(x) = x^2$. Suppose $a_n \to 2$. Use the sequential characterization of continuity to show that $f(a_n) \to 4$.
8. Suppose $a_n \to a$ and $b_n \to b$. Use the algebraic properties of limits to compute $\lim_{n \to \infty} (3a_n - 2b_n)$.
9. Let $a_n = \frac{2n^2+1}{n^2+3}$. Find $\lim_{n \to \infty} a_n$ and justify your answer.
10. Construct a sequence that is bounded but divergent. Justify your answer using subsequences.

## Python-based exercises

Use Python (sympy, numpy, matplotlib) to tackle the following.

1. Use sympy to compute the limit of the following sequences as $n \to \infty$:
   - $a_n = \frac{3n^2 + 2n + 1}{2n^2 + 5}$.
   - $b_n = (1 + \frac{1}{n})^n$.
   - $c_n = \frac{\sin n}{n}$.

   *Hint: Use* `sympy.limit` *with* `n = symbols('n', integer=True)` *and* `oo` *for infinity.*

2. Write a Python function that generates the first $N$ terms of a sequence given by a formula (e. g., $a_n = 1/n$). Plot the sequence and observe its behavior as $n$ increases.
   *Hint: Use* `numpy.arange` *and* `matplotlib.pyplot.plot` *with markers.*

3. Simulate two subsequences of $a_n = (-1)^n$: one for even $n$, one for odd $n$. Plot them separately and verify that they converge to different limits.

4. Implement a numerical check for convergence: for $a_n = \frac{n}{n+1}$, compute $|a_n - 1|$ for $n = 10, 100, 1000, 10000$ and verify that it becomes smaller than any given $\varepsilon > 0$.

5. Use sympy to verify the continuity of $f(x) = \sqrt{x}$ at $x = 4$ via sequences: generate a random sequence $a_n \to 4$ and check numerically that $f(a_n) \to 2$.
   *Hint: Use random perturbations* $a_n = 4 + \frac{1}{n} \cdot \text{rand}()$ *and evaluate* $f(a_n)$.

## 17.4 Limsup and liminf

### 17.4.1 Definitions

This section explores the concepts of limit superior and limit inferior, fundamental tools in real analysis for understanding the long-term behavior of sequences. Unlike simple convergence, which captures a single limiting value, limsup and liminf describe the range of possible limits of subsequences, offering insight into oscillatory and divergent behavior. We develop their definitions, key properties, and characterizations, and present powerful inequalities that govern their interaction under arithmetic operations and transformations. Through theorems, examples, and a comprehensive exercise set, this section provides a robust framework for analyzing sequences beyond pointwise convergence.

Given any sequence, we can construct infinitely many subsequences which may converge (or diverge) to different limits.

Let $a_n$ be a sequence. Define

$$b_n = \sup\{a_n, a_{n+1}, a_{n+2}, \ldots\} \quad \text{and} \quad c_n = \inf\{a_n, a_{n+1}, a_{n+2}, \ldots\}.$$

Obviously, the sequence $b_n$ is decreasing while the sequence $c_n$ is increasing, and moreover,

$$c_n \leq b_n \quad \text{for all } n \in \mathbb{N}.$$

We define the *limit superior* and *limit inferior* of $a_n$ as

$$\limsup_{n\to\infty} a_n = \inf_n b_n \quad \text{and} \quad \liminf_{n\to\infty} a_n = \sup_n c_n,$$

which satisfy

$$\liminf_{n\to\infty} a_n \le \limsup_{n\to\infty} a_n.$$

It can be proved that there exists a subsequence of $a_n$ converging to $\limsup_{n\to\infty} a_n$ and a subsequence converging to $\liminf_{n\to\infty} a_n$. In fact, every convergent (or divergent to $\pm\infty$) subsequence of $a_n$ converges to some number in the interval

$$\left[ \liminf_{n\to\infty} a_n, \quad \limsup_{n\to\infty} a_n \right].$$

For convenience, we give the following definition.

---

**Definition 17.5** (Limit superior and limit inferior). Let $(a_n)_{n\in\mathbb{N}}$ be a sequence of real numbers. We define the lim sup of the sequence as follows:

$$\limsup_{n\to\infty} a_n = \lim_{n\to\infty} \left( \sup_{k\ge n} a_k \right).$$

Alternatively, it is the greatest limit point of the sequence $(a_n)$, or the supremum of the set of all subsequential limits.

We define the lim inf of the sequence as follows:

$$\liminf_{n\to\infty} a_n = \lim_{n\to\infty} \left( \inf_{k\ge n} a_k \right).$$

Alternatively, it is the smallest limit point of the sequence $(a_n)$, or the infimum of the set of all subsequential limits.

---

Clearly, if $a_{n_k}$ is any subsequence of $a_n$, then by definition

$$\lim_{k\to\infty} a_{n_k} \le \limsup_{n\to\infty} a_n.$$

### 17.4.2 Properties of $\limsup_{n\to\infty} a_n$ and $\liminf_{n\to\infty} a_n$

Using Theorem 17.1, we have the following corollary.

---

**Corollary 17.2** (Characterization of convergence via limsup and liminf). *A sequence $a_n$ converges to $a \in \mathbb{R}$ if and only if*

$$\limsup_{n\to\infty} a_n = \liminf_{n\to\infty} a_n = a.$$

*Similarly, $a_n$ diverges to $\pm\infty$ if and only if*

$$\limsup_{n\to\infty} a_n = \liminf_{n\to\infty} a_n = \pm\infty.$$

---

**Theorem 17.3** (Property of limiting behavior with negation). *For any sequence $(a_n)$, the following relationship holds:*

$$\limsup_{n\to\infty}(-a_n) = -\liminf_{n\to\infty} a_n.$$

*Proof.* Let $(a_n)_{n\in\mathbb{N}}$ be a sequence of real numbers. We aim to show that

$$\limsup_{n\to\infty}(-a_n) = -\liminf_{n\to\infty} a_n.$$

By definition,

$$\limsup_{n\to\infty}(-a_n) = \lim_{n\to\infty}\left(\sup_{k\geq n}(-a_k)\right).$$

We now use the identity from real analysis: for any nonempty set $S \subseteq \mathbb{R}$,

$$\sup(-S) = -\inf(S), \quad \text{where } -S = \{-x \mid x \in S\}.$$

Applying this to the set $\{-a_k \mid k \geq n\}$, we have

$$\sup_{k\geq n}(-a_k) = -\left(\inf_{k\geq n} a_k\right).$$

Therefore,

$$\limsup_{n\to\infty}(-a_n) = \lim_{n\to\infty}\left(-\inf_{k\geq n} a_k\right).$$

Since multiplication by $-1$ is continuous, we can bring the limit inside:

$$\lim_{n\to\infty}\left(-\inf_{k\geq n} a_k\right) = -\left(\lim_{n\to\infty}\inf_{k\geq n} a_k\right) = -\liminf_{n\to\infty} a_n.$$

Thus,

$$\limsup_{n\to\infty}(-a_n) = -\liminf_{n\to\infty} a_n,$$

as desired. □

**Theorem 17.4** (Inequalities for sums of sequences with limsup and liminf). *For any two real sequences* $(a_n)$ *and* $(b_n)$, *the following inequalities hold:*

$$\liminf_{n\to\infty} a_n + \liminf_{n\to\infty} b_n \leq \liminf_{n\to\infty}(a_n + b_n),$$

$$\limsup_{n\to\infty}(a_n + b_n) \leq \limsup_{n\to\infty} a_n + \limsup_{n\to\infty} b_n.$$

*Proof.* We prove each inequality separately, using the definitions

$$\liminf_{n\to\infty} a_n = \lim_{n\to\infty}\left(\inf_{k\geq n} a_k\right), \quad \limsup_{n\to\infty} a_n = \lim_{n\to\infty}\left(\sup_{k\geq n} a_k\right).$$

**First inequality:**

$$\liminf_{n\to\infty} a_n + \liminf_{n\to\infty} b_n \leq \liminf_{n\to\infty}(a_n + b_n)$$

For any fixed $n \in \mathbb{N}$, define

$$u_n = \inf_{k\geq n} a_k, \quad v_n = \inf_{k\geq n} b_k, \quad w_n = \inf_{k\geq n}(a_k + b_k).$$

Note that for all $k \geq n$,

$$a_k \geq u_n, \quad b_k \geq v_n \quad \Longrightarrow \quad a_k + b_k \geq u_n + v_n.$$

Therefore, $u_n + v_n$ is a lower bound for the set $\{a_k + b_k \mid k \geq n\}$, so

$$w_n = \inf_{k\geq n}(a_k + b_k) \geq u_n + v_n.$$

Taking the limit as $n \to \infty$,

$$\lim_{n\to\infty} w_n \geq \lim_{n\to\infty}(u_n + v_n) = \lim_{n\to\infty} u_n + \lim_{n\to\infty} v_n,$$

since both $(u_n)$ and $(v_n)$ are monotonic sequences (nondecreasing), and thus their limits exist in the extended real numbers.

Hence,

$$\liminf_{n\to\infty}(a_n + b_n) \geq \liminf_{n\to\infty} a_n + \liminf_{n\to\infty} b_n.$$

**Second inequality:**

$$\limsup_{n\to\infty}(a_n + b_n) \leq \limsup_{n\to\infty} a_n + \limsup_{n\to\infty} b_n$$

Similarly, define

$$x_n = \sup_{k\geq n} a_k, \quad y_n = \sup_{k\geq n} b_k, \quad z_n = \sup_{k\geq n}(a_k + b_k).$$

For all $k \geq n$,

$$a_k \leq x_n, \quad b_k \leq y_n \quad \Longrightarrow \quad a_k + b_k \leq x_n + y_n.$$

Thus, $x_n + y_n$ is an upper bound for the set $\{a_k + b_k \mid k \geq n\}$, so

$$z_n = \sup_{k \geq n}(a_k + b_k) \leq x_n + y_n.$$

Taking the limit as $n \to \infty$,

$$\lim_{n \to \infty} z_n \leq \lim_{n \to \infty}(x_n + y_n) = \lim_{n \to \infty} x_n + \lim_{n \to \infty} y_n,$$

and therefore,

$$\limsup_{n \to \infty}(a_n + b_n) \leq \limsup_{n \to \infty} a_n + \limsup_{n \to \infty} b_n.$$

This completes the proof. $\square$

**Theorem 17.5** (Inequalities for products of nonnegative sequences with limsup and liminf). *If $(a_n)$ and $(b_n)$ are bounded sequences such that $a_n \geq 0$ and $b_n \geq 0$ for all $n$, then the following inequalities hold:*

$$\liminf_{n \to \infty} a_n \cdot \liminf_{n \to \infty} b_n \leq \liminf_{n \to \infty}(a_n b_n) \leq \limsup_{n \to \infty}(a_n b_n) \leq \limsup_{n \to \infty} a_n \cdot \limsup_{n \to \infty} b_n.$$

*Proof.* We prove each inequality in order.

**First inequality:**

$$\liminf_{n \to \infty} a_n \cdot \liminf_{n \to \infty} b_n \leq \liminf_{n \to \infty}(a_n b_n)$$

Let

$$L_a = \liminf_{n \to \infty} a_n, \quad L_b = \liminf_{n \to \infty} b_n.$$

By definition of lim inf, there exists a subsequence $(n_k)$ such that

$$\lim_{k \to \infty}(a_{n_k} b_{n_k}) = \liminf_{n \to \infty}(a_n b_n).$$

Along this subsequence, since $a_{n_k} \geq 0$ and $b_{n_k} \geq 0$, we can extract a further subsequence (still denoted by $(n_k)$ for simplicity) such that both $a_{n_k}$ and $b_{n_k}$ converge in $[0, \infty)$. Let

$$\lim_{k \to \infty} a_{n_k} = A, \quad \lim_{k \to \infty} b_{n_k} = B.$$

Then $A \geq L_a, B \geq L_b$, and since all terms are nonnegative, we have

$$\lim_{k \to \infty}(a_{n_k} b_{n_k}) = A \cdot B \geq L_a \cdot L_b.$$

Therefore,

$$\liminf_{n\to\infty}(a_nb_n) \geq L_a \cdot L_b,$$

which proves the first inequality.

**Middle inequality:**

$$\liminf_{n\to\infty}(a_nb_n) \leq \limsup_{n\to\infty}(a_nb_n)$$

This holds for *any* real sequence: by definition, the limit inferior is always less than or equal to the limit superior (in the extended real numbers). This follows from the fact that

$$\inf_{k\geq n}(a_kb_k) \leq \sup_{k\geq n}(a_kb_k)$$

for all $n$, and taking limits preserves the inequality.

**Right inequality:**

$$\limsup_{n\to\infty}(a_nb_n) \leq \limsup_{n\to\infty} a_n \cdot \limsup_{n\to\infty} b_n$$

Let

$$S_a = \limsup_{n\to\infty} a_n, \quad S_b = \limsup_{n\to\infty} b_n.$$

By definition, there exists a subsequence $(n_j)$ such that

$$\lim_{j\to\infty}(a_{n_j}b_{n_j}) = \limsup_{n\to\infty}(a_nb_n).$$

Along this subsequence, extract a further subsequence (still denoted $(n_j)$) such that both $a_{n_j} \to A'$ and $b_{n_j} \to B'$ in $[0,\infty)$. Then $A' \leq S_a, B' \leq S_b$, and

$$\lim_{j\to\infty}(a_{n_j}b_{n_j}) = A' \cdot B' \leq S_a \cdot S_b.$$

Hence,

$$\limsup_{n\to\infty}(a_nb_n) \leq S_a \cdot S_b,$$

which completes the proof. □

---

**Theorem 17.6** (Comparison of limsup and liminf for ordered sequences). *If two bounded sequences $(a_n)$ and $(b_n)$ satisfy $a_n \leq b_n$ for all n, then*

$$\liminf_{n\to\infty} a_n \leq \liminf_{n\to\infty} b_n, \quad \text{and} \quad \limsup_{n\to\infty} a_n \leq \limsup_{n\to\infty} b_n.$$

*Proof.*

**First inequality:** $\liminf_{n\to\infty} a_n \leq \liminf_{n\to\infty} b_n$

Let $L_a = \liminf_{n\to\infty} a_n$. By definition, there exists a subsequence $(a_{n_k})$ such that

$$\lim_{k\to\infty} a_{n_k} = L_a.$$

Now consider the corresponding subsequence $(b_{n_k})$. Since $a_{n_k} \leq b_{n_k}$ for all $k$, we have

$$a_{n_k} \leq b_{n_k} \quad\Longrightarrow\quad \lim_{k\to\infty} a_{n_k} \leq \liminf_{k\to\infty} b_{n_k}.$$

But since $a_{n_k} \to L_a$, this gives

$$L_a \leq \liminf_{k\to\infty} b_{n_k}.$$

Now, $\liminf_{k\to\infty} b_{n_k}$ is the limit inferior of a subsequence of $(b_n)$, and every subsequential limit inferior (in fact, every subsequential limit) is at least $\liminf_{n\to\infty} b_n$ because the latter is the smallest such limit in $\mathbb{R}$. Therefore

$$\liminf_{k\to\infty} b_{n_k} \geq \liminf_{n\to\infty} b_n.$$

Combining the inequalities,

$$\liminf_{n\to\infty} a_n = L_a \leq \liminf_{k\to\infty} b_{n_k} \geq \liminf_{n\to\infty} b_n,$$

so

$$\liminf_{n\to\infty} a_n \leq \liminf_{n\to\infty} b_n.$$

**Second inequality:** $\limsup_{n\to\infty} a_n \leq \limsup_{n\to\infty} b_n$

Let $S_a = \limsup_{n\to\infty} a_n$. By definition, there exists a subsequence $(a_{n_j})$ such that

$$\lim_{j\to\infty} a_{n_j} = S_a.$$

Along this subsequence, $a_{n_j} \leq b_{n_j}$. Now, extract a further subsequence (still denoted by $n_j$) such that $b_{n_j}$ converges in $\mathbb{R}$. Let

$$\lim_{j\to\infty} b_{n_j} = M.$$

Then $M$ is a subsequential limit of $(b_n)$, so

$$M \leq \limsup_{n\to\infty} b_n.$$

On the other hand, since $a_{n_j} \to S_a$ and $a_{n_j} \le b_{n_j}$, taking limits gives

$$S_a = \lim_{j \to \infty} a_{n_j} \le \lim_{j \to \infty} b_{n_j} = M.$$

Therefore,

$$\limsup_{n \to \infty} a_n = S_a \le M \le \limsup_{n \to \infty} b_n,$$

which proves the second inequality.
This completes the proof. □

---

**Theorem 17.7** (Behavior of lim sup and lim inf under monotone continuous functions). *Let $(a_n) \subset \mathbb{R}$ be a sequence, and let $f$ be a real-valued function that is continuous on a set containing all limit points of $(a_n)$, including $\limsup_{n \to \infty} a_n$ and $\liminf_{n \to \infty} a_n$.*
1. *If $f$ is increasing, then*

$$\limsup_{n \to \infty} f(a_n) = f\left(\limsup_{n \to \infty} a_n\right), \quad \liminf_{n \to \infty} f(a_n) = f\left(\liminf_{n \to \infty} a_n\right).$$

2. *If $f$ is decreasing, then*

$$\limsup_{n \to \infty} f(a_n) = f\left(\liminf_{n \to \infty} a_n\right), \quad \liminf_{n \to \infty} f(a_n) = f\left(\limsup_{n \to \infty} a_n\right).$$

---

*Proof.*
**Part 1: $f$ is increasing**
Let $u_n = \sup_{k \ge n} a_k$, so that $u_n \searrow \limsup_{n \to \infty} a_n =: L$. Since $f$ is increasing, for all $k \ge n$, $a_k \le u_n \implies f(a_k) \le f(u_n)$. Thus, $f(u_n)$ is an upper bound for $\{f(a_k) \mid k \ge n\}$, and therefore

$$\sup_{k \ge n} f(a_k) \le f(u_n).$$

Taking the limit as $n \to \infty$,

$$\limsup_{n \to \infty} f(a_n) = \lim_{n \to \infty} \left(\sup_{k \ge n} f(a_k)\right) \le \lim_{n \to \infty} f(u_n).$$

Since $u_n \to L$ and $f$ is continuous at $L$, we have $\lim_{n \to \infty} f(u_n) = f(L)$. Hence

$$\limsup_{n \to \infty} f(a_n) \le f\left(\limsup_{n \to \infty} a_n\right).$$

For the reverse inequality, note that by definition of lim sup, there exists a subsequence $(a_{n_j})$ such that $a_{n_j} \to L$. Since $f$ is continuous, $f(a_{n_j}) \to f(L)$. Therefore, $f(L)$ is a subsequential limit of $(f(a_n))$, so

$$\limsup_{n \to \infty} f(a_n) \ge f(L) = f\left(\limsup_{n \to \infty} a_n\right).$$

Combining both inequalities,

$$\limsup_{n\to\infty} f(a_n) = f\left(\limsup_{n\to\infty} a_n\right).$$

Now let $l_n = \inf_{k\geq n} a_k$, so $l_n \nearrow \liminf_{n\to\infty} a_n =: l$. Since $f$ is increasing, $a_k \geq l_n \implies f(a_k) \geq f(l_n)$. Thus, $f(l_n)$ is a lower bound for $\{f(a_k) \mid k \geq n\}$, so

$$\inf_{k\geq n} f(a_k) \geq f(l_n).$$

Taking limits,

$$\liminf_{n\to\infty} f(a_n) = \lim_{n\to\infty}\left(\inf_{k\geq n} f(a_k)\right) \geq \lim_{n\to\infty} f(l_n) = f(l),$$

by continuity of $f$ at $l$.

Again, there exists a subsequence $(a_{n_j}) \to l$, so $f(a_{n_j}) \to f(l)$, and thus

$$\liminf_{n\to\infty} f(a_n) \leq f(l),$$

since lim inf is the smallest subsequential limit.

Therefore,

$$\liminf_{n\to\infty} f(a_n) = f\left(\liminf_{n\to\infty} a_n\right).$$

**Part 2: $f$ is decreasing**

Suppose $f$ is decreasing and continuous.

Define $g(x) = -f(x)$. Then $g$ is increasing, and since $f$ is continuous, so is $g$.

Note that

$$\limsup_{n\to\infty} f(a_n) = -\liminf_{n\to\infty}(-f(a_n)) = -\liminf_{n\to\infty} g(a_n).$$

From Part 1, since $g$ is increasing,

$$\liminf_{n\to\infty} g(a_n) = g\left(\liminf_{n\to\infty} a_n\right) = -f\left(\liminf_{n\to\infty} a_n\right).$$

Thus

$$\limsup_{n\to\infty} f(a_n) = -\left(-f\left(\liminf_{n\to\infty} a_n\right)\right) = f\left(\liminf_{n\to\infty} a_n\right).$$

Similarly,

$$\liminf_{n\to\infty} f(a_n) = -\limsup_{n\to\infty} g(a_n) = -g\left(\limsup_{n\to\infty} a_n\right) = -\left(-f\left(\limsup_{n\to\infty} a_n\right)\right) = f\left(\limsup_{n\to\infty} a_n\right).$$

This completes the proof. $\qquad\square$

**Example 17.1.** Consider the sequence

$$a_n = \frac{n^2}{1+n^2} \sin \frac{2\pi n}{3}.$$

Choosing the subsequence $a_{3k}$, we have

$$\lim_{k \to \infty} a_{3k} = 0,$$

so every subsequence of $a_{3k}$ converges to 0. Choosing $n = 3k - 1$, the subsequence $a_{3k-1}$ converges to $-\frac{\sqrt{3}}{2}$, and similarly every subsequence of $a_{3k-1}$ converges to this value. Finally, for $n = 3k - 2$, the subsequence $a_{3k-2}$ converges to $\frac{\sqrt{3}}{2}$, and so do all its subsequences.

Any subsequence of $a_n$ is either a subsequence of one of the three subsequences $a_{3k}, a_{3k-1},$ or $a_{3k-2}$, or a combination of terms from these subsequences. Therefore, no subsequence converges to a limit greater than $\frac{\sqrt{3}}{2}$ or less than $-\frac{\sqrt{3}}{2}$, so

$$\limsup_{n \to \infty} a_n = \frac{\sqrt{3}}{2}, \quad \liminf_{n \to \infty} a_n = -\frac{\sqrt{3}}{2}.$$

## 17.4.3 Exercises

### True or false

Decide whether each statement is **true** or **false**. Justify your answer.

1. If $\limsup_{n \to \infty} a_n = \liminf_{n \to \infty} a_n = a$, then the sequence $(a_n)$ converges to $a$.

   **Answer:** _____

2. For any sequence $(a_n)$, $\limsup_{n \to \infty}(-a_n) = \liminf_{n \to \infty} a_n$.

   **Answer:** _____

3. If $a_n \le b_n$ for all $n$, then $\limsup_{n \to \infty} a_n \le \limsup_{n \to \infty} b_n$.

   **Answer:** _____

4. The sequence $a_n = \sin(\frac{n\pi}{2})$ has $\limsup a_n = 1$ and $\liminf a_n = -1$.

   **Answer:** _____

5. If $f$ is continuous and increasing, then $\limsup f(a_n) = f(\limsup a_n)$.

   **Answer:** _____

### Multiple choice questions

Choose the correct option (A, B, C, or D) for each question.

1. What are lim sup and lim inf of the sequence $a_n = (-1)^n(1 + \frac{1}{n})$?

   A. lim sup = 1, lim inf = −1    B. lim sup = −1, lim inf = 1

   C. lim sup = 1, lim inf = −1    D. lim sup = 1, lim inf = −1

   **Answer:** _____

2. Let $a_n = \frac{n}{n+1}$ and $b_n = \frac{(-1)^n}{n}$. What is $\limsup_{n \to \infty}(a_n + b_n)$?

   A. 0    B. 1

   C. 2    D. Does not exist

   **Answer:** _____

3. For a bounded sequence $(a_n)$, $\limsup_{n \to \infty} a_n$ is equal to

A. the limit of the smallest convergent subsequence

B. the limit of the largest convergent subsequence

C. the supremum of the set $\{a_n : n \in \mathbb{N}\}$

D. the infimum of the set $\{a_n : n \in \mathbb{N}\}$

**Answer:** _____

4. Which of the following inequalities is **always** true for bounded sequences $(a_n)$, $(b_n)$?

A. $\limsup(a_n + b_n) \geq \limsup a_n + \limsup b_n$

B. $\liminf(a_n + b_n) \leq \liminf a_n + \liminf b_n$

C. $\limsup(a_n + b_n) \leq \limsup a_n + \limsup b_n$

D. $\limsup(a_n b_n) = \limsup a_n \cdot \limsup b_n$

**Answer:** _____

5. Let $a_n = n$ and $b_n = -n$. Then $\limsup_{n \to \infty}(a_n + b_n)$ is

A. $+\infty$    B. $-\infty$

C. 0      D. undefined

**Answer:** _____

## Matching exercise

Match each sequence with its lim sup and lim inf.

1. $a_n = \frac{(-1)^n}{n}$     A. lim sup = 1, lim inf = $-1$

2. $a_n = \cos(\frac{n\pi}{3})$    B. lim sup = 0, lim inf = 0

3. $a_n = (-1)^n$     C. lim sup = 1, lim inf = $-1$

4. $a_n = \frac{n^2}{1+n^2}$     D. lim sup = 1, lim inf = 0

5. $a_n = \sin n$     E. lim sup = 1, lim inf = $-1$ (dense in $[-1, 1]$)

**Answers: 1.**_____ 2._____ 3._____ 4._____ 5._____

## Fill in the blank

Fill in each blank with the most appropriate word or phrase.

1. The $\limsup_{n \to \infty} a_n$ is defined as $\inf_n \sup_{k \geq n} a_k$, while $\liminf_{n \to \infty} a_n$ is _____.

**Answer:** _____

2. A sequence $(a_n)$ converges if and only if $\limsup a_n = \liminf a_n = $ _____.

**Answer:** _____

3. If $a_n \leq b_n$ for all $n$, then $\liminf a_n \leq$ _____.

**Answer:** _____

4. For a nonnegative sequence $(a_n)$, $\liminf a_n \cdot \liminf b_n \leq$ _____.

**Answer:** _____

5. If $f$ is continuous and decreasing, then $\limsup f(a_n) = f($_____$)$.

**Answer:** _____

## Theoretical and computational problems

Solve each problem step-by-step.

1. Compute $\limsup_{n \to \infty} a_n$ and $\liminf_{n \to \infty} a_n$ for the sequence $a_n = \frac{n}{n+1} + (-1)^n \frac{2n}{n+2}$. Justify your answer.

2. Prove that for any sequence $(a_n)$, $\limsup_{n\to\infty}(-a_n) = -\liminf_{n\to\infty} a_n$.

3. Let $a_n = \sin(\frac{n\pi}{4})$. Find all possible limits of convergent subsequences. Then compute $\limsup a_n$ and $\liminf a_n$.

4. Prove that if $a_n \le b_n$ for all $n$, then $\limsup a_n \le \limsup b_n$.

5. Let $a_n = n$ and $b_n = \frac{1}{n}$. Compute $\limsup(a_n b_n)$ and compare it with $\limsup a_n \cdot \limsup b_n$.

6. Let $f(x) = e^{-x}$, which is continuous and decreasing. Let $a_n = (-1)^n n$. Compute $\limsup a_n$, $\liminf a_n$, and verify that $\limsup f(a_n) = f(\liminf a_n)$.

7. Prove that $\liminf a_n + \liminf b_n \le \liminf(a_n + b_n)$ for bounded sequences $(a_n)$, $(b_n)$.

8. Suppose $a_n \to a$ and $b_n \to b$. Show that $\limsup(a_n + b_n) = a + b$ and $\liminf(a_n + b_n) = a + b$.

9. Let $a_n = 2 + \frac{(-1)^n}{n}$ and $b_n = 1 + \frac{(-1)^{n+1}}{n}$. Compute $\limsup(a_n b_n)$ and compare it with $\limsup a_n \cdot \limsup b_n$.

10. Use the definition of $\limsup$ and $\liminf$ to prove that for any sequence $(a_n)$, $\liminf a_n \le \limsup a_n$.

### Python-based exercises

Use Python (`sympy`, `numpy`, `matplotlib`) to tackle the following:

1. Write a Python function that approximates $\limsup$ and $\liminf$ of a sequence numerically by computing $\sup_{k\ge n} a_k$ and $\inf_{k\ge n} a_k$ for large $n$. Test it on $a_n = \sin n$ and $a_n = (-1)^n(1 + 1/n)$.

2. Plot the sequence $a_n = \frac{n^2}{1+n^2}\sin(\frac{2\pi n}{3})$ and its upper and lower envelopes (running sup and inf) for $n = 1$ to $50$. Observe the convergence of the envelopes.

3. Simulate multiple subsequences of $a_n = \sin n$ (e. g., using indices where $n \mod 2\pi$ falls in certain intervals) and estimate the $\limsup$ and $\liminf$ numerically.

4. Use sympy to compute $\limsup$ and $\liminf$ symbolically for sequences like $a_n = \frac{n}{n+1}$ and $a_n = (-1)^n$.
   (*Note*: sympy does not have direct `limsup` command; simulate using subsequences.)

5. Implement a check for the inequality $\limsup(a_n + b_n) \le \limsup a_n + \limsup b_n$ using numerical sequences. Test it on $a_n = \sin n$, $b_n = \cos n$.

## 17.5 Some criteria for convergence

The following criterion comes from the theory of series of sequences but applies also for sequences. In that theory one can find also other criteria that can be applied also to sequences.

**Theorem 17.8** (Ratio and root tests). *Let $(a_n)$ be a sequence (not necessarily positive). If*

$$\limsup_{n\to\infty} \frac{|a_{n+1}|}{|a_n|} = p < 1 \quad or \quad \limsup_{n\to\infty} \sqrt[n]{|a_n|} = p < 1,$$

then $a_n \to 0$.
  If

$$\liminf_{n\to\infty} \frac{|a_{n+1}|}{|a_n|} = p > 1 \quad \text{or} \quad \liminf_{n\to\infty} \sqrt[n]{|a_n|} = p > 1,$$

then $|a_n| \to +\infty$.

**Remark 22.** The root test is stronger than the ratio test. For example, the sequence

$$a_n = \frac{1}{3^{(-1)^n + n}}$$

tends to zero and this is seen using the root test but not the ratio test. However, even stronger tests exist (to be seen in series theory). For instance, the sequence

$$a_n = \frac{1}{(\ln n)^a}$$

cannot be analyzed by either test. □

Often, to find the limit of a sequence, it suffices to bound it above and below by sequences with known limits. If those limits coincide, then the intermediate sequence converges to the same limit. This is proved in the following proposition.

**Proposition 17.5** (Squeeze theorem for sequences). *Let $a_n$, $b_n$, $c_n$ be sequences such that*

$$a_n \le b_n \le c_n,$$

*and suppose*

$$a_n \to k, \quad c_n \to k.$$

*Then $b_n \to k$ as well.*

*Proof.* Let $\varepsilon > 0$. Then there exist integers $n_1$, $n_2$ such that

$$|a_n - k| < \frac{\varepsilon}{2} \quad \text{for all } n > n_1, \quad \text{and} \quad |c_n - k| < \frac{\varepsilon}{2} \quad \text{for all } n > n_2.$$

Set $n_0 = \max\{n_1, n_2\}$. For all $n \ge n_0$, we have

$$|b_n - k| \le \max\{|a_n - k|, |c_n - k|\} < \varepsilon. \qquad \square$$

The following theorem is a cornerstone of real analysis, guaranteeing the convergence of monotonic sequences when bounded. It formalizes the intuitive idea that an increasing sequence with an upper bound must approach a least upper limit—the supremum—and similarly for decreasing sequences and infima.

> **Theorem 17.9** (Monotone and bounded sequences converge). *Every monotone and bounded sequence converges. Specifically, if $(a_n)$ is increasing and bounded above, then*
>
> $$\lim_{n \to \infty} a_n = \sup_n a_n = \sup\{a_n : n \in \mathbb{N}\}.$$
>
> *Similarly, for decreasing sequences.*

*Proof.* Assume $a_n$ is increasing. The set $A = \{a_n : n \in \mathbb{N}\}$ is nonempty and bounded above, so it has a least upper bound $a = \sup_n a_n$. Since $a$ is the least upper bound, for every $\varepsilon > 0$ there exists some $a_{n_0}$ such that

$$a - \varepsilon < a_{n_0}.$$

For every $n \geq n_0$, because $(a_n)$ is increasing,

$$|a - a_n| = a - a_n \leq a - a_{n_0} < \varepsilon,$$

so the sequence converges to the supremum. The decreasing case is analogous. $\qquad\square$

Often, it is easier to work with functions to obtain results about the corresponding sequences rather than deal with sequences directly. For example, it is simpler (and more tools are available) to check if a function is increasing, which is useful for applying Theorem 17.9. For this reason, we relate the behavior of a sequence and its corresponding function in the following theorem.

> **Theorem 17.10** (Sequences via functions). *Let $f : \mathbb{R}^+ \to \mathbb{R}$ be a function and $a_n = f(n)$. If*
>
> $$\lim_{x \to \infty} f(x) = a \in \mathbb{R} \cup \{-\infty, +\infty\},$$
>
> *then the sequence $(a_n)$ converges to $a$. If $f$ is increasing (respectively decreasing), then the sequence $(a_n)$ is increasing (respectively decreasing), but the converse is not necessarily true.*

*Proof.* Suppose $a \in (-\infty, \infty)$. Since $f(x) \to a$ as $x \to \infty$, for every $\varepsilon > 0$ there exists $M$ sufficiently large such that

$$|f(x) - a| < \varepsilon \quad \text{for all } x > M.$$

Choosing $N = \lfloor M \rfloor + 1$, it follows that

$$|a_n - a| < \varepsilon \quad \text{for all } n > N,$$

since $a_n = f(n)$. Thus, $a_n \to a$.

If $a = +\infty$, then for every $M$ there exists $M'$ sufficiently large such that

$$f(x) > M \quad \text{for all } x > M'.$$

Choosing $N = \lfloor M' \rfloor + 1$, it follows that $a_n \to +\infty$. Similarly for $a = -\infty$.

If $f(x) \le f(y)$ for every $x < y$ with $x, y \in \mathbb{R}^+$, then

$$f(n) = a_n \le a_{n+1} = f(n+1).$$

Although the sequence $(a_n)$ may be monotone, the corresponding function $f$ may not be. In practice, this means that if the function is not monotone, we cannot conclude that the sequence is not monotone. $\square$

### 17.5.1 Limit of important sequences

This section presents several fundamental limit results involving powers and roots, which frequently appear in calculus and analysis. From the convergence of expressions like $\sqrt[n]{n}$ and $\theta^n$ to the classic exponential limit $(1 + \frac{k}{n})^n \to e^k$, these propositions highlight the behavior of sequences with exponential, logarithmic, and root components. The proofs employ tools such as logarithmic transformations and L'Hospital's rule, and culminate with a striking example that connects root-based expressions to the natural logarithm. These results form a powerful toolkit for evaluating complex sequence limits.

**Proposition 17.6** (Limits of nth roots). *If* $a_n = \sqrt[n]{n}$ *and* $b_n = \sqrt[n]{a}$ *(for* $a > 0$*), then* $a_n, b_n \to 1$.

*Proof.* We will prove that $a_n \to 1$. Define the function

$$f(x) = x^{1/x}$$

and

$$g(x) = \ln f(x) = \frac{\ln x}{x}.$$

We will show

$$\lim_{x \to \infty} g(x) = 0,$$

hence

$$\lim_{x \to \infty} f(x) = 1,$$

which implies $a_n \to 1$ by the previous theorem. We compute

$$\lim_{x \to \infty} \frac{\ln x}{x} = \lim_{x \to \infty} \frac{1/x}{1} = 0,$$

using L'Hospital's rule.

Similarly, for $b_n$, define

$$f(x) = a^{1/x}, \quad g(x) = \ln f(x) = \frac{\ln a}{x}.$$

Clearly,

$$\lim_{x \to \infty} g(x) = 0,$$

thus $\lim_{x \to \infty} f(x) = 1$ and $b_n \to 1$. □

---

**Proposition 17.7** (Limit of powers with base between 0 and 1). *If $0 < \theta < 1$, then $\theta^n \to 0$.*

---

*Proof.* Define $f(x) = \theta^x$ and $g(x) = \ln f(x) = x \ln \theta$. Since $\ln \theta < 0$ (because $\theta \in (0,1)$), we have

$$\lim_{x \to \infty} g(x) = -\infty.$$

Therefore,

$$\lim_{x \to \infty} f(x) = 0,$$

and thus $\theta^n \to 0$. □

---

**Proposition 17.8** (Limit of the sequence $(1 + \frac{k}{n})^n$). *The sequence $a_n = (1 + \frac{1}{n})^n$ converges to the number e, and more generally,*

$$a_n = \left(1 + \frac{k}{n}\right)^n \to e^k$$

*for every $k \in \mathbb{R}$. Moreover, the sequence $a_n = (1 + \frac{k}{n})^n$ is increasing for every $n > -k$. Similarly, if $b_n \to +\infty$, then the sequence*

$$a_n = \left(1 + \frac{k}{b_n}\right)^{b_n}$$

*converges to $e^k$.*

---

*Proof.* Although the sequence $a_n$ is well-defined for every $n \in \mathbb{N}$ and $k \in \mathbb{R}$, the corresponding function

$$f(x) = \left(1 + \frac{k}{x}\right)^x, \quad x \in (-k, \infty)$$

is well-defined only for $x > -k$.
Define

$$g(x) = \ln f(x) = x \ln\left(1 + \frac{k}{x}\right).$$

We compute

$$\lim_{x \to \infty} g(x) = \lim_{x \to \infty} \frac{\ln(1 + \frac{k}{x})}{\frac{1}{x}}.$$

Setting $y = \frac{1}{x}$, this becomes

$$\lim_{y \to 0} \frac{\ln(1 + ky)}{y} = \lim_{y \to 0} \frac{\frac{k}{1+ky}}{1} = k,$$

using L'Hospital's rule.

Therefore,

$$\lim_{x \to \infty} f(x) = e^k,$$

and so $a_n \to e^k$.

Next, we prove $a_n$ is increasing for $n \geq -k$. The derivative of

$$g(x) = x \ln\left(1 + \frac{k}{x}\right)$$

is

$$g'(x) = \ln\left(1 + \frac{k}{x}\right) - \frac{k}{x(1 + \frac{k}{x})}.$$

To show $g$ is increasing, we need $g' \geq 0$.

The second derivative is

$$g''(x) = -\frac{k^2}{x^3(1 + \frac{k}{x})^2},$$

which is negative on $(0, \infty) \cap (-k, \infty)$. Hence, $g'$ is decreasing on this interval and tends to zero as $x \to \infty$, so $g' \geq 0$ there, implying $g$ is increasing and so is $a_n$ for $n > -k$.

Finally, for $a_n = (1 + \frac{k}{b_n})^{b_n}$ with $b_n \to +\infty$, define

$$f(y) = \left(1 + \frac{k}{y}\right)^y,$$

setting $y = b_n$. Since $y \to +\infty$ as $n \to \infty$, the function $f(y)$ converges to $e^k$ as $y \to \infty$. □

**Example 17.2.** Let

$$a_n = n\left(\sqrt[n]{a} - 1\right)$$

with $a > 0$. Show that

$$\lim_{n \to \infty} a_n = \ln a.$$

*Solution.* Consider the function

$$f(x) = x(\sqrt[x]{a} - 1) = \frac{a^y - 1}{y},$$

setting $y = \frac{1}{x}$. Compute the limit as $y \to 0$:

$$\lim_{y \to 0} \frac{a^y - 1}{y}.$$

Using L'Hospital's rule, this limit is $\ln a$. Hence,

$$a_n \to \ln a. \qquad \square$$

## 17.6 Cauchy sequence

This section explores the concept of *Cauchy sequences*, which provide an intrinsic criterion for convergence without reference to a specific limit. We establish the equivalence between convergence and the Cauchy property in the real numbers, and prove that every sequence admits a monotonic subsequence—a key result that underpins the convergence of bounded sequences. These foundational theorems illustrate the completeness of $\mathbb{R}$, a central theme in real analysis.

**Definition 17.6** (Cauchy sequence). A sequence $\{a_n\}$ is called a *Cauchy sequence* if for every $\varepsilon > 0$ there exists a natural number $N$ such that for all $m, n \geq N$,

$$|a_n - a_m| \leq \varepsilon.$$

**Theorem 17.11** (Every sequence contains a monotone subsequence). *Every sequence* $(a_n)$ *contains a monotone subsequence.*

*Proof.* Start from the first term $a_1$. Attempt to construct an increasing subsequence. If impossible, then there exists a greatest term, say $a_{m_1^*}$, in the sequence. Starting from $a_{m_1^*+1}$, attempt again to construct an increasing subsequence. If still impossible, there exists another greatest term $a_{m_2^*}$ after $m_1^*$ with

$$a_{m_1^*} \geq a_{m_2^*}.$$

Repeating this process, either an increasing subsequence is constructed or the terms

$$a_{m_k^*}, \quad k = 1, 2, \ldots,$$

form a decreasing subsequence. $\qquad \square$

> **Theorem 17.12** (Characterization of convergent sequences). *A sequence $\{a_n\}$ converges if and only if it is a Cauchy sequence.*

*Proof.* Suppose $a_n \to a$. For every $\varepsilon > 0$, there exists $N$ such that for all $n, m > N$,

$$|a_m - a| < \frac{\varepsilon}{2} \quad \text{and} \quad |a_n - a| < \frac{\varepsilon}{2}.$$

Therefore,

$$|a_n - a_m| \leq |a_n - a| + |a_m - a| < \varepsilon,$$

so $(a_n)$ is Cauchy.

Conversely, if $(a_n)$ is Cauchy, it is bounded. Indeed, for some $\varepsilon > 0$ and sufficiently large $N$,

$$|a_n - a_m| < \varepsilon \quad \text{for all } n, m > N,$$

which implies

$$a_m - \varepsilon < a_n < a_m + \varepsilon \quad \text{for all } n > N \text{ and some fixed } m > N.$$

Every bounded sequence contains a monotone subsequence which converges (Theorems 17.11 and 17.9), say to $a$. We now show the whole sequence converges to $a$.

For every $\varepsilon > 0$, there exists $N$ such that for all $n, m \geq N$,

$$|a_n - a_m| < \frac{\varepsilon}{2}.$$

Since $a_{m_n} \to a$, there exists $M$ such that

$$m_M > N, \quad |a_{m_n} - a| < \frac{\varepsilon}{2} \quad \text{for all } n \geq M.$$

Then, for all $n \geq N$,

$$|a_n - a| \leq |a_n - a_{m_M}| + |a_{m_M} - a| < \frac{\varepsilon}{2} + \frac{\varepsilon}{2} = \varepsilon.$$

Hence $(a_n)$ converges to $a$. $\qquad\square$

> **Example 17.3** (Arithmetic and geometric means). Let $a_n \to a$ and define the *arithmetic mean*
>
> $$b_n = \frac{a_1 + \cdots + a_n}{n}.$$
>
> We will prove that $b_n \to a$. For simplicity, assume $a = 0$. Since $a_n \to 0$, for every $\varepsilon > 0$ there exists $m$ such that

$$|a_n| < \frac{\varepsilon}{2} \quad \text{for all } n > m.$$

Then

$$|b_n| \leq \frac{|a_1 + \cdots + a_m|}{n} + \frac{|a_{m+1}| + \cdots + |a_n|}{n} < \frac{k}{n} + \frac{\varepsilon}{2},$$

where $k = |a_1 + \cdots + a_m|$. Choosing $N > m$ such that

$$\frac{k}{n} < \frac{\varepsilon}{2} \quad \text{for all } n > N,$$

it follows that $|b_n| < \varepsilon$, so $b_n \to 0$. If $a \neq 0$, apply the same argument to $a_n - a$.

Next, consider the *geometric mean*

$$c_n = \sqrt[n]{a_1 \cdots a_n}.$$

Assume $a_n > 0$ for all $n$ and $a_n \to a \neq 0$. Then

$$\ln c_n = \frac{\ln a_1 + \cdots + \ln a_n}{n}.$$

Since $\ln a_n \to \ln a$, the sequence $\ln c_n$ is the arithmetic mean of $\ln a_n$ and thus $\ln c_n \to \ln a$, or equivalently $c_n \to a$.

---

**Example 17.4.** Suppose $(a_n)$ is monotone and

$$\lim_{n \to \infty} \frac{a_1 + \cdots + a_n}{n} = a.$$

Prove that $\lim_{n \to \infty} a_n = a$.

*Solution.* Assume without loss of generality that $(a_n)$ is increasing and $a_n \geq 0$. Set

$$s_n = \frac{a_1 + \cdots + a_n}{n}.$$

Since $s_n$ converges, it is bounded by some $M > 0$, i. e., $|s_n| \leq M$. Also,

$$2n s_{2n} = a_1 + \cdots + a_{2n} \geq a_n + \cdots + a_{2n} \geq (n+1)a_n \geq na_n,$$

so

$$0 \leq a_n \leq 2s_{2n} \leq 2M.$$

Thus, $(a_n)$ is increasing and bounded above, so converges to some $b \in \mathbb{R}$. If $b \neq a$, then by the previous example,

$$s_n \to b,$$

a contradiction. Hence $b = a$. $\qquad\square$

## 17.6.1 Exercises

### True or false

Decide whether each statement is **true** or **false**. Justify your answer.

1. If $a_n$ is bounded and monotone, then it converges.

    **Answer:** _____

2. If $\lim_{n \to \infty} a_n = 0$, then $a_n$ is bounded.

    **Answer:** _____

3. Every Cauchy sequence is convergent.

    **Answer:** _____

4. The sequence $a_n = \sqrt[n]{n}$ converges to 1.

    **Answer:** _____

5. If $a_n$ is increasing and $b_n$ is decreasing with $a_n \leq b_n$ for all $n$, then both sequences converge to the same limit.

    **Answer:** _____

### Multiple choice questions

Choose the correct option (A, B, C, or D) for each question.

1. What is the value of $\lim_{n \to \infty} (1 + \frac{1}{n})^n$?

    A. 0     B. 1

    C. $e$     D. Does not exist

    **Answer:** _____

2. Let $a_n = n(\sqrt[n]{3} - 1)$. What is $\lim_{n \to \infty} a_n$?

    A. 0     B. $\ln 3$

    C. 3     D. Does not exist

    **Answer:** _____

3. Which of the following sequences diverges to infinity?

    A. $a_n = \frac{n^2}{n+1}$

    B. $a_n = \frac{(-1)^n}{n}$

    C. $a_n = \sin(n)$

    D. $a_n = \sqrt[n]{n}$

    **Answer:** _____

4. Which of the following statements is always true for any two sequences $(a_n)$, $(b_n)$?

    A. If $a_n \to a$ and $b_n \to b$, then $a_n + b_n \to a + b$

    B. If $a_n \to a$, then $|a_n| \to |a|$

    C. If $a_n \leq b_n$ for all $n$, then $\lim a_n \leq \lim b_n$

    D. All of the above

    **Answer:** _____

5. Suppose $a_n$ is increasing and bounded above by 5. What can we say about $\lim_{n \to \infty} a_n$?

    A. It equals 5             B. It is less than 5

    C. It is greater than 5     D. Cannot be determined

    **Answer:** _____

## Matching exercise

Match each sequence with its behavior.

1. $a_n = \theta^n$ with $0 < \theta < 1$    A. Converges to 0
2. $a_n = \sqrt[n]{n}$                    B. Converges to 1
3. $a_n = (1 + \frac{1}{n})^n$       C. Converges to $e$
4. $a_n = \sqrt[n]{a}$ with $a > 0$    D. Converges to 1
5. $a_n = \frac{n}{n+1}$             E. Converges to 1

**Answers: 1._____    2._____    3._____    4._____    5._____**

## Fill in the blanks

Fill in each blank with the most appropriate word or phrase.

1. A sequence $(a_n)$ is called a _____ if for every $\varepsilon > 0$, there exists $N$ such that $|a_n - a_m| < \varepsilon$ for all $n, m > N$.

   **Answer:** _____

2. The sequence $a_n = \frac{1}{n}$ is an example of a sequence that _____ to 0.

   **Answer:** _____

3. If $f(x)$ is continuous and $a_n \to a$, then $f(a_n) \to f(_____)$.

   **Answer:** _____

4. For a function $f(x) = x^{1/x}$, $\lim_{x \to \infty} f(x) = _____$.

   **Answer:** _____

5. If $a_n$ is increasing and bounded above, then $\lim_{n \to \infty} a_n = _____$.

   **Answer:** _____

## Theoretical and computational problems

Solve each problem step-by-step.

1. Prove that the sequence $a_n = \sqrt[n]{n}$ converges to 1 using the squeeze theorem.
2. Show that the sequence $a_n = \theta^n$ with $0 < \theta < 1$ converges to 0.
3. Use the monotone convergence theorem to prove that the sequence $a_n = (1 + \frac{1}{n})^n$ converges.
4. Prove that the arithmetic mean of a convergent sequence also converges to the same limit.
5. Let $a_n = \frac{1}{n}$ and $b_n = (-1)^n$. Compute $\lim \sup(a_n + b_n)$ and $\lim \inf(a_n + b_n)$.
6. Define a function $f(x) = e^{-x}$ and consider the sequence $a_n = (-1)^n n$. Compute $\lim \sup f(a_n)$ and $\lim \inf f(a_n)$, and verify the relation $\lim \sup f(a_n) = f(\lim \inf a_n)$.
7. Prove that for any convergent sequence $(a_n)$, the geometric mean $\sqrt[n]{a_1 a_2 \cdots a_n}$ also converges to the same limit.
8. Let $a_n = \frac{1}{n}$ and $b_n = \frac{1}{n^2}$. Show that the product sequence $a_n b_n$ converges to 0.
9. Let $a_n = \frac{(-1)^n}{n}$. Show that the arithmetic mean $\frac{a_1 + \cdots + a_n}{n}$ converges to 0.
10. Prove that every sequence has a monotonic subsequence.

### Python-based exercises

Use Python (sympy, numpy, matplotlib) to tackle the following:

1. Write a Python script to compute the first 100 terms of the sequence $a_n = (1 + \frac{1}{n})^n$ and plot the values to observe the convergence to $e$.

2. Use sympy to find the limit of the sequence $a_n = \sqrt[n]{n}$ as $n \to \infty$.

3. Plot the sequence $a_n = \frac{n}{n+1}$ and its corresponding function $f(x) = \frac{x}{x+1}$ on the same graph. Observe how the sequence approximates the function.

4. Simulate the sequence $a_n = \sqrt[n]{a}$ for various values of $a > 0$ and show numerically that it converges to 1.

5. Implement a numerical check for the inequality $\lim \sup(a_n + b_n) \le \lim \sup a_n + \lim \sup b_n$ using sequences like $a_n = \sin(n)$ and $b_n = \cos(n)$.

## 17.7 Sequences defined recursively

Recursive sequences are an essential topic in mathematical analysis and appear naturally in many problems of pure and applied mathematics. Rather than being defined explicitly, these sequences are determined by a recurrence relation of the form

$$a_{n+1} = f(a_n),$$

where $f$ is a real-valued function and $a_1 \in \mathbb{R}$ is a given initial value. Despite their seemingly simple structure, recursive sequences exhibit rich behavior and often converge to limits that satisfy fixed-point equations of the form $L = f(L)$.

The primary goal in studying such sequences is to determine whether they converge, and if so, to identify the limit. In many cases, convergence is established by proving that the sequence is monotonic and bounded. Once these properties are shown, powerful theorems from real analysis ensure the existence of a limit.

This section explores common techniques for analyzing recursive sequences, including:

– Identifying potential limits by solving fixed-point equations.
– Proving monotonicity using mathematical induction.
– Establishing bounds to ensure convergence.

We illustrate these ideas through a series of carefully selected examples, including nonlinear iterations, root approximations, and sequences involving radicals and rational functions. These examples highlight both the theoretical principles and practical strategies for analyzing recursive behavior.

A class of sequences is defined recursively, for example

$$a_{n+1} = f(a_n), \quad a_1 = a \in \mathbb{R},$$

where $f(\cdot)$ is a continuous function. Note that if the sequence converges to a limit $L$, then

$$L = f(L),$$

since $f$ is continuous. Typically, we start by assuming convergence, solve

$$f(x) - x = 0$$

to find possible fixed points, and then attempt to prove convergence to one of these fixed points by showing monotonicity and boundedness (above if increasing, below if decreasing). Then the sequence converges to the least upper bound or greatest lower bound accordingly.

The usual steps are:

(i)   Find possible limits $L$ by solving $L = f(L)$.

(ii)  Prove inductively that the sequence is monotone.

(iii) If increasing, find the smallest possible limit bounding the sequence above; if decreasing, find the largest possible limit bounding it below.

---

**Example 17.5.** Consider the sequence defined by the recursion

$$a_{n+1} = (\sqrt{2})^{a_n}, \quad a_0 = 1.$$

We will prove that it is increasing. Indeed, for $n = 1$,

$$a_1 = \sqrt{2} \leq (\sqrt{2})^{\sqrt{2}} = a_2.$$

Assume $a_k \leq a_{k+1}$. Then

$$a_{k+1} = (\sqrt{2})^{a_k} \leq (\sqrt{2})^{a_{k+1}} = a_{k+2}.$$

Next, prove $a_n \leq 2$ for all $n$. For $n = 1$, $\sqrt{2} < 2$. Assume $a_k \leq 2$. Then

$$a_{k+1} = (\sqrt{2})^{a_k} \leq (\sqrt{2})^2 = 2.$$

Let $L$ be the limit of the sequence (so $L \leq 2$). Then

$$\lim_{n \to \infty} a_{n+1} = L, \quad \lim_{n \to \infty} a_n = L,$$

thus

$$L = (\sqrt{2})^L.$$

This implies $L = 2$.

To see this, study

$$f(x) = (\sqrt{2})^x - x,$$

which has a minimum at

$$x = -\frac{\ln \ln \sqrt{2}}{\ln \sqrt{2}}.$$

Moreover, $f$ is decreasing on

$$\left( -\infty, -\frac{\ln \ln \sqrt{2}}{\ln \sqrt{2}} \right),$$

and increasing on

$$\left(-\frac{\ln \ln \sqrt{2}}{\ln \sqrt{2}}, \infty\right).$$

It has zeros at $x = 2$ and $x = 4$. Since the sequence is bounded above by 2, it converges to 2.

**Example 17.6.** Let $0 \le a \le 1$. Define the sequence by

$$a_1 = 0, \quad a_{n+1} = a_n + \frac{1}{2}(a - a_n^2).$$

Prove that $(a_n)$ is increasing and that $\lim a_n = \sqrt{a}$.

*Solution.* We will prove that the sequence is increasing and bounded above, hence convergent.

Indeed,

$$a_{n+1} = \frac{1}{2}((a + 1) - (1 - a_n)^2) \le \frac{a + 1}{2} \le 1,$$

so it is bounded above.

To prove it is increasing, we use induction. Since

$$a_2 = \frac{a}{2} \ge 0 = a_1,$$

the base case holds. Assume $a_1 \le a_2 \le \cdots \le a_n$. Then

$$a_{n+1} - a_n = \left(a_n + \frac{1}{2}(a - a_n^2)\right) - \left(a_{n-1} + \frac{1}{2}(a - a_{n-1}^2)\right)$$

$$= (a_n - a_{n-1}) - \frac{1}{2}(a_n^2 - a_{n-1}^2)$$

$$= (a_n - a_{n-1})\left(1 - \frac{a_n + a_{n-1}}{2}\right) \ge 0.$$

Since the sequence is increasing and bounded above, it converges to some limit $b$. Taking limits in the recursion,

$$b = b + \frac{1}{2}(a - b^2),$$

which implies

$$b = \sqrt{a}. \qquad \square$$

**Example 17.7.** Consider the sequence defined recursively by

$$a_{n+1} = \sqrt{a_n + 1}, \quad a_1 = 1.$$

Prove that
$$\lim a_n = \frac{1 + \sqrt{5}}{2}.$$

*Solution.* First, we prove the sequence is increasing. Since
$$a_2 = \sqrt{2} \geq a_1 = 1,$$

and assuming $a_1 \leq \cdots \leq a_n$, then
$$a_{n+1} - a_n = \sqrt{a_n + 1} - \sqrt{a_{n-1} + 1} = \frac{a_n - a_{n-1}}{\sqrt{a_n + 1} + \sqrt{a_{n-1} + 1}} \geq 0.$$

Next, prove boundedness above. Since the sequence is increasing,
$$a_n \leq \sqrt{a_n + 1} \quad \implies \quad a_n^2 - a_n - 1 \leq 0.$$

Factoring,
$$\left( a_n - \frac{1 - \sqrt{5}}{2} \right)\left( a_n - \frac{1 + \sqrt{5}}{2} \right) \leq 0,$$

so
$$\frac{1 - \sqrt{5}}{2} \leq a_n \leq \frac{1 + \sqrt{5}}{2}.$$

Hence, $(a_n)$ is bounded and increasing, so converges to $b$ satisfying
$$b = \sqrt{b + 1}.$$

Solving,
$$b = \frac{1 + \sqrt{5}}{2}. \qquad \qquad \square$$

**Example 17.8.** Consider the sequence defined by
$$a_{n+1} = \frac{n + 1}{2n} a_n, \quad a_1 = 1.$$

Using the ratio test,
$$\lim_{n \to \infty} \frac{a_{n+1}}{a_n} = \lim_{n \to \infty} \frac{n + 1}{2n} = \frac{1}{2} < 1,$$

hence $a_n \to 0$.

**Example 17.9.** Let $a_{n+1} = 3a_n + 4$ with $a_1 = 1$. Since $a_n > 0$ for all $n$, we apply the ratio test:

$$\liminf_{n\to\infty} \frac{a_{n+1}}{a_n} = 3 + \liminf_{n\to\infty} \frac{4}{a_n} \geq 3.$$

Thus, $a_n \to +\infty$.

**Example 17.10.** Consider

$$a_{n+1} = \frac{1}{2}\left(a_n + \frac{2}{a_n}\right), \quad a_1 = 2.$$

If the limit $L$ exists, it satisfies

$$L = \frac{1}{2}\left(L + \frac{2}{L}\right),$$

which implies $L = \pm\sqrt{2}$.

We check monotonicity: $a_2 = \frac{3}{2} < a_1 = 2$, so try to prove the sequence is decreasing

$$a_{n+1} \leq a_n \iff \frac{1}{2}\left(a_n + \frac{2}{a_n}\right) \leq a_n \iff a_n \geq \sqrt{2}.$$

Inductively, if $a_n \geq \sqrt{2}$, then

$$a_{n+1} = \frac{1}{2}\left(a_n + \frac{2}{a_n}\right) \geq \sqrt{2},$$

since

$$a_n^2 - 2\sqrt{2}a_n + 2 = (a_n - \sqrt{2})^2 \geq 0.$$

Thus, $(a_n)$ is bounded below by $\sqrt{2}$ and decreasing, so converges to $\sqrt{2}$.

**Example 17.11.** Consider

$$a_{n+1} = \frac{a_n + 2}{a_n + 3}, \quad a_1 = 1.$$

Since $0 \leq a_n \leq 1$, the sequence is bounded. Suppose it converges to $L$. Then

$$L = \frac{L+2}{L+3} \implies L = \sqrt{3} - 1,$$

the other root being negative and discarded.

We prove $a_n \geq \sqrt{3} - 1$ by induction. For $n = 1$ true. Assume $a_n \geq \sqrt{3} - 1$, then

$$a_{n+1} \geq \sqrt{3} - 1 \iff a_n + 2 \geq \sqrt{3}a_n + 3\sqrt{3} - a_n - 3,$$

or equivalently,

$$(2 - \sqrt{3})a_n \geq 3\sqrt{3} - 5,$$

which holds by the induction hypothesis.

To prove decreasingness,

$$a_{n+1} \leq a_n \quad \Longleftrightarrow \quad \frac{a_n + 2}{a_n + 3} \leq a_n \quad \Longleftrightarrow \quad a_n^2 + 2a_n - 2 \geq 0,$$

which is true for $a_n \geq \sqrt{3} - 1$.

Therefore, the sequence is decreasing and bounded below, hence convergent to $\sqrt{3} - 1$.

## 17.8 Important lemmas on sequences

The analysis of sequences and series is foundational to real analysis and calculus. Many important results in mathematics, including the theory of integration and approximation, rely on a deep understanding of how sequences behave and under what conditions they converge.

This section presents a range of powerful tools and convergence theorems that are essential when dealing with infinite sums and recursively defined sequences. We explore both classical convergence criteria—such as the ratio and root tests—and deeper analytical results like the dominated convergence theorem, the monotone convergence lemma, and Fatou's lemma.

In addition, we investigate key strategies for proving convergence, including bounding techniques, asymptotic comparisons, and recursive analysis. These methods are applied to a variety of examples that demonstrate how to evaluate limits, determine convergence, and understand the behavior of both explicitly and implicitly defined sequences.

Throughout, we emphasize the logical structure of convergence arguments, including:

- Using *limsup* and *liminf* to capture asymptotic growth.
- Applying convergence tests to sequences involving exponentials, factorials, and logarithms.
- Proving convergence of recursively defined sequences by identifying fixed points and showing monotonicity and boundedness.
- Employing measure-theoretic ideas such as domination and monotonicity to exchange limits and summation.

These theorems and techniques form the analytical backbone for advanced topics in integration, differential equations, and probability theory, making them indispensable for any serious study of mathematics.

**Lemma 17.1** (Dominated convergence theorem). *Let $a_{nk}$ be real numbers for $n, k \in \mathbb{N}$, and let $b_k$ be a nonnegative sequence such that*

$$\sum_{k=1}^{\infty} b_k < \infty$$

and

$$|a_{nk}| \le b_k.$$

If

$$\lim_{n \to \infty} a_{nk} = a_k,$$

then

$$\lim_{n \to \infty} \sum_{k=1}^{\infty} a_{nk} = \sum_{k=1}^{\infty} a_k.$$

*Solution.* Let $\varepsilon > 0$ and $M = M(\varepsilon)$ such that

$$\sum_{k=M+1}^{\infty} b_k < \frac{\varepsilon}{3}.$$

Since $|a_{nk}| \le b_k$, also $|a_k| \le b_k$, and

$$\sum_{k=M+1}^{\infty} |a_{nk}| + \sum_{k=M+1}^{\infty} |a_k| \le \frac{2\varepsilon}{3}.$$

For sufficiently large $n$,

$$\sum_{k=1}^{M} |a_{nk} - a_k| \le \frac{\varepsilon}{3}.$$

Thus,

$$\left| \sum_{k=1}^{\infty} a_{nk} - \sum_{k=1}^{\infty} a_k \right| \le \sum_{k=1}^{M} |a_{nk} - a_k| + \sum_{k=M+1}^{\infty} |a_{nk}| + \sum_{k=M+1}^{\infty} |a_k| \le \varepsilon. \qquad \square$$

**Lemma 17.2** (Monotone convergence lemma). *Let $a_{nk} \ge 0$ for $n, k \in \mathbb{N}$ such that for each $k$, the sequence $(a_{nk})_{n \ge 1}$ is nondecreasing and*

$$\lim_{n \to \infty} a_{nk} = a_k \le \infty.$$

*Then*

$$\lim_{n \to \infty} \sum_{k=1}^{\infty} a_{nk} = \sum_{k=1}^{\infty} a_k.$$

*Solution.* If $\sum_{k=1}^{\infty} a_k < \infty$, this follows from the dominated convergence lemma. If $\sum_{k=1}^{\infty} a_k = \infty$, let $A > 0$ and choose $M = M(A)$ such that

$$\sum_{k=1}^{M} a_k \ge 2A.$$

For sufficiently large $n$,

$$\sum_{k=1}^{\infty}(a_k - a_{nk}) \le A,$$

so

$$\sum_{k=1}^{\infty} a_{nk} \ge \sum_{k=1}^{M} a_{nk} + \sum_{k=1}^{M}(a_k - a_{nk}) \ge 2A - A = A. \qquad \square$$

**Lemma 17.3** (Fatou's lemma). *Let $a_{nk} \ge 0$ for $n, k \in \mathbb{N}$. Then*

$$\sum_{k=1}^{\infty} \liminf_{n\to\infty} a_{nk} \le \liminf_{n\to\infty} \sum_{k=1}^{\infty} a_{nk}.$$

*Solution.* Define

$$z_{nk} = \inf\{a_{nk}, a_{n+1,k}, \ldots\}.$$

Then $z_{nk}$ is an increasing sequence converging to $\liminf_{n\to\infty} a_{nk}$. By the monotone convergence lemma,

$$\sum_{k=1}^{\infty} \liminf_{n\to\infty} a_{nk} = \lim_{n\to\infty} \sum_{k=1}^{\infty} z_{nk}.$$

Since $z_{nk} \le a_{nk}$,

$$\sum_{k=1}^{\infty} z_{nk} \le \sum_{k=1}^{\infty} a_{nk},$$

and taking limits,

$$\lim_{n\to\infty} \sum_{k=1}^{\infty} z_{nk} \le \liminf_{n\to\infty} \sum_{k=1}^{\infty} a_{nk}. \qquad \square$$

## 17.9 Python and AI

---

**AI Request**

Compute the lim inf and lim sup of the sequence

$$a_n = \frac{n^2}{1+n^2}\sin\frac{2\pi n}{3}.$$

Give me the Latex code of the solution step-by-step. Write a Python code to determine the above so I can compare the results.

---

## 17.10 Exercises

### True or false
Decide whether each statement is **true** or **false**. Justify your answer.
1.  Every recursive sequence defined by $a_{n+1} = f(a_n)$ converges if $f$ is continuous.
    **Answer:** _____
2.  If a sequence $(a_n)$ is bounded and satisfies $a_{n+1} = f(a_n)$ with $f$ increasing, then it converges.
    **Answer:** _____
3.  A recursive sequence always has a fixed point.
    **Answer:** _____
4.  The sequence $a_{n+1} = \sqrt{a_n + 1}$ with $a_1 = 1$ converges to $\frac{1+\sqrt{5}}{2}$.
    **Answer:** _____
5.  If $|a_{n+1}| < |a_n|$ for all $n$, then $a_n \to 0$.
    **Answer:** _____

### Multiple choice questions
Choose the correct option (A, B, C, or D) for each question.
1.  What is the limit of the sequence $a_{n+1} = (\sqrt{2})^{a_n}$ with $a_1 = 1$?
    A. 1    B. $\sqrt{2}$
    C. 2    D. Does not converge
    **Answer:** _____
2.  Let $a_{n+1} = \frac{1}{2}(a_n + \frac{2}{a_n})$ with $a_1 = 2$. What is the limit?
    A. 1    B. $\sqrt{2}$
    C. 2    D. Does not exist
    **Answer:** _____
3.  Which of the following tests can be used to show that $a_n = \frac{n!}{n^n} \to 0$?
    A. Ratio test
    B. Root test
    C. Comparison test
    D. All of the above
    **Answer:** _____
4.  For which value of $k$ does the sequence $a_n = (1 + \frac{k}{n})^n$ converge to $e^3$?
    A. $k = 1$    B. $k = 2$
    C. $k = 3$    D. $k = e$
    **Answer:** _____
5.  Suppose $a_n \to L$ and $b_n = \frac{a_1 + \cdots + a_n}{n}$. Then $b_n \to$
    A. $L$    B. 0
    C. 1    D. Does not exist
    **Answer:** _____

## Matching exercise

Match each recursive sequence with its behavior or limit.

1. $a_{n+1} = \sqrt{a_n + 1}$, $a_1 = 1$     A. Converges to $\frac{1+\sqrt{5}}{2}$
2. $a_{n+1} = \frac{1}{2}(a_n + \frac{2}{a_n})$, $a_1 = 2$     B. Converges to $\sqrt{2}$
3. $a_{n+1} = (\sqrt{2})^{a_n}$, $a_1 = 1$     C. Converges to 2
4. $a_{n+1} = \frac{a_n+2}{a_n+3}$, $a_1 = 1$     D. Converges to $\sqrt{3} - 1$
5. $a_{n+1} = 3a_n + 4$, $a_1 = 1$     E. Diverges to $+\infty$

**Answers: 1.**_____   **2.**_____   **3.**_____   **4.**_____   **5.**_____

## Fill in the blank

Fill in each blank with the most appropriate word or phrase.

1. A recursive sequence $a_{n+1} = f(a_n)$ may converge to a _____ of the function $f$.

   **Answer:** _____

2. The sequence $a_n = (1 + \frac{1}{n})^n$ converges to _____.

   **Answer:** _____

3. If $a_{n+1} = \frac{a_n+2}{a_n+3}$ with $a_1 = 1$, then the sequence is _____ and _____.

   **Answer:** _____, _____

4. The sequence $a_n = \frac{n}{2^n}$ converges to _____.

   **Answer:** _____

5. A sequence defined recursively by $a_{n+1} = \frac{n+1}{2n} a_n$ with $a_1 = 1$ converges to _____.

   **Answer:** _____

## Theoretical and computational problems

Solve each problem step-by-step.

1. Consider the sequence defined recursively by $a_{n+1} = \sqrt{a_n + 1}$ with $a_1 = 1$. Prove that it is increasing and bounded above, and find its limit.
2. Let $a_{n+1} = \frac{1}{2}(a_n + \frac{2}{a_n})$ with $a_1 = 2$. Show that the sequence is decreasing and bounded below by $\sqrt{2}$, and compute its limit.
3. Use the ratio test to prove that $a_n = \frac{n!}{n^n} \to 0$ as $n \to \infty$.
4. Define $a_n = (1 + \frac{1}{n^2})^n$. Prove that $a_n \to 1$ using the fact that $\lim_{n\to\infty}(1 + \frac{1}{n^2})^{n^2} = e$.
5. Let $a_n = \frac{c_k n^k + \cdots + c_1 n + c_0}{b_r n^r + \cdots + b_1 n + b_0}$. Find the limit as $n \to \infty$ depending on the degrees $k$ and $r$.
6. Study the convergence of the sequence $a_n = \sqrt{n+1} - \sqrt{n}$. Justify your answer.
7. Prove that the arithmetic mean of a convergent sequence also converges to the same limit.
8. Use the squeeze theorem to prove that $a_n = \frac{\sin(n!e^n)}{n+1} \to 0$.
9. Show that the sequence $a_n = \ln(n + 1) - \ln n$ converges to 0.
10. Prove that the sequence $a_n = \frac{a^n}{n!} \to 0$ for any real number $a$.

11. Prove the following limits:

$$\lim_{n\to\infty} (1 + n + n^2)^{\frac{1}{n}} = 1, \quad \lim_{n\to\infty} (2^n + 3^n)^{\frac{1}{n}} = 3, \quad \lim_{n\to\infty} \frac{\ln n}{n} = 0.$$

12. Prove that for every $a \in \mathbb{R}$

$$\lim_{n\to\infty} \frac{a^n}{n!} = 0.$$

13. Determine convergence of the sequences

$$\frac{n}{2^n}, \quad \frac{2^n}{n^2}, \quad \frac{n!}{n^n}.$$

14. Study the sequence defined recursively by

$$0 < a_1 < 1, \quad a_{n+1} = 1 - \sqrt{1 - a_n}.$$

15. Find the limit of the sequence

$$a_n = \left(1 + \frac{1}{n^2}\right)^n.$$

16. Compute the limit of the sequence

$$a_n = \frac{c_k n^k + \cdots + c_1 n + c_0}{b_r n^r + \cdots + b_1 n + b_0},$$

where $c_0, \ldots, c_k, b_0, \ldots, b_r \in \mathbb{R}$ for all $r, k \in \mathbb{N}$.

17. Find the limits of the sequences

$$a_n = \sqrt{n+1} - \sqrt{n}, \quad a_n = \ln(n+1) - \ln n, \quad a_n = \frac{\sqrt{n}\sin(n!e^n)}{n+1}.$$

## Python-based exercises

Use Python (sympy, numpy, matplotlib) to tackle the following:

1. Write a Python script to simulate the recursive sequence $a_{n+1} = \sqrt{a_n + 1}$ with $a_1 = 1$ and plot the first 50 terms. Observe the convergence behavior.
2. Plot the sequence $a_n = (1 + \frac{1}{n})^n$ for $n = 1$ to 100 and observe how it approaches $e$.
3. Simulate the sequence $a_n = \frac{1}{2}(a_n + \frac{2}{a_n})$ starting from $a_1 = 2$ and observe its convergence to $\sqrt{2}$.
4. Use sympy to compute the limit of the sequence $a_n = \frac{n!}{n^n}$ as $n \to \infty$.
5. Implement a numerical check for the inequality $\limsup(a_n + b_n) \le \limsup a_n + \limsup b_n$ using sequences like $a_n = \sin(n)$ and $b_n = \cos(n)$.

# 18 Indefinite integral

## Contents

The indefinite integral, or antiderivative, is the reverse process of differentiation. Given a function $f(x)$, an antiderivative $F(x)$ satisfies $F'(x) = f(x)$. All antiderivatives of $f$ differ by a constant, so the general indefinite integral is written as $\int f(x)\, dx = F(x) + c$. Finding indefinite integrals is a key tool in calculus, enabling us to solve problems involving motion, area, and accumulation. Throughout this chapter, we will develop the main rules and techniques for integration—including substitution, integration by parts, and the decomposition of rational functions—supported by examples and exercises to build both understanding and skill.

## 18.1 Introduction

The concept of the indefinite integral, or antiderivative, is a cornerstone of calculus and mathematical analysis. While the derivative measures the instantaneous rate of change of a function, the indefinite integral performs the opposite task: it reconstructs a function whose rate of change is given. More precisely, given a function $f(x)$, an antiderivative is a function $F(x)$ such that $F'(x) = f(x)$. The collection of all antiderivatives of $f$ forms a family differing only by an additive constant, reflecting the fact that the derivative of any constant is zero. This is why, whenever we write $\int f(x)\, dx$, we represent not a single function but a whole family, written as $F(x) + c$ for $c \in \mathbb{R}$.

The process of finding an indefinite integral is known as integration, and it serves as a fundamental tool for solving problems in mathematics, physics, engineering, probability, and beyond. For instance, knowing an object's velocity function, one can find its displacement via integration. Likewise, the area under a curve, accumulation of quantities, and solutions to a wide variety of differential equations all rely on the properties of indefinite integrals.

A crucial property of the indefinite integral is its linearity: the integral of a sum is the sum of the integrals, and constants can be factored out. This property, together with a set of basic integration rules and techniques, forms the basis for almost all calculations in integral calculus. Some of the most important techniques include substitution (the

https://doi.org/10.1515/9783112228289-018

reverse of the chain rule for derivatives), integration by parts (derived from the product rule), and the decomposition of rational functions into partial fractions.

However, not every function can be integrated in closed form in terms of elementary functions; the antiderivative of $e^{-x^2}$, for example, is famously nonelementary. This reality leads to the development of special functions and numerical methods for integration.

Throughout this chapter, we will explore the fundamental principles of the indefinite integral, develop essential techniques for computing antiderivatives, and solve a wide variety of examples and exercises. Special attention will be paid to the role of substitutions, integration by parts, the treatment of rational functions, and the link between indefinite and definite integrals—a connection that culminates in the fundamental theorem of calculus. The goal is to build both conceptual understanding and practical fluency, preparing you to tackle real-world problems and more advanced mathematical theory.

## 18.2 Definitions

In this chapter, we will work on the problem of finding an *antiderivative* of a function $f$, that is, a function $F$ such that $F' = f$.

If we assume there exist two such functions, $F_1$ and $F_2$, then by Corollary 8.1, it follows that $F_1 = F_2 + c$, where $c \in \mathbb{R}$ is an arbitrary constant. Therefore, the set of all antiderivatives of a function $f$ is actually of the form $\{F + c \mid c \in \mathbb{R}\}$, where $F$ is any one antiderivative.

We will see below, when we study the definite integral, how such an $F$ (apart from being an antiderivative) relates to $f$.

We denote the set $\{F + c \mid c \in \mathbb{R}\}$ of all antiderivatives of a function $f$ by

$$\int f(x)\,dx$$

and we will explain (once we state and prove the first fundamental theorem of calculus) why we use the integral symbol. We will call this the `indefinite integral` of $f$.

Next, we present an important theorem regarding the `linearity of the indefinite integral`, which is extremely useful in computations.

**Theorem 18.1** (Linearity of indefinite integral). *Let $f$ and $g$ be continuous functions and let $a, b \in \mathbb{R}$. Then the following holds:*

$$\int \big(af(x) + bg(x)\big)\,dx = a \int f(x)\,dx + b \int g(x)\,dx.$$

*Proof.* Let $H$ be an antiderivative of $h(x) = af(x) + bg(x)$, and let $F$ and $G$ be antiderivatives of $f$ and $g$, respectively. Then we have

$$H'(x) = af(x) + bg(x) = aF'(x) + bG'(x) = (aF(x) + bG(x))',$$

where we used the linearity of differentiation.

Then we have

$$H(x) = aF(x) + bG(x) + c, \quad \text{where } c \in \mathbb{R} \text{ is an arbitrary constant.}$$

Therefore, in terms of indefinite integrals, we can write

$$\int (af(x) + bg(x))\, dx = a \int f(x)\, dx + b \int g(x)\, dx. \qquad \square$$

As we will see below, every continuous function has an antiderivative, but we do not necessarily know its functional form in closed form (if such a form even exists). A classic example is the function $e^{-x^2}$, which is continuous, but its antiderivative does not have a "closed-form" expression.

### 18.2.1 Integration by parts

In this section, we will describe a method for finding an antiderivative of a function $f$, or equivalently, for computing the indefinite integral of $f$. The technique we will develop is called integration by parts, and it is based on the following theorem (the integration-by-parts theorem).

---

**Theorem 18.2** (Integration by parts). *If $f$ and $g$ are continuously differentiable on an interval $I \subseteq \mathbb{R}$, then*

$$\int f(x)g'(x)\, dx = f(x)g(x) - \int f'(x)g(x)\, dx$$

*on $I$.*

---

*Proof.* Since $f$ and $g$ are differentiable, so is their product, and in particular $(fg)' = f'g + fg'$.

Therefore, an antiderivative of $(fg)'$ is $fg$, which means it is also an antiderivative of $f'g + fg'$. Let $H_1(x)$ be an antiderivative of $f'g$, and let $H_2(x)$ be an antiderivative of $fg'$. Then we have

$$(fg)'(x) = f'(x)g(x) + f(x)g'(x) = H_1'(x) + H_2'(x) = (H_1(x) + H_2(x))'$$

where we used the linearity of differentiation.

We conclude that

$$(fg)(x) = H_1(x) + H_2(x) + c$$

or, in terms of integrals,

$$\int f(x)g'(x)\, dx = f(x)g(x) - \int f'(x)g(x)\, dx. \qquad \square$$

**Example 18.1** (The derivative of ln $|f(x)|$). Let $f(x)$ be a function that does not vanish, and suppose that ln $|f(x)|$ is well-defined. Then the derivative of the latter is $\frac{f'(x)}{f(x)}$. Therefore, the integral of $\frac{f'(x)}{f(x)}$ is the family of functions ln $|f(x)| + c$, that is,

$$\int \frac{f'(x)}{f(x)} = \ln|f(x)| + c.$$

**Example 18.2.** Compute the following indefinite integrals:

$$\int \ln x\, dx, \quad \int xe^x\, dx, \quad \int x^2 e^x\, dx, \quad \int x \sin x\, dx.$$

*Solution.*

- $\int \ln x\, dx = \int \ln x \cdot (x)'\, dx = x \ln x - \int (\ln x)' \cdot x\, dx = x \ln x - \int dx = x \ln x - x + c.$
- $\int xe^x\, dx = \int x(e^x)'\, dx = xe^x - \int e^x \cdot (x)'\, dx = xe^x - e^x + c.$
- $\int x^2 e^x\, dx = \int x^2 (e^x)'\, dx = x^2 e^x - 2 \int xe^x\, dx = x^2 e^x - 2xe^x + 2e^x + c.$
- $\int x \sin x\, dx = -\int x(\cos x)'\, dx = -x \cos x + \int \cos x\, dx = -x \cos x + \sin x + c.$

□

**Example 18.3.** Prove the following identities:

$$I = \int \cos(ax + b)e^{kx}\, dx$$

$$= \frac{k \cos(ax + b) + a \sin(ax + b)}{a^2 + k^2} e^{kx} + c, \quad a^2 + k^2 \neq 0,$$

$$J = \int \sin(ax + b)e^{kx}\, dx$$

$$= \frac{k \sin(ax + b) - a \cos(ax + b)}{a^2 + k^2} e^{kx} + c, \quad a^2 + k^2 \neq 0.$$

*Solution.* We apply integration by parts twice:

$$I = \frac{1}{k} \int \cos(ax + b)(e^{kx})'\, dx$$

$$= \frac{1}{k} \cos(ax + b)e^{kx} - \frac{1}{k} \int (\cos(ax + b))' e^{kx}\, dx$$

$$= \frac{1}{k} \cos(ax + b)e^{kx} + \frac{a}{k^2} \int \sin(ax + b)(e^{kx})'\, dx$$

$$= \frac{1}{k} \cos(ax + b)e^{kx} + \frac{a}{k^2} \sin(ax + b)e^{kx} - \frac{a}{k^2} \int (\sin(ax + b))' e^{kx}\, dx$$

$$= \frac{k \cos(ax + b) + a \sin(ax + b)}{k^2} e^{kx} - \frac{a^2}{k^2} I.$$

Solving for $I$, we obtain the desired result. The computation for $J$ is similar and is left as an exercise.                                                    □

## 18.3  Integration by substitution

In this section, we will study the method of substitution. Our goal is to transform an integral with respect to $x$ into an integral with respect to a new variable $u$, by setting $u = \phi(x)$. This new integral should be easier to compute. Symbolically, we write $du = \phi'(x)\,dx$ and perform the substitution accordingly.

**Example 18.4.** Compute the integral $\int x \cos(x^2 + 1)\,dx$.

*Solution.* Let $u = x^2 + 1$, so (symbolically) $du = 2x\,dx$. Then the original integral becomes

$$\frac{1}{2}\int \cos(u)\,du.$$

Thus,

$$\frac{1}{2}\int \cos u\,du = \frac{1}{2}\sin u + c.$$

Substituting back $u = x^2 + 1$, we obtain

$$\int x \cos(x^2 + 1)\,dx = \frac{1}{2}\sin(x^2 + 1) + c.$$

We have made the substitution $du = 2x\,dx$, but we have not defined rigorously the $du$. Therefore one has to verify the result by differentiating the function $\frac{1}{2}\sin(x^2 + 1) + c$. $\qquad\square$

**Example 18.5.** Compute the integral $\int \sin^2 x \cos x\,dx$.

*Solution.* Let $u = \sin x$, so that $du = \cos x\,dx$. Then the original integral becomes

$$\int u^2\,du = \frac{u^3}{3} + c.$$

Substituting back $u = \sin x$, we get

$$\int \sin^2 x \cos x\,dx = \frac{\sin^3 x}{3} + c.$$

Verify the result or give a rigorous mathematical meaning to $du$! $\qquad\square$

**Example 18.6.** Compute the integral $\int \cos^2 x\,dx$.

*Solution.* From trigonometry, we know that $\cos^2 x = \frac{1+\cos 2x}{2}$. Therefore,

$$\int \cos^2 x \, dx = \frac{1}{2} \int dx + \frac{1}{2} \int \cos(2x) \, dx = \frac{x}{2} + \frac{1}{2} \int \cos(2x) \, dx.$$

Let $u = 2x$, so that $du = 2dx$. Then

$$\int \cos(2x) \, dx = \frac{1}{2} \int \cos u \, du = \frac{1}{2} \sin u + c.$$

Substituting back $u = 2x$, we get

$$\int \cos^2 x \, dx = \frac{x}{2} + \frac{1}{4} \sin(2x) + c.$$

Verify the result. □

In the following examples, we set $t = \phi(x)$ and solve for $x$ in order to proceed.

**Example 18.7.** Compute the integral

$$\int \frac{dx}{1 + \sqrt{x}}.$$

*Solution.* Let $t = \sqrt{x}$, so that $x = t^2$. Then (symbolically), $dx = 2t \, dt$. Therefore, the original integral becomes

$$\int \frac{2t}{1+t} \, dt.$$

We simplify the integrand

$$\int \frac{2t}{1+t} \, dt = 2 \int \frac{1+t-1}{1+t} \, dt$$
$$= 2 \int dt - 2 \int \frac{1}{1+t} \, dt$$
$$= 2t - 2 \ln|1 + t| + c$$
$$= G(t) + c.$$

Substituting back $t = \sqrt{x}$, we obtain the final result

$$\int \frac{dx}{1 + \sqrt{x}} = 2\sqrt{x} - 2\ln(1 + \sqrt{x}) + c, \quad \text{for } x \geq 0.$$

Verify the result. □

**Example 18.8.** Compute the integral

$$\int \sqrt{4 - x^2} \, dx.$$

*Solution.* Let $x = \phi(t) = 2 \sin t$, so that $dx = 2 \cos t \, dt$. Then $t = \phi^{-1}(x) = \arcsin(\frac{x}{2})$. Substituting, we have

$$\int \sqrt{4 - x^2} \, dx = \int \sqrt{4 - (2 \sin t)^2} \cdot 2 \cos t \, dt$$

$$= \int \sqrt{4 - 4 \sin^2 t} \cdot 2 \cos t \, dt$$

$$= \int 2 \cos t \cdot 2 \cos t \, dt$$

$$= \int 4 \cos^2 t \, dt.$$

Using the identity $\cos^2 t = \frac{1 + \cos(2t)}{2}$, we get

$$\int 4 \cos^2 t \, dt = 4 \int \frac{1 + \cos(2t)}{2} \, dt$$

$$= 2 \int 1 \, dt + 2 \int \cos(2t) \, dt$$

$$= 2t + \sin(2t) + c.$$

Now, recall that $t = \arcsin(\frac{x}{2})$, and

$$\sin(2t) = 2 \sin t \cos t = 2 \sin\left(\arcsin\left(\frac{x}{2}\right)\right) \cos\left(\arcsin\left(\frac{x}{2}\right)\right) = \frac{x}{2} \sqrt{4 - x^2},$$

since

$$\cos\left(\arcsin\left(\frac{x}{2}\right)\right) = \sqrt{1 - \left(\frac{x}{2}\right)^2} = \sqrt{1 - \frac{x^2}{4}}.$$

Therefore, the original integral becomes

$$\int \sqrt{4 - x^2} \, dx = 2 \arcsin\left(\frac{x}{2}\right) + \frac{x}{2} \sqrt{4 - x^2} + c.$$

Verify the result. □

### 18.3.1 Exercises

### True or false

Decide whether each statement is **true** or **false**. Justify your answer.

1. If $F$ is an antiderivative of $f$, then so is $F + 5$.

   **Answer:** _____

2. The indefinite integral $\int f(x)dx$ represents a unique function.

   **Answer:** _____

3. The linearity of integration states that $\int (f(x) + g(x))dx = \int f(x)dx + \int g(x)dx$.

   **Answer:** _____

4. Integration by parts is derived from the product rule for differentiation.

   **Answer:** _____

5. The integral $\int e^{-x^2} dx$ has a closed-form antiderivative in terms of elementary functions.

   **Answer:** _____

### Multiple choice questions

Choose the correct option (A, B, C, or D) for each question.

1. Which of the following is *not* a valid antiderivative of $2x$?

   A. $x^2 + 1$   B. $x^2 - 3$

   C. $2x^2$      D. $x^2 + \pi$

   **Answer:** _____

2. What is $\int 3e^x dx$?

   A. $3e^x + C$   B. $e^{3x} + C$

   C. $\frac{3}{2}e^{2x} + C$   D. $3xe^x + C$

   **Answer:** _____

3. Which method should be used to compute $\int x \cos x dx$?

   A. Substitution         B. Linearity

   C. Integration by parts   D. Direct formula

   **Answer:** _____

4. If we let $u = x^2 + 1$, what is $du$ in terms of $dx$?

   A. $du = 2xdx$   B. $du = xdx$

   C. $du = 2dx$    D. $du = 2x^2 dx$

   **Answer:** _____

5. What is the correct expression for $\int \frac{f'(x)}{f(x)} dx$, assuming $f(x) > 0$?

   A. $\ln |f'(x)| + C$   B. $\frac{1}{f(x)} + C$

   C. $\ln |f(x)| + C$     D. $f(x) \ln |f(x)| + C$

   **Answer:** _____

### Matching exercise

Match each integration technique or concept with its correct description.

1. Indefinite integral      A. $\int f(g(x))g'(x)dx = \int f(u)du$
2. Integration by parts     B. Represents the family of all antiderivatives
3. $u$-Substitution          C. $\int u dv = uv - \int v du$
4. Linearity                D. $\int (af(x) + bg(x))dx = a \int f(x)dx + b \int g(x)dx$
5. Antiderivative            E. A function $F$ such that $F' = f$

Answers: 1._____   2._____   3._____   4._____   5._____

## Fill in the blanks

Fill in each blank with the most appropriate word or phrase.

1. The constant of integration is usually denoted by _____.

   Answer: _____

2. The indefinite integral of $x^n$ (for $n \neq -1$) is _____.

   Answer: _____

3. The formula $\int f'(x)g(x)dx = f(x)g(x) - \int f(x)g'(x)dx$ is known as _____.

   Answer: _____

4. When using substitution, if $u = \sin x$, then $du = $ _____.

   Answer: _____

5. The integral $\int \cos(3x)dx$ can be computed using _____, and the re-

   sult is _____.

   Answer: _____

## Theoretical and computational problems

Solve each problem step-by-step.

1. Prove that if $F$ and $G$ are both antiderivatives of $f$ on an interval, then $F(x) = G(x) + C$ for some constant $C$.
2. Use the linearity of integration to compute

$$\int (4x^3 - 2\sin x + e^x)dx.$$

3. Compute $\int x \ln x dx$ using integration by parts. Clearly state your choice of $u$ and $dv$.
4. Compute $\int xe^{2x} dx$ using integration by parts.
5. Evaluate $\int \frac{2x}{x^2+1} dx$ using the method of substitution.
6. Evaluate $\int \sin^3 x \cos x dx$ using substitution.
7. Compute $\int x^2 \cos x dx$ using integration by parts (you may need to apply it twice).
8. Verify that the function $F(x) = \frac{1}{2}\sin(2x) + x$ is an antiderivative of $f(x) = \cos(2x) + 1$.
9. Derive the formula $\int \frac{f'(x)}{f(x)} dx = \ln|f(x)| + C$ by considering the derivative of $\ln|f(x)|$.
10. Explain why the function $e^{-x^2}$ has an antiderivative (by a fundamental theorem of calculus), but we cannot express it in terms of elementary functions.

## Python-based exercises

Use Python (sympy, numpy, matplotlib) to tackle the following:

1. Use sympy to symbolically compute the indefinite integrals of $x^2$, $\sin(2x)$, and $\frac{1}{x}$. Print the results and include the constant of integration.
2. Write a Python function that uses sympy to verify the result of integration by parts for $\int xe^x dx$. Differentiate the result and confirm it equals $xe^x$.
3. Plot the function $f(x) = e^{-x^2}$ and its numerical antiderivative $F(x) = \int_0^x e^{-t^2} dt$ on the interval $[-2, 2]$ using numpy and matplotlib. Use scipy.integrate.quad for numerical integration.
4. Use sympy to compute $\int \cos^2 x dx$ and confirm that the result matches $\frac{x}{2} + \frac{\sin(2x)}{4} + C$.
5. Write a Python script that takes a user-input expression and attempts to compute its indefinite integral using sympy. If the integral cannot be found, print an appropriate message.

## 18.4 Integration of rational functions

In this section, we explore the method of partial fraction decomposition, a powerful technique for integrating rational functions—expressions formed by the ratio of two polynomials. When the degree of the numerator is strictly less than that of the denominator, such functions can be broken down into sums of simpler fractions, called partial fractions, which can be integrated term by term. This algebraic tool plays a central role in calculus, complex analysis, and applications such as the Fourier transform. Through definitions, propositions, and detailed examples, we demonstrate how to perform these decompositions effectively.

Let the rational function $R(x) = \frac{P(x)}{Q(x)}$, where $P$ and $Q$ are polynomials, and suppose that the degree of $P$ is less than the degree of $Q$.

### 18.4.1 Polynomial division

Polynomial division is a method used to divide one polynomial by another. The process is similar to long division of numbers. Given two polynomials $P(x)$ (the dividend) and $D(x)$ (the divisor), the goal is to find two polynomials $Q(x)$ (the quotient) and $R(x)$ (the remainder) such that

$$P(x) = D(x) \cdot Q(x) + R(x),$$

where the degree of $R(x)$ is less than the degree of $D(x)$.

The steps for polynomial division are as follows:
1. Arrange both polynomials in descending order of powers.
2. Divide the leading term of the dividend by the leading term of the divisor to get the first term of the quotient.
3. Multiply the entire divisor by this term and subtract the result from the dividend.
4. Repeat the process with the new polynomial obtained until the degree of the remainder is less than the degree of the divisor.

## Step-by-step polynomial division with diagrams

In this section, we will demonstrate the step-by-step process of dividing the polynomial $P(x) = x^3 + 2x^2 - 3x + 4$ by $D(x) = x - 1$. The division is illustrated with diagrams at each step.

### Step 1: Divide the leading terms

We divide the leading term of the dividend $(x^3)$ by the leading term of the divisor $(x)$:

$$\frac{x^3}{x} = x^2.$$

The first term of the quotient is $x^2$. Multiply $x^2$ by the divisor $x - 1$ and subtract the result from the dividend.

$$
\begin{array}{r|l}
 & x^2 \\
\hline
x - 1 & x^3 + 2x^2 - 3x + 4 \\
 & \quad -\quad x^3 - x^2 \\
 & \overline{\phantom{xxxxxxxx}}
\end{array}
$$

### Step 2: Subtract and bring down the next term

Subtract $x^3 - x^2$ from $x^3 + 2x^2$, which gives

$$(x^3 + 2x^2) - (x^3 - x^2) = 3x^2.$$

Bring down the next term $(-3x)$.

$$
\begin{array}{r|l}
 & x^2 \\
\hline
x - 1 & x^3 + 2x^2 - 3x + 4 \\
 & \quad -\quad x^3 - x^2 \\
 & \overline{\phantom{xxxxxx}} \\
 & \quad 3x^2 - 3x \\
 & \qquad +4
\end{array}
$$

### Step 3: Divide the new leading terms

Divide the leading term of the new dividend $(3x^2)$ by the leading term of the divisor $(x)$:

$$\frac{3x^2}{x} = 3x.$$

The next term of the quotient is $3x$. Multiply $3x$ by the divisor $x - 1$ and subtract the result.

$$
\begin{array}{r|l}
 & x^2 + 3x \\
\hline
x - 1 & x^3 + 2x^2 - 3x + 4 \\
 & -\quad x^3 - x^2 \\
\hline
 & 3x^2 - 3x \\
 & +4 \\
 & -\quad 3x^2 - 3x \\
\hline
\end{array}
$$

### Step 4: Subtract and finalize the remainder

Subtract $3x^2 - 3x$ from $3x^2 - 3x$, which gives

$$(3x^2 - 3x) - (3x^2 - 3x) = 0.$$

Bring down the final term (+4). Since there are no more terms to divide, the remainder is 4.

$$
\begin{array}{r|l}
 & x^2 + 3x \\
\hline
x - 1 & x^3 + 2x^2 - 3x + 4 \\
 & -\quad x^3 - x^2 \\
\hline
 & 3x^2 - 3x \\
 & +4 \\
 & -\quad 3x^2 - 3x \\
\hline
 & 4 \ \text{Remainder: } 4 \\
\end{array}
$$

### Final result

The quotient is $Q(x) = x^2 + 3x$, and the remainder is $R(x) = 4$. Thus, the result of the division can be written as

$$x^3 + 2x^2 - 3x + 4 = (x - 1)(x^2 + 3x) + 4.$$

### 18.4.2 Decomposing rational functions

In this section, we will see how to integrate such expressions. The key idea is to decompose them into a sum of simpler fractions and then integrate term by term.

**Definition 18.1** (Simple fractions). Rational functions of the form

$$\frac{A}{(x-a)^k}, \quad \frac{Ax+B}{(ax^2+bx+c)^k}, \quad k \in \mathbb{N}, \quad b^2 - 4ac < 0,$$

are called simple fractions or partial fractions.

With the following proposition, which we state without proof, we will see how rational functions can be decomposed into simple fractions.

**Proposition 18.1** (Decomposing rational fractions). *Every rational function, as described above, can be uniquely decomposed into a sum of simple fractions.*
  *Specifically,*
(1)  *If $a$ is a real root of $Q(x)$ of multiplicity $k$, i. e., $Q(x) = (x-a)^k Q_1(x)$ with $Q_1(a) \neq 0$, then*

$$\frac{P(x)}{Q(x)} = \frac{A_k}{(x-a)^k} + \frac{A_{k-1}}{(x-a)^{k-1}} + \cdots + \frac{A_1}{x-a} + \frac{P_1(x)}{Q_1(x)},$$

  *where $A_1, A_2, \ldots, A_k$ are real constants.*
(2)  *If $a + bi$ is a complex root of $Q(x)$ of multiplicity $\mu$, i. e., $Q(x) = [(x-a)^2 + b^2]^\mu Q_2(x)$ with $Q_2(a+bi) \neq 0$, then*

$$\frac{P(x)}{Q(x)} = \frac{A_\mu x + B_\mu}{[(x-a)^2 + b^2]^\mu} + \frac{A_{\mu-1}x + B_{\mu-1}}{[(x-a)^2 + b^2]^{\mu-1}} + \cdots + \frac{A_1 x + B_1}{(x-a)^2 + b^2} + \frac{P_2(x)}{Q_2(x)}.$$

A particularly useful result, especially in the context of the Fourier transform and the computation of constants is the following (see [5]).

**Proposition 18.2.** *If $P(x)$ and $Q(x)$ are as above, then*

$$\frac{P(x)}{Q(x)} = \frac{a_{11}}{x - x_1} + \frac{a_{12}}{(x - x_1)^2} + \cdots + \frac{a_{1n_1}}{(x - x_1)^{n_1}}$$

$$+ \frac{a_{21}}{x - x_2} + \frac{a_{22}}{(x - x_2)^2} + \cdots + \frac{a_{2n_2}}{(x - x_2)^{n_2}}$$

$$+ \cdots$$

$$+ \frac{a_{m1}}{x - x_m} + \frac{a_{m2}}{(x - x_m)^2} + \cdots + \frac{a_{mn_m}}{(x - x_m)^{n_m}},$$

*where $x_1, x_2, \ldots, x_m$ are the $m$ distinct (generally complex) roots of $Q(x)$, with multiplicities $n_1, n_2, \ldots, n_m$, respectively, and $a_{ij}$ are unique coefficients.*

**Remark 23.** If we want to decompose the rational function $\frac{P(x)}{Q(x)}$, where $P(x)$ and $Q(x)$ are polynomials, we must first perform polynomial division if the degree of the numerator is greater than that of the denominator. Thus, we will have

$$\frac{P(x)}{Q(x)} = \pi(x) + \frac{v(x)}{Q(x)},$$

where $\pi(x)$ is some polynomial and $v(x)$ is the remainder of the division, which will have degree less than the degree of $Q(x)$.

If the polynomial $Q(x)$ has real coefficients, then it can be written in the form

$$Q(x) = (x - x_1)^{r_1}(x - x_2)^{r_2} \cdots (x - x_m)^{r_m}(x - z_1)^{l_1}(x - \bar{z}_1)^{l_1} \cdots (x - z_k)^{l_k}(x - \bar{z}_k)^{l_k},$$

where $x_1, \ldots, x_m$ are real roots of multiplicity $r_1, \ldots, r_m$, and $z_1, \ldots, z_k$ are complex roots of multiplicity $l_1, \ldots, l_k$, respectively. Note that complex roots come in conjugate pairs with the same multiplicity.

According to the previous theorem, there exist appropriate constants such that

$$\frac{v(x)}{Q(x)} = \frac{a_{1,1}}{(x - x_1)^{r_1}} + \cdots + \frac{a_{1,r_1}}{x - x_1}$$

$$+ \frac{a_{2,1}}{(x - x_2)^{r_2}} + \cdots + \frac{a_{2,r_2}}{x - x_2}$$

$$\vdots$$

$$+ \frac{a_{m+1,1}}{(x - z_1)^{l_1}} + \cdots + \frac{a_{m+k,l_1}}{x - z_1}$$

$$+ \frac{a_{m+1,1}}{(x - \bar{z}_1)^{l_1}} + \cdots + \frac{a_{m+k,l_1}}{x - \bar{z}_1}$$

$$\vdots$$

We can compute the coefficients (in red) that are above the highest-degree terms (e. g., above $(x - x_1)^{r_1}$) by multiplying both sides by this denominator and substituting $x = x_1$ (or taking the limit as $x \to x_1$). In this way, we can compute all the coefficients above the highest powers, such as $a_{1,1}, a_{2,1}, \ldots$. Then we move those fractions to the left-hand side and compute the constants over the remaining highest powers, and so on.

**Example 18.9.** We will decompose the rational function

$$\frac{1}{x^4 + 1}$$

into partial fractions.

First, we compute the roots of the denominator. Since $x^4 = -1$, we have $x^2 = \pm i$, and therefore the roots are

$$x_{1,2} = \pm\sqrt{i}, \quad x_{3,4} = \pm\sqrt{-i}.$$

Since $e^{i\pi/2} = i$, we have $\sqrt{i} = e^{i\pi/4} = \cos(\pi/4) + i\sin(\pi/4) = \frac{1}{\sqrt{2}} + i\frac{1}{\sqrt{2}}$. Therefore, the denominator factors as

$$x^4 + 1 = \left(x - \frac{1}{\sqrt{2}} - i\frac{1}{\sqrt{2}}\right)\left(x - \frac{1}{\sqrt{2}} + i\frac{1}{\sqrt{2}}\right)$$

$$\cdot \left(x + \frac{1}{\sqrt{2}} + i\frac{1}{\sqrt{2}}\right)\left(x + \frac{1}{\sqrt{2}} - i\frac{1}{\sqrt{2}}\right)$$

$$= \left( \left( x - \frac{1}{\sqrt{2}} \right)^2 + \frac{1}{2} \right) \left( \left( x + \frac{1}{\sqrt{2}} \right)^2 + \frac{1}{2} \right).$$

Thus, the partial fraction decomposition takes the form

$$\frac{1}{x^4 + 1} = \frac{Ax + B}{(x - \frac{1}{\sqrt{2}})^2 + \frac{1}{2}} + \frac{Cx + D}{(x + \frac{1}{\sqrt{2}})^2 + \frac{1}{2}}.$$

**Example 18.10.** Decompose the rational function into partial fractions:

$$\frac{P(x)}{Q(x)} = \frac{2x - 1}{(x - 1)(x + 2)(x - 3)}.$$

*Solution.* Since the denominator has three distinct real roots, there exist constants $A$, $B$, and $\Gamma$ such that

$$\frac{2x - 1}{(x - 1)(x + 2)(x - 3)} = \frac{A}{x - 1} + \frac{B}{x + 2} + \frac{\Gamma}{x - 3},$$

for every $x \in \mathbb{R}$.

To compute the constants $A$, $B$, and $\Gamma$, we proceed as follows. Multiply both sides of the equation by $(x - 1)$, then evaluate the limit as $x \to 1$ to isolate $A$:

$$A = \frac{2 \cdot 1 - 1}{(1 + 2)(1 - 3)} = -\frac{1}{6}.$$

Similarly, we find

$$B = -\frac{1}{3}, \quad \Gamma = \frac{1}{2}.$$

Hence,

$$\frac{2x - 1}{(x - 1)(x + 2)(x - 3)} = -\frac{1}{6}\frac{1}{x - 1} - \frac{1}{3}\frac{1}{x + 2} + \frac{1}{2}\frac{1}{x - 3}. \qquad \square$$

**Example 18.11.** Decompose the following rational function into partial fractions:

$$\frac{x + 2}{(x + 1)(x^2 + 1)^2}.$$

*Solution.* The function has one real root and one complex root with multiplicity two (since $x^2 + 1$ has two complex conjugate roots, and the factor $(x^2 + 1)^2$ appears).

Thus, the function decomposes as

$$\frac{x + 2}{(x + 1)(x^2 + 1)^2} = \frac{A}{x + 1} + \frac{Bx + \Gamma}{(x^2 + 1)^2} + \frac{\Delta x + E}{x^2 + 1}.$$

To find the constants, we make all terms have the same denominator (i. e., multiply through by the denominator on the left-hand side), and then equate numerators.

Matching coefficients from both sides of the resulting polynomial identity will yield a system of five equations with five unknowns.

Compute the partial fraction decomposition using Proposition 18.2.  □

**Example 18.12.** Let us decompose the following into partial fractions:

$$\frac{1}{(x-1)(x^2+1)}.$$

The denominator has one real root, $x = 1$, and two complex conjugate roots, $x = i$ and $x = -i$, each with multiplicity one. Thus,

$$\frac{1}{(x-1)(x^2+1)} = \frac{A}{x-1} + \frac{B}{x-i} + \frac{C}{x+i}.$$

Multiplying both sides by $(x - 1)$ and letting $x = 1$, we find

$$A = \frac{1}{2}.$$

To find $B$ and $C$, we multiply both sides by $(x - i)$ and $(x + i)$ respectively and substitute $x = i$ and $x = -i$. We obtain

$$B = \frac{1}{(i-1)\cdot 2i}, \quad C = \frac{1}{(i+1)\cdot 2i}.$$

Thus,

$$\frac{1}{(x-1)(x^2+1)} = \frac{1}{2}\cdot\frac{1}{x-1} + \frac{1}{(i-1)2i(x-i)} + \frac{1}{(i+1)2i(x+i)}.$$

Combining the last two terms, we get

$$\frac{1}{(x-1)(x^2+1)} = \frac{1}{2}\cdot\frac{1}{x-1} + \frac{-\frac{1}{2}x-\frac{1}{2}}{x^2+1}.$$

**Example 18.13.** Evaluate the integral

$$\int \frac{3}{2x^2+4x+10}\,dx.$$

*Solution.* We factor out the constant $\frac{3}{2}$, reducing the problem to the integral of $\frac{1}{x^2+2x+5}$. The discriminant of the denominator is $\Delta = 2^2 - 4\cdot 1\cdot 5 = -16$, so the quadratic has no real roots but two complex conjugates: $-1 \pm 2i$.

Our goal is to rewrite the quadratic in the form $(x - a)^2 + b^2$. Completing the square

$$x^2 + 2x + 5 = (x+1)^2 + 2^2.$$

We then use the substitution $x + 1 = 2t$, so that $dx = 2dt$.

Therefore,

$$\int \frac{3}{2x^2 + 4x + 10}\,dx = \frac{3}{4}\arctan\left(\frac{x+1}{2}\right) + c.$$

Verify the result. □

**Example 18.14.** Evaluate the integral

$$I = \int \frac{2x^2 + 2x - 7}{(x-2)(x^2+1)^2}\,dx.$$

*Solution.* We begin by decomposing the rational function into partial fractions:

$$\int \frac{2x^2 + 2x - 7}{(x-2)(x^2+1)^2}\,dx = \int \left( \frac{1}{5} \cdot \frac{1}{x-2} + \frac{x+4}{(x^2+1)^2} - \frac{1}{5} \cdot \frac{x+2}{x^2+1} \right) dx$$

$$= \frac{1}{5}\ln|x-2| - \frac{1}{2} \cdot \frac{1}{x^2+1} + 4\int \frac{dx}{(x^2+1)^2} - \frac{1}{10}\ln(x^2+1) - \frac{2}{5}\arctan x$$

$$= \frac{1}{5}\ln|x-2| + \frac{4x-1}{2(x^2+1)} - \frac{1}{10}\ln(x^2+1) + \frac{8}{5}\arctan x + c$$

$$= \frac{4x-1}{2(x^2+1)} + \frac{1}{10}\ln\left(\frac{(x-2)^2}{x^2+1}\right) + \frac{8}{5}\arctan x + c.$$

Verify the result. □

## 18.4.3 Exercises

### True or false
Decide whether each statement is **true** or **false**. Justify your answer.

1. Every rational function $\frac{P(x)}{Q(x)}$ where $\deg(P) < \deg(Q)$ can be decomposed into a sum of partial fractions.
   Answer: _____

2. The partial fraction decomposition of a rational function with a repeated complex root in the denominator will include terms with denominators raised to increasing powers.
   Answer: _____

3. The expression $\frac{Ax+B}{x^2+1}$ is a valid partial fraction, even though $x^2 + 1$ has no real roots.
   Answer: _____

4. To find the coefficient $A$ in the term $\frac{A}{(x-a)^k}$ for a root of multiplicity $k$, you can multiply both sides of the equation by $(x-a)^k$ and then substitute $x = a$.
   Answer: _____

5. The integral $\int \frac{1}{x^2+1}\,dx$ is equal to $\ln|x^2+1| + C$.
   Answer: _____

## Multiple choice questions

Choose the correct option (A, B, C, or D) for each question.

1. Which of the following is the correct form of the partial fraction decomposition for $\frac{1}{(x-1)(x+2)}$?

A. $\frac{A}{x-1} + \frac{B}{x+2}$   B. $\frac{A}{x-1} + \frac{B}{x-1} + \frac{C}{x+2}$

C. $\frac{A}{(x-1)(x+2)}$   D. $\frac{Ax+B}{(x-1)(x+2)}$

Answer: _____

2. What is the correct decomposition form for $\frac{x}{(x-1)^2(x^2+4)}$?

A. $\frac{A}{x-1} + \frac{B}{(x-1)^2} + \frac{Cx+D}{x^2+4}$   B. $\frac{A}{x-1} + \frac{B}{x^2+4}$

C. $\frac{A}{x-1} + \frac{B}{(x-1)^2} + \frac{C}{x^2+4}$   D. $\frac{A}{(x-1)^2} + \frac{Bx+C}{x^2+4}$

Answer: _____

3. After decomposing $\frac{3x}{(x-1)(x+1)}$, what is the value of the coefficient for $\frac{1}{x-1}$?

A. 1   B. –1

C. $\frac{3}{2}$   D. $-\frac{3}{2}$

Answer: _____

4. To evaluate $\int \frac{1}{x^2+4x+8}dx$, which method should you use first?

A. Partial fractions with real roots   B. Polynomial division

C. Completing the square   D. Integration by parts

Answer: _____

5. What is $\int \frac{2x}{x^2+1}dx$?

A. $\ln(x^2+1) + C$   B. $2\ln|x| + \arctan x + C$

C. $\frac{1}{2}\ln(x^2+1) + C$   D. $2\arctan x + C$

Answer: _____

## Matching exercise

Match each type of factor in the denominator with its corresponding partial fraction term(s).

1. $(x - a)$, multiplicity 1   A. $\frac{Ax+B}{(x^2+bx+c)^2} + \frac{Cx+D}{x^2+bx+c}$

2. $(x - a)^2$   B. $\frac{A}{x-a}$

3. $(x^2 + bx + c), b^2 - 4c < 0$   C. $\frac{A}{(x-a)^2} + \frac{B}{x-a}$

4. $(x^2 + bx + c)^2, b^2 - 4c < 0$   D. $\frac{Ax+B}{x^2+bx+c}$

5. $(x - a)(x^2 + bx + c)$   E. $\frac{A}{x-a} + \frac{Bx+C}{x^2+bx+c}$

Answers: 1._____   2._____   3._____   4._____   5._____

## Fill in the blanks

Fill in each blank with the most appropriate word or phrase.

1. Before performing partial fraction decomposition, if the degree of the numerator is greater than or equal to the degree of the denominator, you must first perform

_____.

Answer: _____

2. For a rational function with a complex conjugate pair of roots $a \pm bi$ of multiplicity one, the corresponding partial fraction term is of the form _____.

   **Answer:** _____

3. The method of _____ is used to rewrite a quadratic $x^2 + bx + c$ with no real roots into the form $(x + d)^2 + e^2$.

   **Answer:** _____

4. The integral $\int \frac{1}{(x-2)^3} dx$ is equal to _____.

   **Answer:** _____

5. To find the coefficient $A$ in the term $\frac{A}{x-3}$ for a simple root, you can multiply the rational function by $(x - 3)$ and evaluate the limit as $x \to$ _____.

   **Answer:** _____

## Theoretical and computational problems

Solve each problem step-by-step.

1. Explain why complex roots of polynomials with real coefficients always come in conjugate pairs, and why this property is crucial for partial fraction decomposition over the real numbers.

2. Decompose the following rational function into partial fractions:

$$\frac{1}{(x - 1)(x + 3)}.$$

3. Decompose the following rational function into partial fractions:

$$\frac{x}{(x - 2)^2}.$$

4. Decompose the following rational function into partial fractions:

$$\frac{1}{x^2 + 4}.$$

5. Evaluate the integral

$$\int \frac{1}{x^2 - 1} dx.$$

   (*Hint*: Factor the denominator and use partial fractions.)

6. Evaluate the integral

$$\int \frac{2x + 3}{x^2 + 4x + 5} dx.$$

   (*Hint*: Split the numerator and complete the square in the denominator.)

7. Evaluate the integral

$$\int \frac{1}{(x-1)(x^2+1)}\,dx.$$

(Use the decomposition from the provided text as a starting point.)

8. Evaluate the integral

$$\int \frac{1}{x^2+6x+10}\,dx.$$

9. Find the partial fraction decomposition of

$$\frac{3x^2+2x-1}{(x+1)^2(x-2)}.$$

10. Evaluate the integral

$$\int \frac{x}{(x^2+1)^2}\,dx.$$

**Python-based exercises**

Use Python (sympy, numpy, matplotlib) to tackle the following:

1. Use sympy to symbolically compute the partial fraction decomposition of $\frac{1}{x^2-1}$ and $\frac{x}{(x-2)^2}$. Print the results.

2. Use sympy to compute the indefinite integral of $\frac{2x+3}{x^2+4x+5}$. Verify the result by differentiating it.

3. Write a Python function that takes a rational function (as a sympy expression) and returns its partial fraction decomposition. Test it on $\frac{3x^2+2x-1}{(x+1)^2(x-2)}$.

4. Plot the function $f(x) = \frac{1}{x^2+1}$ and its antiderivative $F(x) = \arctan(x)$ on the interval $[-5, 5]$. Comment on the relationship between the graph of $f$ and the slope of $F$.

5. Use sympy to decompose $\frac{1}{x^4+1}$ into partial fractions, as shown in the example. Compare your result to the one provided in the text.

## 18.5 Integration of rational functions via substitutions

This section presents a comprehensive toolkit of substitution techniques used to evaluate integrals involving algebraic, trigonometric, exponential, and root expressions. By transforming complex integrands into simpler or rational forms, these substitutions—ranging from Weierstrass and Euler to hyperbolic and root-based methods—enable efficient integration and highlight the interplay between different function classes. Through examples, theory, and practice problems (including Python-assisted exploration), learners build a deep understanding of strategic integral transformations (Table 18.1).

Table 18.1: Common substitutions for indefinite integrals.

| Integral type | Substitution | Purpose |
|---|---|---|
| $\int R(x, \sqrt{a^2 - x^2})dx$ | $x = a\sin t$ or $x = a\cos t$ | Eliminate the square root |
| $\int R(x, \sqrt{a^2 + x^2})dx$ | $x = a\tan t$ or $x = a\sinh t$ | Eliminate the square root |
| $\int R(x, \sqrt{x^2 - a^2})dx$ | $x = a\sec t$ or $x = a\cosh t$ | Eliminate the square root |
| $\int R(\sin x, \cos x)dx$ | $t = \tan(\frac{x}{2})$ | Convert to rational function |
| $\int R(x, \sqrt[n]{\frac{ax+b}{cx+d}})dx$ | $t = \sqrt[n]{\frac{ax+b}{cx+d}}$ | Convert to rational function |
| $\int R(e^x)dx$ | $t = e^x$ | Convert to rational function |
| $\int x^n(ax^m + b)^p dx$ | $u = ax^m + b$ | Simplify the power |
| $\int f(g(x))g'(x)dx$ | $u = g(x)$ | Simplify (reverse chain rule) |
| $\int \frac{f'(x)}{f(x)} dx$ | (Direct application) | Result: $\ln|f(x)| + c$ |

In certain types of integrals, we apply a specific substitution that transforms the original integral into a more manageable form.

Note that we are not seeking the optimal substitution for evaluating a particular indefinite integral, but rather the substitution that will reliably lead to the solution of this entire class of integrals. In this way, we can develop appropriate software that will always yield a solution once the type of indefinite integral has been identified.

In the integrals below, we denote by $R(x, y)$ a rational function in two variables, i. e., a function that is rational in $x$ and $y$.

## 18.5.1  $\int R(x, \sqrt[n]{\frac{ax+b}{cx+e}})dx$

In integrals of the form

$$I = \int R\left(x, \sqrt[n]{\frac{ax+b}{cx+e}}\right)dx,$$

we make the substitution

$$t = \sqrt[n]{\frac{ax+b}{cx+e}},$$

which gives

$$x = \phi(t) = \frac{et^n - b}{-ct^n + a}.$$

Next, we compute $dx = \phi'(t) dt$ and substitute $x$ and $dx$ into the original integral, yielding a new integral with respect to $t$.

More generally, in integrals of the form

$$I = \int R\left(x, \left(\frac{ax+b}{cx+e}\right)^{a_1}, \ldots, \left(\frac{ax+b}{cx+e}\right)^{a_k}\right)dx,$$

where each exponent $a_i = \frac{m_i}{n_i}$ with $m_i, n_i \in \mathbb{N}$ for $i = 1, \ldots, k$, we set

$$t = \sqrt[n]{\frac{ax + b}{cx + e}},$$

where $n$ is the least common multiple of $n_1, \ldots, n_k$.

**Example 18.15.** Evaluate the integral

$$\int \frac{dx}{\sqrt[4]{(x+1)^3(x-1)}}.$$

*Solution.* We begin by rewriting the integrand as

$$\int \frac{dx}{\sqrt[4]{(x+1)^3(x-1)}} = \int \frac{dx}{(x+1) \cdot \sqrt[4]{\frac{x-1}{x+1}}}.$$

Now, we make the substitution

$$t = \sqrt[4]{\frac{x-1}{x+1}} \quad \Rightarrow \quad x = \phi(t) = \frac{1+t^4}{1-t^4}.$$

Then,

$$x + 1 = \frac{2}{1-t^4}, \quad dx = \phi'(t)\, dt = \frac{8t^3}{(t^4-1)^2}\, dt.$$

Substituting into the original integral, we obtain

$$\int \frac{dx}{(x+1) \cdot \sqrt[4]{\frac{x-1}{x+1}}} = \int \frac{8t^3}{\frac{2}{1-t^4} \cdot t} \cdot \frac{1}{(1-t^4)^2}\, dt = \int \frac{4t^2}{1-t^4}\, dt.$$

This rational integrand can be decomposed into partial fractions

$$\frac{4t^2}{1-t^4} = \frac{A}{1-t} + \frac{B}{1+t} + \frac{Ct+D}{1+t^2}$$

for suitable constants $A, B, C,$ and $D$. Solving this yields

$$\int \frac{4t^2}{1-t^4}\, dt = \ln \frac{|1+t|}{|1-t|} - 2\arctan t + c.$$

Returning to the original variable:

$$\int \frac{dx}{\sqrt[4]{(x+1)^3(x-1)}} = \ln \left| \frac{1 + \sqrt[4]{\frac{x-1}{x+1}}}{1 - \sqrt[4]{\frac{x-1}{x+1}}} \right| - 2\arctan\left( \sqrt[4]{\frac{x-1}{x+1}} \right) + C.$$

Since we do not have formally define $dx$, the **proof** that the indefinite integral is indeed the above will come by **verification**! □

## 18.5.2 $\int R(\cos x, \sin x)\, dx$

In integrals of the form

$$\int R(\cos x, \sin x)\, dx,$$

we use the substitution

$$t = \tan\left(\frac{x}{2}\right),$$

which implies

$$x = \phi(t) = 2\arctan t, \quad \phi'(t) = \frac{2}{1+t^2}, \quad \sin x = \frac{2t}{1+t^2}, \quad \cos x = \frac{1-t^2}{1+t^2}.$$

In some special cases, we can apply simpler substitutions depending on the symmetry of the function $R$:

– B1. If $R(\sin x, -\cos x) = -R(\sin x, \cos x)$, then let

$$\sin x = t \quad \Rightarrow \quad x = \arcsin t, \quad \frac{dx}{dt} = \frac{1}{\sqrt{1-t^2}}.$$

– B2. If $R(-\sin x, \cos x) = -R(\sin x, \cos x)$, then let

$$\cos x = t \quad \Rightarrow \quad x = \arccos t, \quad \frac{dx}{dt} = -\frac{1}{\sqrt{1-t^2}}.$$

– B3. If $R(-\sin x, -\cos x) = R(\sin x, \cos x)$, then let

$$\tan x = t \quad \Rightarrow \quad x = \arctan t, \quad \frac{dx}{dt} = \frac{1}{1+t^2}.$$

**Example 18.16.** Evaluate the integrals

$$I_1 = \int \frac{\cos x}{2 + \sin x}\, dx, \quad I_2 = \int \frac{\cos x + 1}{3\sin x + \sin^3 x}\, dx.$$

*Solution.*

– For $I_1$, we notice that the integrand satisfies $R(\sin x, -\cos x) = -R(\sin x, \cos x)$, so we make the substitution

$$\sin x = t \quad \Rightarrow \quad x = \arcsin t, \quad dx = \frac{1}{\sqrt{1-t^2}}\, dt.$$

We also have $\cos x = \sqrt{1-t^2}$. Substituting

$$\int \frac{\cos x}{2 + \sin x}\, dx = \int \frac{\sqrt{1-t^2}}{2+t} \cdot \frac{1}{\sqrt{1-t^2}}\, dt = \int \frac{1}{2+t}\, dt = \ln|2 + t| + c.$$

Returning to $x$, we get

$$\int \frac{\cos x}{2 + \sin x} dx = \ln |2 + \sin x| + c.$$

- For $I_2$, observe that $R(-\sin x, \cos x) = -R(\sin x, \cos x)$, so we use

$$\cos x = t \quad \Rightarrow \quad x = \arccos t, \quad dx = -\frac{1}{\sqrt{1 - t^2}} dt.$$

Now, $\sin x = \sqrt{1 - t^2}$ and

$$\sin^3 x = (1 - t^2)^{3/2}, \quad \cos x + 1 = t + 1.$$

But instead of substituting all expressions symbolically, we notice from the original derivation

$$\int \frac{\cos x + 1}{3 \sin x + \sin^3 x} dx = \int \frac{-1}{(t - 1)(t - 2)(t + 2)} dt.$$

By partial fraction decomposition,

$$= \frac{1}{3} \ln |1 - t| - \frac{1}{4} \ln |2 - t| - \frac{1}{12} \ln |t + 2| + c.$$

Returning to $\cos x = t$,

$$\int \frac{\cos x + 1}{3 \sin x + \sin^3 x} dx = \frac{1}{3} \ln |1 - \cos x| - \frac{1}{4} \ln |2 - \cos x| - \frac{1}{12} \ln |\cos x + 2| + c. \quad \square$$

Verify the result or give a rigorous mathematical meaning to $dx$!

### 18.5.3 $\int R(x, \sqrt{a^2 - x^2}) \, dx$

In integrals of the form

$$\int R(x, \sqrt{a^2 - x^2}) \, dx,$$

we use the substitution

$$x = \phi(t) = a \sin t \quad \Rightarrow \quad t = \arcsin \frac{x}{a}, \quad dx = \phi'(t) \, dt = a \cos t \, dt.$$

**Example 18.17.** Evaluate the integral

$$\int \frac{1}{\sqrt{2 - x^2}} \, dx.$$

*Solution.* We set $x = \phi(t) = \sqrt{2}\sin t$, so that

$$dx = \phi'(t)\,dt = \sqrt{2}\cos t\,dt.$$

Substituting into the integral

$$\int \frac{1}{\sqrt{2-x^2}}\,dx = \int \frac{1}{\sqrt{2-2\sin^2 t}} \cdot \sqrt{2}\cos t\,dt = \int 1\,dt = t.$$

Thus,

$$\int \frac{1}{\sqrt{2-x^2}}\,dx = \arcsin\left(\frac{x}{\sqrt{2}}\right) + c.$$

Verify the result. ☐

### 18.5.4 $\int R(x, \sqrt{bx^2 - a})\,dx$

In integrals of the form

$$\int R(x, \sqrt{bx^2 - a})\,dx,$$

we use the substitution

$$x = \phi(t) = \frac{\sqrt{a}}{\sqrt{b}}\cosh t \quad \Rightarrow \quad t = \cosh^{-1}\left(\frac{\sqrt{b}}{\sqrt{a}}x\right), \quad dx = \phi'(t)\,dt = \frac{\sqrt{a}}{\sqrt{b}}\sinh t\,dt.$$

**Example 18.18.** Evaluate the indefinite integral

$$\int \frac{1}{\sqrt{2x^2 - 1}}\,dx.$$

*Solution.* We make the substitution

$$x = \phi(t) = \frac{1}{\sqrt{2}}\cosh t = \frac{1}{\sqrt{2}} \cdot \frac{e^t + e^{-t}}{2}, \quad t \in (0, +\infty).$$

Then,

$$dx = \phi'(t)\,dt = \frac{1}{\sqrt{2}}\sinh t\,dt, \quad \text{and} \quad t = \ln(\sqrt{2}x + \sqrt{2x^2 - 1}).$$

Substituting into the integral

$$\int \frac{1}{\sqrt{2x^2 - 1}}\,dx = \int \frac{1}{\sqrt{2}}\,dt = \frac{t}{\sqrt{2}}.$$

Therefore

$$\int \frac{1}{\sqrt{2x^2-1}}\,dx = \frac{1}{\sqrt{2}}\ln(\sqrt{2}x + \sqrt{2x^2-1}) + c.$$

If instead we considered $t \in (-\infty, 0)$, then the inverse hyperbolic identity gives

$$t = \ln(\sqrt{2}x - \sqrt{2x^2-1}),$$

and

$$\int \frac{1}{\sqrt{2x^2-1}}\,dx = -\frac{1}{\sqrt{2}}\ln(\sqrt{2}x - \sqrt{2x^2-1}) + c.$$

The minus sign appears because for $t < 0$, the derivative $\sinh t < 0$, so $\sqrt{\sinh^2 t} = -\sinh t$. Note that both antiderivatives are valid and differ by a constant, as expected. Verify the result. □

### 18.5.5 $\int R(x, \sqrt{bx^2 + a})\,dx$

In integrals of the form

$$\int R(x, \sqrt{bx^2 + a})\,dx,$$

we use the substitution

$$x = \phi(t) = \frac{\sqrt{a}}{\sqrt{b}}\sinh t \quad \Rightarrow \quad t = \sinh^{-1}\!\left(\frac{\sqrt{b}}{\sqrt{a}}x\right), \quad t \in \mathbb{R}.$$

**Example 18.19.** Evaluate the indefinite integral

$$\int \frac{1}{\sqrt{2x^2 + 1}}\,dx.$$

*Solution.* We set $x = \phi(t) = \frac{1}{\sqrt{2}}\sinh t$, so

$$dx = \phi'(t)\,dt = \frac{1}{\sqrt{2}}\cosh t\,dt.$$

Substituting into the integral

$$\int \frac{1}{\sqrt{2x^2 + 1}}\,dx = \int \frac{1}{\sqrt{2}}\,dt = \frac{t}{\sqrt{2}}.$$

Returning to the variable $x$, we have

$$\int \frac{1}{\sqrt{2x^2 + 1}}\,dx = \frac{1}{\sqrt{2}}\sinh^{-1}(\sqrt{2}x) = \frac{1}{\sqrt{2}}\ln\!\left(\sqrt{2}x + \sqrt{2x^2 + 1}\right) + c, \quad x \in \mathbb{R}.$$

Verify the result. ☐

### 18.5.6 $\int R(x, \sqrt{ax^2 + bx + c})\,dx$

In integrals of the form

$$\int R(x, \sqrt{ax^2 + bx + c})\,dx,$$

we distinguish the following three cases (see also: Euler substitutions):

– If $a > 0$, we set

$$\sqrt{ax^2 + bx + c} = t \pm \sqrt{a}x.$$

Solving for $x$, we obtain

$$x = \phi(t) = \frac{t^2 - c}{b \mp 2t\sqrt{a}}.$$

If the discriminant is negative, we transform the square root into the form $\sqrt{kt^2 + d}$ and proceed as in earlier cases.

– If $a < 0$ and $c > 0$, we set

$$\sqrt{ax^2 + bx + c} = tx \pm \sqrt{c},$$

leading to

$$x = \phi(t) = \frac{b \mp 2t\sqrt{c}}{t^2 - a}.$$

The inverse is

$$t = \phi^{-1}(x) = \frac{\sqrt{ax^2 + bx + c} \mp \sqrt{c}}{x},$$

which must be extended continuously at $x = 0$. The continuous extension is

$$t = \phi^{-1}(x) = \begin{cases} \frac{\sqrt{ax^2+bx+c} \mp \sqrt{c}}{x}, & x \neq 0, \\[2mm] \frac{b}{2\sqrt{c}}, & x = 0. \end{cases}$$

– If $b^2 - 4ac > 0$ and $r_1$ is a root of the quadratic, we set

$$\sqrt{ax^2 + bx + c} = t(x - r_1).$$

Then,

$$ax^2 + bx + c = a(x - r_1)(x - r_2),$$

and solving for $x$, we find

$$x = \phi(t) = \frac{ar_2 - t^2 r_1}{a - t^2}.$$

**Example 18.20.** Evaluate the indefinite integrals

$$\int \frac{dx}{x\sqrt{x^2 - x + 1}}, \quad \int \frac{dx}{\sqrt{-2x^2 + x + 1}}, \quad \int \frac{x + 1}{\sqrt{x^2 - 3x + 2}}\, dx.$$

*Solution.*

–  For the first integral, we set

$$t = \sqrt{x^2 - x + 1} + x \quad \Rightarrow \quad x = \phi(t) = \frac{1 - t^2}{1 - 2t}, \quad dx = 2 \cdot \frac{t^2 - t + 1}{(1 - 2t)^2}\, dt.$$

Substituting into the integral, we get

$$\int -\frac{2}{1 - t^2}\, dt = -\ln\left|\frac{t + 1}{t - 1}\right|.$$

Back-substituting:

$$\int \frac{dx}{x\sqrt{x^2 - x + 1}} = \ln\left|\frac{\sqrt{x^2 - x + 1} + x - 1}{\sqrt{x^2 - x + 1} + x + 1}\right| + c.$$

What would the result be if we had chosen $t = \sqrt{x^2 - x + 1} - x$?

–  For the second integral, we set

$$tx + 1 = \sqrt{-2x^2 + x + 1} \quad \Rightarrow \quad x = \phi(t) = \frac{1 - 2t}{t^2 + 2}, \quad dx = \frac{2t^2 - 2t - 4}{(t^2 + 2)^2}\, dt.$$

Then the integral becomes

$$\int -\frac{2}{t^2 + 2}\, dt = -\frac{2}{\sqrt{2}} \arctan\left(\frac{t}{\sqrt{2}}\right).$$

Back-substituting:

$$\int \frac{dx}{\sqrt{-2x^2 + x + 1}} = \sqrt{2}\arctan\left(\frac{1 - \sqrt{-2x^2 + x + 1}}{x\sqrt{2}}\right) + c.$$

What would the result be if we had set $tx - 1 = \sqrt{-2x^2 + x + 1}$?

An alternative method for this integral is to rewrite the integrand as

$$f(x) = \frac{1}{\sqrt{-2x^2 + x + 1}} = \frac{2\sqrt{2}}{3} \cdot \frac{1}{\sqrt{1 - (\frac{4}{3}x - \frac{1}{3})^2}}.$$

This implies that

$$F(x) = \frac{1}{\sqrt{2}} \arcsin\left(\frac{4}{3}x - \frac{1}{3}\right)$$

is an antiderivative of $f$, valid for $x \in (-\frac{1}{2}, 1)$.

Since both expressions are antiderivatives of the same function on the same interval, they must differ by a constant. (What is the value of this constant?)

– We now set $t(x - 2) = \sqrt{x^2 - 3x + 2}$, which implies

$$x = \phi(t) = \frac{1 - 2t^2}{1 - t^2}, \quad \phi'(t) = -\frac{2t}{(1 - t^2)^2}, \quad dx = -\frac{2t}{(1 - t^2)^2} dt.$$

Substituting $x$ and $dx$ into the original integral

$$\int \frac{x + 1}{\sqrt{x^2 - 3x + 2}} dx \quad \Rightarrow \quad \int \frac{4 - 6t^2}{(1 - t^2)^2} dt.$$

This simplifies to

$$\int \frac{4 - 6t^2}{(1 - t^2)^2} dt = \frac{t}{t^2 - 1} + \frac{5}{2} \ln\left|\frac{t + 1}{t - 1}\right| + c.$$

Back-substituting $t = \frac{\sqrt{x^2 - 3x + 2}}{x - 2}$, we find

$$\int \frac{x + 1}{\sqrt{x^2 - 3x + 2}} dx = \sqrt{x^2 - 3x + 2} + \frac{5}{2} \ln\left|2x - 3 + 2\sqrt{x^2 - 3x + 2}\right| + c.$$

What would the result be if we had chosen the substitution $\sqrt{x^2 - 3x + 2} = t(x - 1)$? Verify the results! □

## 18.5.7 $\int R(e^x)\, dx$

For integrals of the form

$$I = \int R(e^x)\, dx,$$

where $R(\cdot)$ is a rational function, we set

$$x = \ln t \quad \Rightarrow \quad dx = \frac{1}{t} dt.$$

**Example 18.21.** Evaluate the integral

$$\int \frac{dx}{\cosh x + 1},$$

where $\cosh x = \frac{e^x + e^{-x}}{2}$.

*Solution.* We first rewrite the integrand using the identity

$$\cosh x + 1 = \frac{e^x + e^{-x}}{2} + 1 = \frac{e^{2x} + 2e^x + 1}{2e^x}.$$

So the integrand becomes

$$\frac{dx}{\cosh x + 1} = \frac{2e^x}{e^{2x} + 2e^x + 1} dx.$$

Now apply the substitution $x = \ln t$, so $e^x = t$, $dx = \frac{1}{t} dt$. Then

$$\int \frac{2t}{t^2 + 2t + 1} \cdot \frac{1}{t} dt = \int \frac{2}{(t + 1)^2} dt.$$

This integrates to

$$\int \frac{2}{(t + 1)^2} dt = -\frac{2}{t + 1} + c.$$

Returning to the variable $x$, with $t = e^x$, we obtain

$$\int \frac{dx}{\cosh x + 1} = -\frac{2}{e^x + 1} + c.$$

Verify the result! □

### 18.5.8 Exercises

**True or false**
Decide whether each statement is **true** or **false**. Justify your answer.
1. The Weierstrass substitution $t = \tan(x/2)$ can be used to integrate any rational function of $\sin x$ and $\cos x$.
   **Answer:** _____
2. For the integral $\int R(x, \sqrt[n]{\frac{ax+b}{cx+e}}) dx$, the substitution $t = \sqrt[n]{\frac{ax+b}{cx+e}}$ will always transform the integrand into a rational function of $t$.
   **Answer:** _____
3. The substitution $x = a \sin t$ is appropriate for integrals involving $\sqrt{a^2 + x^2}$.
   **Answer:** _____

4. The substitution $x = \frac{\sqrt{a}}{\sqrt{b}} \sinh t$ is used when the integrand contains $\sqrt{bx^2 + a}$ with $a, b > 0$.

   **Answer:** _____

5. The substitution $x = \ln t$ is used to transform integrals of the form $\int R(e^x)dx$ into integrals of rational functions.

   **Answer:** _____

## Multiple choice questions

Choose the correct option (A, B, C, or D) for each question.

1. Which substitution is most appropriate for evaluating $\int \frac{1}{1+\cos x}dx$?

   A. $x = \ln t$    B. $t = \tan(x/2)$

   C. $x = \sin t$    D. $t = \sqrt{\frac{x-1}{x+1}}$

   **Answer:** _____

2. What is the correct substitution for the integral $\int \frac{dx}{\sqrt{3x^2-2}}$?

   A. $x = \frac{\sqrt{2}}{\sqrt{3}} \sinh t$    B. $x = \frac{\sqrt{2}}{\sqrt{3}} \cosh t$

   C. $x = \frac{\sqrt{3}}{\sqrt{2}} \cos t$    D. $x = \frac{\sqrt{3}}{\sqrt{2}} \sin t$

   **Answer:** _____

3. After applying the Weierstrass substitution $t = \tan(x/2)$, what is the expression for $\sin x$?

   A. $\frac{1-t^2}{1+t^2}$    B. $\frac{t}{1+t^2}$

   C. $\frac{2t}{1+t^2}$    D. $\frac{1}{1+t^2}$

   **Answer:** _____

4. For the integral $\int \frac{dx}{\sqrt[3]{x+1}\sqrt{x-1}}$, what is the appropriate substitution?

   A. $t = \sqrt[3]{\frac{x+1}{x-1}}$    B. $t = \sqrt{\frac{x-1}{x+1}}$

   C. $t = \sqrt[6]{\frac{x+1}{x-1}}$    D. $t = \sqrt[5]{\frac{x+1}{x-1}}$

   **Answer:** _____

5. What is $\int \frac{1}{\sqrt{x^2+4}}dx$?

   A. $\ln|x + \sqrt{x^2 + 4}| + C$    B. $\frac{1}{2}\ln|x + \sqrt{x^2 + 4}| + C$

   C. $\arcsin(\frac{x}{2}) + C$    D. $\frac{1}{2}\arctan(\frac{x}{2}) + C$

   **Answer:** _____

## Matching exercise

Match each type of integral with the most appropriate substitution technique.

1. $\int R(\cos x, \sin x)dx$    A. $x = a \sin t$
2. $\int R(x, \sqrt{a^2 - x^2})dx$    B. $x = \frac{\sqrt{a}}{\sqrt{b}} \sinh t$
3. $\int R(x, \sqrt{bx^2 + a})dx$    C. $t = \tan(x/2)$
4. $\int R(x, \sqrt{bx^2 - a})dx$    D. $x = \frac{\sqrt{a}}{\sqrt{b}} \cosh t$
5. $\int R(e^x)dx$    E. $x = \ln t$

**Answers: 1._____ 2._____ 3._____ 4._____ 5._____**

## Fill in the blanks

Fill in each blank with the most appropriate word or phrase.

1. The substitution $t = \tan(x/2)$ is also known as the _____ substitution.

   **Answer:** _____

2. For the integral $\int \frac{dx}{\sqrt{9-x^2}}$, the trigonometric substitution is $x = $ _____.

   **Answer:** _____

3. The derivative $dx$ when using the substitution $x = \ln t$ is $dx = $ _____.

   **Answer:** _____

4. To integrate $\int \frac{1}{\sqrt{4x^2+1}} dx$, use the substitution $x = \frac{1}{2}$_____.

   **Answer:** _____

5. For the integral $\int R(x, \sqrt[n]{\frac{ax+b}{cx+e}}) dx$, if the exponents are $\frac{1}{3}$ and $\frac{1}{4}$, the substitution should be $t = $ _____ $\sqrt{\frac{ax+b}{cx+e}}$.

   **Answer:** _____

## Theoretical and computational problems

Solve each problem step-by-step.

1. Explain why the substitution $t = \tan(x/2)$ is effective for rational functions of $\sin x$ and $\cos x$. Derive the formulas for $\sin x$, $\cos x$, and $dx$ in terms of $t$.

2. Use the substitution $t = \sqrt[3]{\frac{x-1}{x+1}}$ to evaluate the integral $\int \frac{dx}{\sqrt[3]{(x-1)^2(x+1)}}$.

3. Evaluate the integral $\int \frac{1}{1+\sin x} dx$ using the Weierstrass substitution $t = \tan(x/2)$.

4. Evaluate the integral $\int \frac{1}{\sqrt{3x^2-1}} dx$ using the hyperbolic substitution $x = \frac{1}{\sqrt{3}} \cosh t$.

5. Evaluate the integral $\int \frac{1}{\sqrt{x^2+6x+10}} dx$ by first completing the square and then using an appropriate trigonometric or hyperbolic substitution.

6. Use the substitution $x = \ln t$ to evaluate the integral $\int \frac{1}{e^x+e^{-x}} dx$.
   (*Hint:* Recognize the integrand as sech $x$.)

7. Evaluate the integral $\int \frac{1}{x\sqrt{4x^2-1}} dx$ using the trigonometric substitution $2x = \sec\theta$.

8. Evaluate the integral $\int \frac{1}{\sqrt{-x^2+4x-3}} dx$. Identify the appropriate method and justify your choice.

9. Use the Euler substitution $\sqrt{ax^2+bx+c} = t \pm \sqrt{a}x$ (for $a > 0$) to evaluate $\int \frac{1}{\sqrt{x^2+2x+2}} dx$.

10. Show that the two different substitutions $t = \sqrt{x^2-x+1}+x$ and $t = \sqrt{x^2-x+1}-x$ for the integral $\int \frac{dx}{x\sqrt{x^2-x+1}}$ lead to antiderivatives that differ by a constant.

## Python-based exercises

Use Python (sympy, numpy, matplotlib) to tackle the following:

1. Use sympy to symbolically compute the integral $\int \frac{1}{1+\sin x} dx$. Compare the result to the one obtained using the Weierstrass substitution.

2. Write a Python function that applies the substitution $x = \ln t$ to an integral of the form $\int R(e^x) dx$ and returns the new integrand in terms of $t$. Test it on $\int \frac{1}{e^x+1} dx$.

3. Use sympy to verify the result of the integral $\int \frac{1}{\sqrt{2x^2+1}}dx$ by differentiating the antiderivative $\frac{1}{\sqrt{2}}\sinh^{-1}(\sqrt{2}x)$.

4. Plot the function $f(x) = \frac{1}{\sqrt{x^2+1}}$ and its antiderivative $F(x) = \sinh^{-1}(x)$ on the interval $[-5, 5]$. Comment on the behavior of $F(x)$ as $x \to \pm\infty$.

5. Use sympy to perform the Weierstrass substitution on the integral $\int \frac{1}{2+\cos x}dx$ and compute the resulting rational integral.

## 18.6 Python and AI

**AI Request**

Evaluate the indefinite integral

$$\int \frac{1}{\sqrt{2x^2 - 1}} dx$$

using the substitution $x = \phi(t) = \frac{1}{\sqrt{2}}\cosh t$. Write a Python code in order to compare the above result.

## 18.7 Review exercises

### True or false

Decide whether each statement is **true** or **false**. Justify your answer.

1. If $F(x)$ is an antiderivative of $f(x)$, then $F(x) + 5$ is also an antiderivative of $f(x)$.
   Answer: _____

2. The linearity property of indefinite integrals states that $\int [f(x)g(x)]dx = \int f(x)dx \cdot \int g(x)dx$.
   Answer: _____

3. Integration by substitution is essentially the reverse process of the chain rule for differentiation.
   Answer: _____

4. To integrate $\int \frac{1}{x^2+1}dx$, one can use the substitution $x = \tan t$.
   Answer: _____

5. Integration by parts is derived from the product rule of differentiation.
   Answer: _____

### Multiple choice questions

Choose the correct option (A, B, C, or D) for each question.

1. Which of the following is the correct application of integration by parts to $\int x \cos x \, dx$?
   A. $x \sin x - \int \sin x \, dx$      B. $x \sin x + \int \sin x \, dx$
   C. $\frac{x^2}{2} \sin x - \int \frac{x^2}{2} \sin x \, dx$    D. $-\frac{x^2}{2} \sin x + \int \frac{x^2}{2} \sin x \, dx$
   Answer: _____

2. To evaluate $\int \frac{1}{\sqrt{1-x^2}} dx$, the most appropriate substitution is

   A. $x = \sin t$    B. $x = \tan t$

   C. $x = \sec t$    D. $x = \cosh t$

   **Answer:** _____

3. For integrating $\int \frac{2x-1}{(x-1)(x+2)} dx$, the form of partial fraction decomposition is

   A. $\frac{A}{x-1} + \frac{B}{x+2}$    B. $\frac{A}{(x-1)^2} + \frac{B}{x+2}$

   C. $\frac{Ax+B}{(x-1)(x+2)}$    D. $\frac{A}{x-1} + \frac{Bx+C}{x+2}$

   **Answer:** _____

4. When using the substitution $u = x^2 + 1$ in $\int x \cos(x^2 + 1)dx$, the integral becomes

   A. $\int \cos u \, du$    B. $\frac{1}{2} \int \cos u \, du$

   C. $2 \int \cos u \, du$    D. $\int x \cos u \, du$

   **Answer:** _____

5. The integral $\int e^x \sin x \, dx$ requires

   A. only substitution              B. only partial fractions

   C. integration by parts twice    D. trigonometric substitution

   **Answer:** _____

## Matching exercise

Match each integral with the most appropriate method for solving it.

1. $\int xe^x dx$        A. Trigonometric substitution $x = a \sin t$

2. $\int \frac{1}{\sqrt{4-x^2}} dx$    B. Integration by parts

3. $\int \frac{x+1}{x^2+x-2} dx$    C. Substitution $u = x^2 + x - 2$

4. $\int \frac{1}{x^2+4x+5} dx$    D. Partial fraction decomposition

5. $\int \frac{x}{\sqrt{x^2+1}} dx$    E. Completing the square and substitution

**Answers: 1.** _____    **2.** _____    **3.** _____    **4.** _____    **5.** _____

## Fill in the blanks

Fill in each blank with the most appropriate word or phrase.

1. The process of finding an antiderivative is called _____.

   **Answer:** _____

2. The linearity property allows us to split $\int (3f(x)-2g(x))dx$ into _____.

   **Answer:** _____

3. In integration by parts, if we choose $u = \ln x$, then $dv$ should contain _____.

   **Answer:** _____

4. For integrals involving $\sqrt{a^2 - x^2}$, we typically use the substitution _____.

   **Answer:** _____

5. When decomposing $\frac{P(x)}{Q(x)}$ where $Q(x)$ has a repeated linear factor $(x-a)^3$, the partial fractions include terms _____.

   **Answer:** _____

## Theoretical and computational problems

Solve each problem step-by-step.

1. Prove the linearity property of indefinite integrals: $\int [af(x)+bg(x)]dx = a\int f(x)dx + b\int g(x)dx$.

2. Using the substitution $u = x^2 + 1$, evaluate $\int x(x^2 + 1)^5 dx$.

3. Decompose $\frac{3x-2}{(x+1)(x-2)}$ into partial fractions and hence evaluate $\int \frac{3x-2}{(x+1)(x-2)}dx$.

4. Apply integration by parts to find $\int x^2 e^x dx$.

5. Evaluate $\int \frac{1}{\sqrt{9-x^2}}dx$ using the appropriate trigonometric substitution.

6. Show that $\int \ln x \, dx = x \ln x - x + C$ using integration by parts.

7. Find $\int \frac{dx}{x^2+4x+13}$ by completing the square in the denominator.

8. Use integration by parts twice to evaluate $\int e^x \cos x \, dx$.

9. Evaluate $\int \frac{x^2+1}{x^3+x}dx$ using partial fraction decomposition.

10. Verify by differentiation that your answer to $\int \frac{1}{\sqrt{a^2-x^2}}dx$ is correct.

## Python-based exercises

Use Python (sympy, numpy, matplotlib) to tackle the following:

1. Use sympy to compute the following integrals symbolically and verify the results by differentiation:
   - $\int x \sin(x^2)dx$
   - $\int \frac{\ln x}{x}dx$
   - $\int \frac{1}{x^2+2x+2}dx$

   Hint: Use sympy.integrate and sympy.diff.

2. Implement a Python function that performs integration by parts for simple cases. Test it on $\int xe^x dx$ and $\int x \cos x dx$.

   Hint: The function should take expressions for $u$ and $dv$, compute $du$ and $v$, and return $uv - \int v du$.

3. Write a Python script that performs partial fraction decomposition for rational functions with linear factors. Apply it to decompose $\frac{2x+3}{(x-1)(x+2)}$.

   Hint: Use sympy.apart to check your results.

4. Plot the function $f(x) = \frac{1}{\sqrt{1-x^2}}$ and its antiderivative $F(x) = \arcsin(x)$ on the interval $(-1, 1)$. Comment on their relationship.

   Hint: Use numpy for function evaluation and matplotlib.pyplot for plotting.

# 19 Definite integral

## Contents

The definite integral is a central concept in mathematical analysis, providing a rigorous way to measure the area under a curve and the accumulation of quantities. It formalizes the idea of summing infinitely many small parts, moving from discrete sums to continuous accumulation. Through the definitions of partitions, Riemann sums, and Darboux sums, we explore what it means for a function to be integrable. This chapter develops the basic properties and techniques of definite integration, such as linearity, additivity, and inequalities, as well as the mean value theorem. The fundamental theorem of calculus reveals the deep connection between integration and differentiation. Through theoretical results, computational examples, and Python-based exercises, we gain insight into the power and applications of the definite integral across mathematics and science.

## 19.1 Introduction

The definite integral is one of the foundational concepts in mathematical analysis, serving as a bridge between discrete sums and continuous accumulation. It originated as a method for calculating areas, lengths, and volumes, but its scope extends far beyond simple geometry. Today, the definite integral is central not only to mathematics but also to the physical and social sciences, engineering, and beyond, wherever the idea of "accumulating" quantities is essential.

At its heart, the definite integral formalizes the intuitive process of summing infinitely many infinitesimal quantities, such as the area under a curve or the total change of a function over an interval. This process, rigorously established by Riemann and Darboux, leads to the precise definition of the integral via partitions and sums. By dividing an interval into subintervals, evaluating the function on each, and letting the number of subintervals increase without bound, we transition from the discrete to the continuous, capturing the notion of total accumulation.

https://doi.org/10.1515/9783112228289-019

This chapter begins with the basic definitions of partitions, Riemann sums, and the Darboux approach, providing a dual perspective on integrability. We explore the essential conditions for a function to be Riemann integrable, including monotonicity, continuity, and the nature of discontinuities. Through concrete examples—ranging from continuous and piecewise-defined functions to the notorious Dirichlet function—we illustrate the subtle interplay between function behavior and integrability.

We then develop the fundamental properties of the definite integral: linearity, additivity over intervals, and order-preserving inequalities. These properties not only simplify computations but also reveal the integral's algebraic structure and its connection to function spaces. The chapter proceeds to mean value theorems for integrals, demonstrating that the integral can be interpreted as an "average value" of a function over an interval, both in classical and weighted senses.

A pivotal point is the fundamental theorem of calculus, which unites the processes of differentiation and integration. We show how the definite integral defines new functions (antiderivatives) and how, under suitable conditions, every integrable function possesses an antiderivative whose difference at the endpoints gives the value of the integral. Techniques such as integration by parts and substitution further extend our computational toolkit, enabling the evaluation of a wide array of definite integrals, including those that appear in probability, physics, and engineering.

Beyond computation, we connect the integral with infinite sums and limits via Riemann sums, culminating in classical results such as Wallis's and Stirling's formulas. These not only showcase the power of integration in asymptotic analysis but also build a bridge to deeper areas such as series and probability theory.

Throughout the chapter, an array of examples, theoretical and computational problems, and Python-based exercises provide opportunities to apply the concepts in both symbolic and numerical settings. These problems are designed to strengthen intuition, develop proof skills, and foster computational fluency—preparing students for more advanced topics in real and complex analysis.

By mastering the definite integral, the reader gains not only a crucial analytical tool but also a deeper appreciation for the unity of mathematics, where geometry, algebra, and analysis converge to describe and solve a vast spectrum of problems.

## 19.2 Definitions and basic results

**Definition 19.1** (Partition). Let $[a, b] \subset \mathbb{R}$ be a closed interval. A partition of $[a, b]$ is a finite set

$$P = \{x_0, x_1, \ldots, x_n\}, \quad a = x_0 < x_1 < \cdots < x_n = b.$$

The subintervals are $[x_{i-1}, x_i]$. The *mesh* (norm) of $P$ is

$$\|P\| := \max_{1 \le i \le n}(x_i - x_{i-1}).$$

**Definition 19.2** (Tagged partition and Riemann sum). A *tagged partition* is a pair $(P, \{\xi_i\}_{i=1}^n)$ where $P = \{x_0, \ldots, x_n\}$ is a partition of $[a, b]$ and each tag $\xi_i \in [x_{i-1}, x_i]$. If $f : [a, b] \to \mathbb{R}$ is bounded, the corresponding *Riemann sum* is

$$R\left(f; P, \{\xi_i\}\right) := \sum_{i=1}^n f(\xi_i)(x_i - x_{i-1}).$$

**Definition 19.3** (Lower and upper Darboux sums). Let $f : [a, b] \to \mathbb{R}$ be bounded. Given a partition $P = \{x_0, \ldots, x_n\}$, define

$$s(P) := \sum_{i=1}^n m_i(x_i - x_{i-1}), \quad m_i := \inf_{x \in [x_{i-1}, x_i]} f(x),$$

$$S(P) := \sum_{i=1}^n M_i(x_i - x_{i-1}), \quad M_i := \sup_{x \in [x_{i-1}, x_i]} f(x).$$

These are the lower and upper Darboux sums. For any tagged partition $(P, \{\xi_i\})$,

$$s(P) \le R\left(f; P, \{\xi_i\}\right) \le S(P).$$

See Figures 19.1, 19.2, 19.3 for the Darboux sums.

**Definition 19.4** (Lower and upper integrals). Let $f : [a, b] \to \mathbb{R}$ be bounded. Define

$$I := \sup_P s(P), \quad J := \inf_P S(P),$$

where the supremum and infimum are taken over all partitions of $[a, b]$. Then $I$ is the lower integral and $J$ is the upper integral. Always

$$s(P) \le I \le J \le S(P)$$

for every partition $P$; in particular $I \le J$.

**Definition 19.5** (Riemann integrability). A bounded function $f : [a, b] \to \mathbb{R}$ is *Riemann integrable* on $[a, b]$ if there exists a number $A$ such that for every $\varepsilon > 0$ there is $\delta > 0$ with the property that for every tagged partition $(P, \{\xi_i\})$ satisfying $\|P\| < \delta$ we have

$$\left| R\left(f; P, \{\xi_i\}\right) - A \right| < \varepsilon.$$

In that case we write

$$\int_a^b f(x)\, dx := A.$$

**Theorem 19.1** (Refinement inequality). *If $Q$ is a refinement of a partition $P$, then*

$$s(P) \le s(Q) \le S(Q) \le S(P).$$

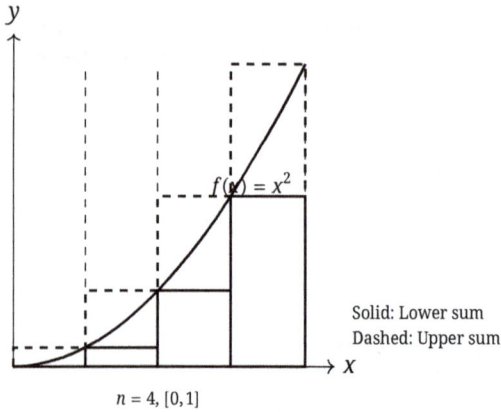

Figure 19.1: Darboux sums for $f(x) = x^2$ on $[0, 1]$ with $n = 4$ (solid: lower, dashed: upper rectangles).

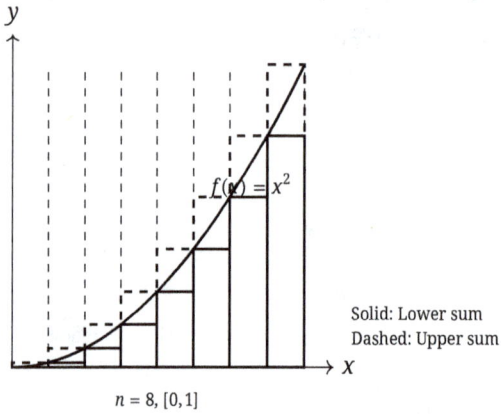

Figure 19.2: Darboux sums for $f(x) = x^2$ on $[0, 1]$ with $n = 8$ (solid: lower, dashed: upper rectangles).

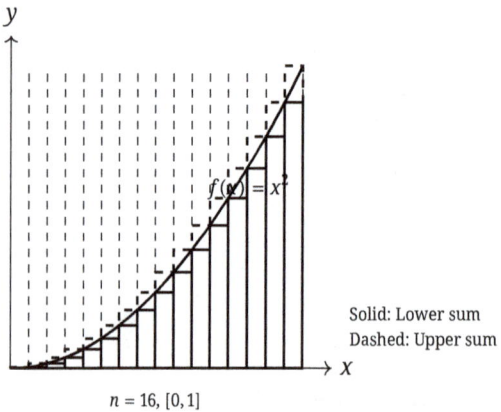

Figure 19.3: Darboux sums for $f(x) = x^2$ on $[0, 1]$ with $n = 16$ (solid: lower, dashed: upper rectangles).

*Proof.* Refining $P$ means subdividing some of its subintervals. On a smaller subinterval the infimum can only increase (or stay the same), so each summand in $s(Q)$ is at least the corresponding combined contribution in $s(P)$; hence $s(P) \leq s(Q)$. Similarly, the supremum on a smaller subinterval can only decrease, so $S(Q) \leq S(P)$. The middle inequality $s(Q) \leq S(Q)$ is clear. Combining these yields the asserted chain. □

**Proposition 19.1** (Comparison of lower and upper Darboux sums). *For any two partitions $P$ and $Q$ of $[a, b]$, it holds that*

$$s(P) \leq S(Q).$$

*Proof.* Let $R$ be a common refinement of $P$ and $Q$. Then, by the refinement inequality,

$$s(P) \leq s(R) \leq S(R) \leq S(Q),$$

so $s(P) \leq S(Q)$. □

**Lemma 19.1** (Control of the Darboux gap by the mesh). *Assume $I = J =: A$. Then for every $\varepsilon > 0$ there exists $\delta > 0$ such that for every partition $P$ of $[a, b]$ with $\|P\| < \delta$ one has*

$$S(P) - s(P) < \varepsilon.$$

*Proof.* Suppose to the contrary that there exist $\varepsilon_0 > 0$ and a sequence of partitions $P_n$ with $\|P_n\| < 1/n$ but $S(P_n) - s(P_n) \geq \varepsilon_0$ for all $n$. Since $I = J = A$, there exists a partition $R$ such that $S(R) - s(R) < \varepsilon_0$. For each $n$, let $R_n$ be the common refinement of $R$ and $P_n$. Then $R_n$ refines $R$, so

$$S(R_n) - s(R_n) \leq S(R) - s(R) < \varepsilon_0,$$

and $R_n$ also refines $P_n$, so by the refinement inequality

$$S(P_n) - s(P_n) \geq S(R_n) - s(R_n).$$

Thus $S(P_n) - s(P_n) < \varepsilon_0$, contradicting the assumption $S(P_n) - s(P_n) \geq \varepsilon_0$. Hence the assumption is false and the lemma holds. □

**Theorem 19.2** (Darboux criterion for Riemann integrability). *Let $f : [a, b] \to \mathbb{R}$ be bounded. Then $f$ is Riemann integrable on $[a, b]$ if and only if $I = J$, and in that case*

$$\int_a^b f(x)\, dx = I = J.$$

*Proof.* **(Only if.)** Suppose $f$ is Riemann integrable with $\int_a^b f = A$. Given $\varepsilon > 0$ there exists $\delta > 0$ such that for every tagged partition $(P, \{\xi_i\})$ with $\|P\| < \delta$ one has

$$\left| R(f; P, \{\xi_i\}) - A \right| < \varepsilon.$$

For a fixed partition $P$ with $\|P\| < \delta$, the set of all Riemann sums $R(f; P, \{\xi_i\})$ as the tags vary lies inside the interval $(A - \varepsilon, A + \varepsilon)$. Therefore, the infimum of that set, which is $s(P)$, satisfies $s(P) \geq A - \varepsilon$, and the supremum $S(P)$ satisfies $S(P) \leq A + \varepsilon$. Thus

$$A - \varepsilon \leq s(P) \leq S(P) \leq A + \varepsilon.$$

Taking the supremum over such $P$ gives $I \geq A - \varepsilon$, and taking the infimum gives $J \leq A + \varepsilon$. Since $\varepsilon$ was arbitrary, $I = J = A$.

**(If.)** Now assume $I = J = A$. Let $\varepsilon > 0$. By the previous lemma choose $\delta > 0$ such that any partition $P$ with $\|P\| < \delta$ satisfies

$$S(P) - s(P) < \varepsilon.$$

Then for such $P$ we have $s(P) \leq A \leq S(P)$, hence $|S(P) - A| < \varepsilon$ and $|A - s(P)| < \varepsilon$. Therefore for any tagged partition $(P, \{\xi_i\})$ with $\|P\| < \delta$

$$\left| R(f; P, \{\xi_i\}) - A \right| \leq \max\{S(P) - A, A - s(P)\} < \varepsilon,$$

showing that $f$ is Riemann integrable with integral $A$. □

**Remark 24** (Uniqueness of the integral). If a bounded function $f : [a, b] \to \mathbb{R}$ satisfies the Riemann integrability condition with two values $A$ and $B$, then for every $\varepsilon > 0$ sufficiently fine tagged partitions force both $|R(f; P, \{\xi_i\}) - A| < \varepsilon$ and $|R(f; P, \{\xi_i\}) - B| < \varepsilon$, hence $|A - B| < 2\varepsilon$. Since $\varepsilon$ is arbitrary, $A = B$.

**Theorem 19.3** (Monotone functions are integrable). *If $f : [a, b] \to \mathbb{R}$ is bounded and monotone on $[a, b]$, then $f$ is Riemann integrable on $[a, b]$.*

*Proof.* Assume $f$ is increasing (the decreasing case is analogous). Then for a partition $P = \{x_0, \ldots, x_n\}$,

$$m_k = f(x_{k-1}), \quad M_k = f(x_k),$$

so

$$S(P) - s(P) = \sum_{k=1}^{n} (f(x_k) - f(x_{k-1}))(x_k - x_{k-1}).$$

If $\|P\| < \delta$, then

$$S(P) - s(P) \le \delta \sum_{k=1}^{n} (f(x_k) - f(x_{k-1})) = \delta(f(b) - f(a)).$$

Given $\varepsilon > 0$, choose $\delta < \varepsilon/(f(b) - f(a))$. Then $S(P) - s(P) < \varepsilon$, so $I = J$ and $f$ is integrable. □

---

**Theorem 19.4** (Continuity on a closed interval implies uniform continuity (see Definition 5.4)). *If $f$ : $[a, b] \to \mathbb{R}$ is continuous on the compact interval $[a, b]$, then $f$ is uniformly continuous on $[a, b]$.*

---

*Proof.* Suppose not. Then there exist $\varepsilon_0 > 0$ and sequences $x_n, y_n \in [a, b]$ with $|x_n - y_n| < 1/n$ but $|f(x_n) - f(y_n)| \ge \varepsilon_0$. Since $[a, b]$ is closed, passing to subsequences we may assume $x_n \to x^*$, and hence $y_n \to x^*$. Continuity of $f$ at $x^*$ implies $|f(x_n) - f(y_n)| \to 0$, contradiction. Thus $f$ is uniformly continuous. □

---

**Theorem 19.5** (Continuous functions are integrable). *If $f : [a, b] \to \mathbb{R}$ is continuous on the compact interval $[a, b]$, then $f$ is Riemann integrable on $[a, b]$.*

---

*Proof.* By the previous theorem, $f$ is uniformly continuous. Given $\varepsilon > 0$, choose $\delta > 0$ such that if $|x - y| < \delta$ then $|f(x) - f(y)| < \varepsilon/(b - a)$. For any partition $P$ with $\|P\| < \delta$, on each subinterval $[x_{k-1}, x_k]$ we have $M_k - m_k < \varepsilon/(b - a)$, so

$$S(P) - s(P) = \sum_{k=1}^{n} (M_k - m_k)(x_k - x_{k-1}) < \frac{\varepsilon}{b - a} \sum_{k=1}^{n} (x_k - x_{k-1}) = \varepsilon.$$

Thus $I = J$ and $f$ is integrable. □

**Remark 25.** The inequalities $s(P) \le I \le J \le S(P)$ show that the lower integral $I$ is always less than or equal to the upper integral $J$; integrability is equivalent to their equality.

---

**Definition 19.6.**
- We set $\int_a^a f(x)\, dx = 0$.
- If $a > b$ and $f$ is integrable on $[b, a]$, then define $\int_a^b f := -\int_b^a f$.

---

**Example 19.1** (Integrability of a piecewise function). Show that the function $f : \mathbb{R} \to \mathbb{R}$ defined by

$$f(x) = \begin{cases} x, & 0 \le x \le 1, \\ x + 1, & 1 < x \le 2, \end{cases}$$

is integrable on $[0, 2]$.

---

*Solution.* We use Theorem 19.3 and observe that $f$ is monotone and bounded on $[0, 2]$. Clearly, $0 \le f(x) \le 3$ for all $x \in [0, 2]$. Moreover, $f$ is increasing: for any $x, y \in [0, 2]$ with $y \ge x$, we have $f(y) \ge f(x)$. The monotonicity holds on $[0, 1]$ and on $(1, 2]$ separately.

If $x \in [0,1]$ and $y \in (1,2]$, then $f(y) - f(x) = y + 1 - x = (y - x) + 1 \geq 1 \geq 0$. Hence, $f$ is monotone and bounded on $[0,2]$ and thus integrable. Observe f that $f$ is not continuous at $x = 1$ since

$$\lim_{x \to 1^+} f(x) = 2 \neq 1 = \lim_{x \to 1^-} f(x).$$

This shows that the converse of Theorem 19.5 does not hold. □

**Example 19.2** (Dirichlet function is not integrable). Consider the function $f : \mathbb{R} \to \mathbb{R}$ defined by

$$f(x) = \begin{cases} 1, & x \text{ rational,} \\ -1, & x \text{ irrational.} \end{cases}$$

Show that $f$ is not integrable on any interval $[a,b]$.

*Solution.* Using the Darboux sums, it is clear that the lower sums $s_n$ and upper sums $S_n$ never approach each other.

Indeed, for any partition,

$$S_n = \sum_{k=1}^{n} M_k(x_k - x_{k-1}) = b - a,$$

because $M_k = 1$ on every subinterval (rationals are dense). Also,

$$s_n = \sum_{k=1}^{n} m_k(x_k - x_{k-1}) = -(b - a),$$

because $m_k = -1$ (irrationals are dense).

Therefore,

$$S_n - s_n = 2(b - a) \neq 0,$$

and by the necessary condition for integrability (Theorem 19.2), $f$ is not integrable. □

**Theorem 19.6** (Integrability with finitely many discontinuities). *If $f$ is bounded and has finitely many discontinuities in $[a,b]$ then it is integrable on $[a,b]$.*

*Proof.* Suppose $f$ has finitely many discontinuities, say at points $x_0$. Consider a partition such that $x_0$ is one of the partition points, say $x_0 = x_j$ for some $j$. Then the integral splits as

$$S_n = \sum_{i=1}^{j} M_i(x_i - x_{i-1}) + \sum_{i=j+1}^{n} M_i(x_i - x_{i-1}) \to \int_a^{x_0} f(x)\,dx + \int_{x_0}^{b} f(x)\,dx,$$

where $f$ is continuous on $(a, x_0)$ and $(x_0, b)$.

Similarly for the lower sums $s_n$. Since $x_0 \in (x_{j-1}, x_j)$, the difference

$$S_n = \sum_{i=1}^{j-1} M_i(x_i - x_{i-1}) + \sum_{i=j+1}^{n} M_i(x_i - x_{i-1}) + M_j(x_j - x_{j-1}) \rightarrow \int_a^{x_0} f(x)\,dx + \int_{x_0}^{b} f(x)\,dx,$$

where the added term $M_j(x_j - x_{j-1})$ tends to zero as the partition is refined because $M_i$ are bounded.

Therefore, $f$ is integrable despite finitely many discontinuities.  □

## 19.2.1 Exercises

### True or false

Decide whether each statement is **true** or **false**. Justify your answer briefly.
1.  Every bounded function is Riemann integrable.
    **Answer:** _____
2.  Every monotone and bounded function is integrable.
    **Answer:** _____
3.  The Dirichlet function (defined as 1 at rationals and −1 at irrationals) is integrable.
    **Answer:** _____
4.  If $f$ is continuous on $[a, b]$, then it is uniformly continuous on $[a, b]$.
    **Answer:** _____
5.  Adding points to a partition increases the lower Darboux sum and decreases the upper Darboux sum.
    **Answer:** _____

### Multiple choice questions

Choose the correct option (A, B, C, or D).
1.  For a given function $f$, the lower Darboux sums are always
    A. decreasing   B. increasing
    C. constant     D. oscillating
    **Answer:** _____
2.  If the upper and lower Darboux sums converge to the same limit, then
    A. $f$ is continuous   B. $f$ is differentiable
    C. $f$ is integrable    D. $f$ is monotone
    **Answer:** _____
3.  Uniform continuity on a closed interval $[a, b]$ implies
    A. integrability    B. differentiability
    C. monotonicity    D. none of these
    **Answer:** _____

4. A function with finitely many discontinuities on $[a, b]$ is
   A. never integrable    B. sometimes integrable
   C. always integrable    D. integrable only if continuous
   **Answer:** _____
5. Which of the following functions is NOT integrable on any interval?
   A. Constant function    B. Dirichlet function
   C. Monotone function    D. Polynomial
   **Answer:** _____

## Matching exercise

Match each concept/theorem to the correct description or property.
1. Partition                      A. Limit independent of $\xi_i$
2. Riemann sum                    B. Infimum of upper sums
3. Lower Darboux sum    C. Finite set of points in an interval
4. Upper integral ($J$)          D. Uses infimum on subintervals
5. Integrability                  E. Uses supremum on subintervals
**Answers:** 1._____    2._____    3._____    4._____    5._____

## Fill in the blanks

Complete each sentence with the most appropriate word or phrase.
1. The mesh size of a partition tends to _____ for regular partitions.
   **Answer:** _____
2. A continuous function on a closed interval is always _____ continuous.
   **Answer:** _____
3. A function with finitely many discontinuities on a closed interval is always _____.
   **Answer:** _____
4. The lower Darboux sum is always less than or equal to the _____
   sum.
   **Answer:** _____
5. If upper and lower Darboux sums differ by less than any $\varepsilon > 0$, the function is
   _____.
   **Answer:** _____

## Theoretical and computational problems

Solve each problem step-by-step.
1. Prove that the function $f(x) = x^2$ is integrable on $[0, 1]$ by using Darboux sums explicitly.

2. Show that a monotone bounded function is integrable using the Darboux criterion.
3. Provide a detailed proof that if $f$ is continuous on $[a, b]$, then it is integrable on $[a, b]$.
4. Prove that the Dirichlet function is not integrable on any interval using lower and upper sums.
5. Demonstrate explicitly that a function with a finite number of discontinuities on a closed interval is integrable.

### Python-based exercises

Use Python (`sympy`, `numpy`, `matplotlib`) to perform the following tasks:
1. Write Python code to approximate the integral of $f(x) = x^2$ on $[0, 1]$ using upper and lower Darboux sums. Plot your approximations for different partitions (e. g., $n = 4, 8, 16$).
2. Use Python to visualize upper and lower Darboux sums for the absolute value function $f(x) = |x|$ on $[-1, 1]$.
3. Implement a Python function to test integrability by numerically checking if upper and lower Darboux sums converge to the same limit for given partitions.
4. Numerically illustrate that the Dirichlet function does not converge to a unique integral value using a Python simulation.
5. Using Python, investigate the uniform continuity of $f(x) = \sin(x)$ on $[0, \pi]$ by plotting $\Delta y$ against small intervals $\Delta x$.

## 19.3 Properties of the definite integral

**Definition 19.7** (Oscillation on a subinterval). Let $f : [a, b] \to \mathbb{R}$ be bounded and let $I \subseteq [a, b]$ be a nonempty interval. The *oscillation* of $f$ on $I$ is

$$\mathrm{osc}_I f := \sup_{x \in I} f(x) - \inf_{x \in I} f(x) = \sup\{|f(x) - f(y)| : x, y \in I\}.$$

Equivalently, writing $M_I = \sup_{x \in I} f(x)$ and $m_I = \inf_{x \in I} f(x)$, we have $\mathrm{osc}_I f = M_I - m_I$.

**Theorem 19.7** (Integrability of $f^+, f^-, |f|$, sums and products). *Let $f, g : [a, b] \to \mathbb{R}$ be bounded and Riemann integrable. Then the functions $f^+, f^-, |f|, f + g$, and $fg$ are also Riemann integrable on $[a, b]$. Moreover, their restrictions to any subinterval $[c, d] \subseteq [a, b]$ are integrable as well.*

*Proof.* We proceed in steps.

(i) **Integrability of $f^+, f^-$, and $|f|$.** Since $f$ is bounded, the maps $\phi(t) = \max(t, 0)$, $\psi(t) = \max(-t, 0)$, and $\chi(t) = |t|$ are Lipschitz with constant 1. Hence, for any subinterval $[x_{i-1}, x_i]$,

$$\sup_{[x_{i-1}, x_i]} \phi(f) - \inf_{[x_{i-1}, x_i]} \phi(f) \le \sup_{[x_{i-1}, x_i]} f - \inf_{[x_{i-1}, x_i]} f,$$

and similarly for $\psi(f)$ and $\chi(f)$. Therefore,

$$S(P, \phi \circ f) - s(P, \phi \circ f) \le S(P, f) - s(P, f),$$

so as $\|P\| \to 0$ the Darboux gap for each of $f^+, f^-$, and $|f|$ tends to zero, and they are integrable.

**(ii) Integrability of $f + g$.** For any partition $P$,

$$\mathrm{osc}_{[x_{i-1}, x_i]}(f + g) \le \mathrm{osc}_{[x_{i-1}, x_i]} f + \mathrm{osc}_{[x_{i-1}, x_i]} g,$$

so

$$S(P, f + g) - s(P, f + g) \le (S(P, f) - s(P, f)) + (S(P, g) - s(P, g)).$$

As $\|P\| \to 0$, the right-hand side tends to zero, hence $f + g$ is integrable.

**(iii) Integrability of $f^2$ (and similarly $g^2$).** If $|f| \le M$ on $[a, b]$, then for any $x, y$,

$$\left| f(x)^2 - f(y)^2 \right| = |f(x) - f(y)| |f(x) + f(y)| \le 2M |f(x) - f(y)|.$$

Thus the oscillation of $f^2$ on each subinterval is at most $2M$ times that of $f$, so

$$S(P, f^2) - s(P, f^2) \le 2M(S(P, f) - s(P, f)) \to 0,$$

showing $f^2$ is integrable.

**(iv) Integrability of $fg$.** We use the identity

$$fg = \frac{(f + g)^2 - f^2 - g^2}{2}.$$

From (ii) $f + g$ is integrable, so $(f + g)^2$ is integrable by (iii), and $f^2, g^2$ are integrable. Linear combinations of integrable functions are integrable, hence $fg$ is integrable.

**(v) Subintervals.** If a function is Riemann integrable on $[a, b]$, then its restriction to any subinterval $[c, d] \subseteq [a, b]$ is integrable, since any partition of $[c, d]$ can be extended to $[a, b]$, and the Darboux gap on $[c, d]$ is controlled by that on $[a, b]$.

Therefore all claimed functions are integrable on $[a, b]$ and on every subinterval. $\qquad\qquad\square$

---

**Theorem 19.8** (Linearity of the Riemann integral). *Let $f, g : [a, b] \to \mathbb{R}$ be bounded and Riemann integrable with $\int_a^b f = A$ and $\int_a^b g = B$, and let $\alpha, \beta \in \mathbb{R}$. Then $\alpha f + \beta g$ is Riemann integrable on $[a, b]$ and*

$$\int_a^b \left( \alpha f(x) + \beta g(x) \right) dx = \alpha A + \beta B.$$

---

*Proof.* Let $\varepsilon > 0$. Since $f$ and $g$ are integrable, there exists $\delta > 0$ such that for every tagged partition $(P, \{\xi_i\})$ with $\|P\| < \delta$ we have

$$|R(f;P,\{\xi_i\}) - A| < \frac{\varepsilon}{2|\alpha|} \quad \text{(or trivially if } \alpha = 0\text{)},$$

$$|R(g;P,\{\xi_i\}) - B| < \frac{\varepsilon}{2|\beta|} \quad \text{(or trivially if } \beta = 0\text{)}.$$

Then

$$R(\alpha f + \beta g; P, \{\xi_i\}) = \alpha R(f; P, \{\xi_i\}) + \beta R(g; P, \{\xi_i\}),$$

so

$$|R(\alpha f + \beta g; P, \{\xi_i\}) - (\alpha A + \beta B)| \leq |\alpha| |R(f; P, \{\xi_i\}) - A| + |\beta| |R(g; P, \{\xi_i\}) - B| < \frac{\varepsilon}{2} + \frac{\varepsilon}{2} = \varepsilon.$$

This shows that $\alpha f + \beta g$ is integrable and its integral is $\alpha A + \beta B$. □

---

**Theorem 19.9** (Integrability of composite functions). *Let $f : [a,b] \to \mathbb{R}$ be bounded and Riemann integrable. Denote*

$$m_f := \inf_{x \in [a,b]} f(x), \quad M_f := \sup_{x \in [a,b]} f(x),$$

*and suppose $\phi$ is continuous on the compact interval $[m_f, M_f]$. Then the composite $h := \phi \circ f$ is Riemann integrable on $[a,b]$.*

---

*Proof.* Since $\phi$ is continuous on the compact set $[m_f, M_f]$, it is uniformly continuous and bounded. Let $K := \max_{y \in [m_f, M_f]} |\phi(y)| < \infty$.

Fix $\varepsilon > 0$. By the uniform continuity of $\phi$, there exists $\delta > 0$ such that

$$|s - t| < \delta \quad \Longrightarrow \quad |\phi(s) - \phi(t)| < \frac{\varepsilon}{2(b-a)}.$$

Since $f$ is integrable, choose a partition $P = \{x_0, \ldots, x_n\}$ of $[a,b]$ with

$$S(P,f) - s(P,f) < \frac{\delta \varepsilon}{4K}.$$

For each $i$ write $M_i^f = \sup_{[x_{i-1}, x_i]} f$, $m_i^f = \inf_{[x_{i-1}, x_i]} f$, and similarly define $M_i^{\phi \circ f}, m_i^{\phi \circ f}$. Let $A := \{i : M_i^f - m_i^f < \delta\}$ and $B := \{1, \ldots, n\} \setminus A$. Then

$$S(P, \phi \circ f) - s(P, \phi \circ f) = \sum_{i=1}^n (M_i^{\phi \circ f} - m_i^{\phi \circ f})(x_i - x_{i-1}) = \sum_{i \in A} \cdots + \sum_{i \in B} \cdots.$$

If $i \in A$, the oscillation of $f$ on $[x_{i-1}, x_i]$ is $< \delta$, so by uniform continuity $M_i^{\phi \circ f} - m_i^{\phi \circ f} < \frac{\varepsilon}{2(b-a)}$, hence

$$\sum_{i \in A} (M_i^{\phi \circ f} - m_i^{\phi \circ f})(x_i - x_{i-1}) < \frac{\varepsilon}{2(b-a)} \sum_{i \in A} (x_i - x_{i-1}) \leq \frac{\varepsilon}{2(b-a)}(b-a) = \frac{\varepsilon}{2}.$$

If $i \in B$, then $M_i^f - m_i^f \geq \delta$, so

$$\sum_{i \in B}(x_i - x_{i-1}) \leq \frac{1}{\delta}\sum_{i \in B}(M_i^f - m_i^f)(x_i - x_{i-1}) \leq \frac{1}{\delta}(S(P,f) - s(P,f)) < \frac{\varepsilon}{4K}.$$

But trivially $M_i^{\phi \circ f} - m_i^{\phi \circ f} \leq 2K$, so

$$\sum_{i \in B}(M_i^{\phi \circ f} - m_i^{\phi \circ f})(x_i - x_{i-1}) \leq 2K\sum_{i \in B}(x_i - x_{i-1}) < 2K \cdot \frac{\varepsilon}{4K} = \frac{\varepsilon}{2}.$$

Combining the two parts,

$$S(P, \phi \circ f) - s(P, \phi \circ f) < \frac{\varepsilon}{2} + \frac{\varepsilon}{2} = \varepsilon.$$

Thus the Darboux gap for $\phi \circ f$ can be made arbitrarily small, so $\phi \circ f$ is Riemann integrable on $[a, b]$. □

**Theorem 19.10** (Integral of a nonnegative function is nonnegative). *If $f$ is bounded and integrable on $[a, b]$ and $f(x) \geq 0$ for all $x \in [a, b]$, then*

$$\int_a^b f(x)\, dx \geq 0.$$

*Proof.* Since $f$ is integrable, consider any tagged partition. The lower Darboux sums satisfy

$$s_n = \sum_{i=1}^n m_i(x_i - x_{i-1}) \geq 0,$$

because each $m_i \geq 0$. Passing to the limit completes the proof. □

**Remark 26.** If $f(x) \leq g(x)$ for all $x \in [a, b]$, then

$$\int_a^b f(x)\, dx \leq \int_a^b g(x)\, dx,$$

since the function $h(x) = g(x) - f(x) \geq 0$ is integrable.

**Proposition 19.2** (Integral zero implies function is zero at continuity points). *If $f(x) \geq 0$ on $[a, b]$ and*

$$\int_a^b f(x)\, dx = 0,$$

*then $f(x) = 0$ for all $x \in [a, b]$ at which $f$ is continuous.*

*Proof.* Assume that there exists $x_0 \in [a, b]$ such that $f$ is continuous at $x_0$ and $f(x_0) > 0$. Then by continuity, there exists $\delta > 0$ such that

$$f(x) > \varepsilon := \frac{f(x_0)}{2} > 0 \quad \text{for all } x \in (x_0 - \delta, x_0 + \delta).$$

Therefore,

$$\int_{x_0-\delta}^{x_0+\delta} f(x)\, dx > 2\varepsilon\delta > 0,$$

contradicting $\int_a^b f(x)\, dx = 0$.

Hence, for every point of continuity, $f(x) = 0$. □

**Theorem 19.11** (Absolute value inequality for integrals). *If $f$ is integrable on $[a, b]$, then*

$$\left| \int_a^b f(x)\, dx \right| \le \int_a^b |f(x)|\, dx.$$

*Proof.* Since for all $x \in [a, b]$,

$$-|f(x)| \le f(x) \le |f(x)|,$$

by Theorem 19.10 it follows that

$$-\int_a^b |f(x)|\, dx \le \int_a^b f(x)\, dx \le \int_a^b |f(x)|\, dx,$$

and thus

$$\left| \int_a^b f(x)\, dx \right| \le \int_a^b |f(x)|\, dx. \qquad □$$

**Theorem 19.12** (Additivity of the integral over subintervals). *If $f$ is bounded and integrable on $[a, b]$ and $a < c < b$, then $f$ is integrable on $[a, c]$ and $[c, b]$ and*

$$\int_a^b f(x)\, dx = \int_a^c f(x)\, dx + \int_c^b f(x)\, dx.$$

*Proof.* Since $f$ is integrable on both subintervals (Theorem 19.7), for partitions refining the point $c$ and mesh tending to zero,

$$S_n = \sum_{i=1}^{j} M_i(x_i - x_{i-1}) + \sum_{i=j+1}^{n} M_i(x_i - x_{i-1}) \to \int_a^c f + \int_c^b f. \qquad □$$

## 19.4 Mean value theorems for integrals

**Theorem 19.13** (Mean value theorem for integrals (bounded version)). *If $f$ is integrable on $[a, b]$ and there exist constants $m, M \in \mathbb{R}$ such that*

$$m \le f(x) \le M \quad \text{for all } x \in [a, b],$$

*then there exists $\mu \in [m, M]$ such that*

$$\int_a^b f(x)\, dx = \mu(b - a).$$

*Proof.* By integration and inequalities,

$$m(b - a) \le \int_a^b f(x)\, dx \le M(b - a).$$

Dividing by $b - a > 0$,

$$m \le \frac{1}{b - a} \int_a^b f(x)\, dx \le M,$$

so we can pick $\mu$ in $[m, M]$ satisfying the statement. □

**Theorem 19.14** (Mean value theorem for integrals). *If $f$ is continuous on $[a, b]$, then there exists $\xi \in [a, b]$ such that*

$$\int_a^b f(x)\, dx = f(\xi)(b - a).$$

*Proof.* By Theorem 19.13, there exists $\mu \in [m, M]$, where $m$ and $M$ are the minimum and maximum of $f$ on $[a, b]$. By the intermediate value theorem for continuous functions, there exists $\xi \in [a, b]$ with $\mu = f(\xi)$. □

**Theorem 19.15** (Weighted mean value theorem for integrals). *Let $f, g$ be integrable on $[a, b]$ and suppose $m \le f(x) \le M$ for all $x \in [a, b]$. If $g(x) \ge 0$ or $g(x) \le 0$ on $[a, b]$, then there exists $\mu \in [m, M]$ such that*

$$\int_a^b f(x)g(x)\, dx = \mu \int_a^b g(x)\, dx.$$

*Proof.* Assume $g(x) \ge 0$. Then for all $x \in [a, b]$,

$$mg(x) \le f(x)g(x) \le Mg(x).$$

Integrating, we get

$$m \int_a^b g(x)\, dx \le \int_a^b f(x)g(x)\, dx \le M \int_a^b g(x)\, dx.$$

If $\int_a^b g(x)\, dx = 0$, then $\int_a^b f(x)g(x)\, dx = 0$ trivially. If $\int_a^b g(x)\, dx > 0$, divide the inequalities to get

$$m \le \frac{\int_a^b f(x)g(x)\, dx}{\int_a^b g(x)\, dx} \le M.$$

The case $g(x) \le 0$ is similar. □

---

**Theorem 19.16** (Mean value theorem for integrals with non-changing sign weight). *Let $f$ be continuous on $[a, b]$ and $g$ integrable on $[a, b]$ such that for all $x \in [a, b]$, either $g(x) \ge 0$ or $g(x) \le 0$. Then there exists $\xi \in [a, b]$ such that*

$$\int_a^b f(x)g(x)\, dx = f(\xi) \int_a^b g(x)\, dx.$$

---

*Proof.* The proof follows from the intermediate value theorem and the integrability of $f$. □

## 19.4.1 Exercises

### True or false

Decide whether each statement is **true** or **false**. Justify briefly.

1. If $f$ and $g$ are integrable on $[a, b]$, then $f + g$ is integrable on $[a, b]$.
   **Answer:** _____
2. If $f$ is integrable, then $|f|$ is always integrable.
   **Answer:** _____
3. If $f(x) \ge 0$ and $\int_a^b f(x)\, dx = 0$, then $f(x) = 0$ for all $x \in [a, b]$.
   **Answer:** _____
4. $|\int_a^b f(x)\, dx| \le \int_a^b |f(x)|\, dx.$
   **Answer:** _____
5. If $f(x) \le g(x)$ for all $x \in [a, b]$, then $\int_a^b f(x)\, dx \le \int_a^b g(x)\, dx$.
   **Answer:** _____

## Multiple choice questions

Choose the correct option (A, B, C, or D).

1. If $f$ is integrable on $[a, b]$ and $c \in \mathbb{R}$, then
   A. $cf$ is always integrable          B. $cf$ is never integrable
   C. $cf$ is integrable only if $c > 0$   D. $cf$ is integrable only if $c \geq 0$
   **Answer:** _____

2. If $f$ is integrable on $[a, b]$, then which of these is always integrable?
   A. $f^+$   B. $f^-$
   C. $|f|$   D. All of the above
   **Answer:** _____

3. If $f$ is continuous on $[a, b]$, then
   A. it may not be integrable          B. it is always integrable
   C. it is integrable only if monotone   D. none of the above
   **Answer:** _____

4. The integral $\int_a^b f(x)\, dx$ equals zero implies
   A. $f(x) = 0$ for all $x$   B. $f(x) = 0$ at continuity points
   C. $f(x) \geq 0$          D. $f(x)$ is negative
   **Answer:** _____

5. The function $f(x) = \sqrt{|x|}$ is integrable on $[-1, 1]$ because
   A. it is continuous          B. it is monotone
   C. it has finitely many discontinuities   D. none of these
   **Answer:** _____

## Matching exercise

Match each statement with the correct property or theorem.

1. $\int_a^b (f + g) = \int_a^b f + \int_a^b g$          A. Integrability of composite functions

2. $\int_a^b cf = c \int_a^b f$                          B. Nonnegative integrals property

3. $\int_a^b |f| \geq |\int_a^b f|$                       C. Linearity of integral

4. $f(x) \geq 0 \Rightarrow \int_a^b f(x)\, dx \geq 0$    D. Integral absolute value inequality

5. $h = \phi \circ f$ integrable if $\phi$ continuous   E. Integral scaling

**Answers: 1.**_____   **2.**_____   **3.**_____   **4.**_____   **5.**_____

## Fill in the blanks

Complete each sentence with the most appropriate word or phrase.

1. The function $f^+$ represents the _____ part of $f$.
   **Answer:** _____

2. If $f(x) \leq g(x)$, the integral inequality is _____.
   **Answer:** _____

3. The integral of a nonnegative function is always _____.

 **Answer:** _____

4. If $f(x) \geq 0$ and $\int_a^b f(x)\,dx = 0$, then $f(x)$ equals zero at its _____ points.

 **Answer:** _____

5. The integral over an interval can be split into integrals over _____.

 **Answer:** _____

## 19.5 Fundamental theorem of calculus

Let $f$ be integrable on $[a, b]$ and assume that for every $x \in [a, b]$, the restriction of $f$ to $[a, x]$ is integrable. Define the function $F : [a, b] \to \mathbb{R}$ by

$$F(x) = \int_a^x f(t)\,dt.$$

**Theorem 19.17** (Continuity of the indefinite integral). *If $f$ is integrable on $[a, b]$, then the function*

$$F(x) = \int_a^x f(t)\,dt$$

*is continuous on $[a, b]$.*

*Proof.* For any $\xi \in [a, b]$, we have

$$F(x) = \int_a^x f(t)\,dt = \int_a^\xi f(t)\,dt + \int_\xi^x f(t)\,dt = F(\xi) + \int_\xi^x f(t)\,dt,$$

hence

$$F(x) - F(\xi) = \int_\xi^x f(t)\,dt.$$

Since $f$ is bounded on $[a, b]$ by $M = \sup_{x \in [a,b]} |f(x)|$, it follows that

$$|F(x) - F(\xi)| = \left| \int_\xi^x f(t)\,dt \right| \leq M|x - \xi|.$$

Therefore, $F$ is continuous. $\square$

The following theorem states the fundamental theorem of calculus, linking differentiation and integration.

**Theorem 19.18** (Fundamental theorem of calculus (differentiation of integrals)). *If $f$ is integrable on $[a,b]$ and*

$$F(x) = \int_a^x f(t)\, dt,$$

*then F is differentiable and for every $x \in (a,b)$ at which f is continuous,*

$$F'(x) = f(x).$$

*Proof.* Fix $x \in (a,b)$ such that $f$ is continuous at any $x \in (a,b)$. For $h > 0$,

$$F(x+h) - F(x) = \int_x^{x+h} f(s)\, ds.$$

Given $\varepsilon > 0$, by the continuity of $f$ at $x$ there exists $h$ small enough such that for all $s \in (x, x+h)$,

$$f(x) - \varepsilon \le f(s) \le f(x) + \varepsilon.$$

Hence,

$$\int_x^{x+h} (f(x) - \varepsilon)\, ds \le \int_x^{x+h} f(s)\, ds \le \int_x^{x+h} (f(x) + \varepsilon)\, ds,$$

and dividing by $h$,

$$\left| \frac{F(x+h) - F(x)}{h} - f(x) \right| < \varepsilon.$$

This shows $F'(x) = f(x)$. $\qquad\qquad\square$

**Theorem 19.19** (Second fundamental theorem of calculus (evaluation theorem)). *If $f$ is integrable on $[a,b]$ and $G$ is an antiderivative of $f$ on $[a,b]$, then*

$$\int_a^b f(t)\, dt = G(b) - G(a).$$

*Proof.* Define $F(x) = \int_a^x f(t)\,dt$ and $G(x) = c + \int_a^x f(t)\,dt$ for some constant $c$. Then

$$G(a) = c + \int_a^a f(t)\,dt = c,$$

$$G(b) = c + \int_a^b f(t)\,dt,$$

so

$$G(b) - G(a) = \int_a^b f(t)\,dt. \qquad \square$$

---

**Theorem 19.20 (Integration by parts).** *Let $f, g$ be differentiable functions on $[a, b]$. Then*

$$\int_a^b f'(x)g(x)\,dx = f(x)g(x)\Big|_a^b - \int_a^b f(x)g'(x)\,dx.$$

---

*Proof.* Let $f, g$ be differentiable functions on $[a, b]$. By the product rule, we have

$$(fg)'(x) = f'(x)g(x) + f(x)g'(x).$$

Integrating both sides over the interval $[a, b]$ yields

$$\int_a^b (fg)'(x)\,dx = \int_a^b f'(x)g(x)\,dx + \int_a^b f(x)g'(x)\,dx.$$

Using the fundamental theorem of calculus for the left-hand side, we obtain

$$\int_a^b (fg)'(x)\,dx = f(x)g(x)\Big|_a^b = f(b)g(b) - f(a)g(a).$$

Thus,

$$f(x)g(x)\Big|_a^b = \int_a^b f'(x)g(x)\,dx + \int_a^b f(x)g'(x)\,dx.$$

Solving for the first integral gives

$$\int_a^b f'(x)g(x)\,dx = f(x)g(x)\Big|_a^b - \int_a^b f(x)g'(x)\,dx,$$

which is the desired result. $\qquad \square$

**Theorem 19.21** (Change of variable: first form). *Let f be continuous (and thus integrable) on $[a, b]$ and suppose $\phi : [c, d] \to [a, b]$ and $g : [c, d] \to \mathbb{R}$ satisfy*

$$f(\phi(t))\phi'(t) = g(t),$$

*where $\phi$ is continuously differentiable at $[c, d]$ so that g is integrable on $[c, d]$. Then*

$$\int_c^d g(t)\, dt = \int_{\phi(c)}^{\phi(d)} f(x)\, dx.$$

*Proof.* Consider the function

$$F(x) = \int_a^x f(u)\, du.$$

By the fundamental theorem of calculus, $F'(x) = f(x)$ for all $x \in [a, b]$.
Define

$$H(t) = F(\phi(t)).$$

By the chain rule,

$$H'(t) = F'(\phi(t))\, \phi'(t) = f(\phi(t))\, \phi'(t) = g(t).$$

Integrating $H'(t) = g(t)$ over $[c, d]$, we obtain

$$\int_c^d g(t)\, dt = \int_c^d H'(t)\, dt = H(d) - H(c) = F(\phi(d)) - F(\phi(c)).$$

Using the definition of $F$, this becomes

$$\int_c^d g(t)\, dt = \int_a^{\phi(d)} f(x)\, dx - \int_a^{\phi(c)} f(x)\, dx = \int_{\phi(c)}^{\phi(d)} f(x)\, dx. \qquad \square$$

**Theorem 19.22** (Change of variables: second form). *Let $f : [a, b] \to \mathbb{R}$ be continuous and thus integrable. Let $\phi$ be any strictly increasing (or decreasing) continuously differentiable bijection defined on the interval $[m, M]$, where $m = \min\{\phi^{-1}(a), \phi^{-1}(b)\}$ and $M = \max\{\phi^{-1}(a), \phi^{-1}(b)\}$. Then we have*

$$\int_a^b f(x)\, dx = \int_{\phi^{-1}(a)}^{\phi^{-1}(b)} f(\phi(t))\, \phi'(t)\, dt.$$

*Proof.* Note that $f(\phi(t))\,\phi'(t)$ is a continuous function and thus integrable on the interval $[m, M]$.

Since $f$ is integrable on $[a, b]$, we can define the function $F(x) = \int_a^x f(u)\,du$ for any $x \in [a, b]$. By the fundamental theorem of calculus,

$$\int_a^b f(x)\,dx = F(b) - F(a).$$

Consider the substitution $x = \phi(t)$, where $t$ ranges from $\phi^{-1}(a)$ to $\phi^{-1}(b)$. Then

$$\frac{d}{dt}F(\phi(t)) = F'(\phi(t)) \cdot \phi'(t) = f(\phi(t))\,\phi'(t).$$

If $\phi(t)$ is strictly increasing then $\phi^{-1}$ is also strictly increasing, and thus $\phi^{-1}(a) \le \phi^{-1}(b)$. Therefore,

$$\int_{\phi^{-1}(a)}^{\phi^{-1}(b)} f(\phi(t))\,\phi'(t)\,dt = F(\phi(\phi^{-1}(b))) - F(\phi(\phi^{-1}(a))) = F(b) - F(a).$$

Hence,

$$\int_a^b f(x)\,dx = \int_{\phi^{-1}(a)}^{\phi^{-1}(b)} f(\phi(t))\,\phi'(t)\,dt.$$

This completes the proof.

In the case where $\phi$ is strictly decreasing note that $\phi' < 0$ and

$$\int_{\phi^{-1}(a)}^{\phi^{-1}(b)} f(\phi(t))\,\phi'(t)\,dt = - \int_{\phi^{-1}(b)}^{\phi^{-1}(a)} f(\phi(t))\,\phi'(t)\,dt. \qquad \square$$

**Remark 27** (How to compute a definite integral). In general, when we want to compute a definite integral, we can first determine its indefinite integral and then apply the second fundamental theorem of integral calculus (Theorem 19.19), without resorting to Theorem 19.21 or Theorem 19.22. However, there are functions whose antiderivatives cannot be expressed in closed form, see Table 19.1. In such cases, if a substitution is required, the only available method is to use Theorem 19.21 or Theorem 19.22.

**Table 19.1:** Examples of functions without an elementary antiderivative.

| Function | Notes |
|---|---|
| $e^{-x^2}$ | Related to the error function erf($x$); no closed-form antiderivative |
| $\frac{\sin x}{x}$ | Known as the sinc function; antiderivative defines the sine integral Si($x$) |
| $\frac{\cos x}{x}$ | Antiderivative defines the cosine integral Ci($x$) |
| $(\ln x)^n$ | Requires special functions for $n \neq -1$ |
| $e^{x^2}$ | Leads to the Dawson integral or other special functions |
| $\frac{1}{\ln x}$ | Defines the logarithmic integral Li($x$) |

**Example 19.3.** The function $f(t) = e^{-t^2}$ is not elementary integrable, but the integral

$$\frac{2}{\sqrt{\pi}} \int_0^x e^{-t^2}\, dt$$

defines the *error function* erf($x$), a special function widely used in probability and statistics. From standard Gaussian integral properties,

$$\int_0^\infty e^{-t^2}\, dt = \frac{\sqrt{\pi}}{2} \quad \text{and} \quad \int_{-\infty}^\infty e^{-t^2}\, dt = \sqrt{\pi}.$$

Consider now the integral

$$\frac{2}{\sqrt{\pi}} \int_0^x e^{-(t-2)^2}\, dt.$$

Using the substitution $\phi(t) = t - 2$, with $f(x) = e^{-x^2}$ and $g(t) = f(\phi(t))\phi'(t)$, by Theorem 19.21 we have

$$\frac{2}{\sqrt{\pi}} \int_0^x g(t)\, dt = \frac{2}{\sqrt{\pi}} \int_{\phi(0)}^{\phi(x)} f(r)\, dr = \text{erf}(x-2) - \text{erf}(-2).$$

More generally, for parameters $m, \sigma > 0$,

$$\frac{2}{\sqrt{\pi}} \int_0^x e^{-\frac{(t-m)^2}{2\sigma^2}}\, dt = \sigma\sqrt{2}\big(\text{erf}(\phi(x)) - \text{erf}(\phi(0))\big),$$

where $\phi(t) = \frac{t-m}{\sigma\sqrt{2}}$.
Moreover, the Gaussian integral over $\mathbb{R}$ satisfies

$$\int_{-\infty}^\infty e^{-\frac{(t-m)^2}{2\sigma^2}}\, dt = \sigma\sqrt{2\pi}.$$

As an application of substitution, evaluate

$$\int_1^2 \frac{e^{-x}}{2\sqrt{x}}\, dx.$$

Using substitution $t = \sqrt{x}$ so that $x = t^2$, we get

$$\int\limits_1^2 \frac{e^{-x}}{2\sqrt{x}}dx = \int\limits_1^{\sqrt{2}} e^{-t^2} dt = \frac{\sqrt{\pi}}{2}\left(\mathrm{erf}(\sqrt{2}) - \mathrm{erf}(1)\right).$$

In the last step we applied the substitution

$$t = \sqrt{x}, \quad x = t^2,$$

which maps the interval $[1, 2]$ onto $[1, \sqrt{2}]$. Notice that $\phi(t) = t^2$ is strictly increasing on $[1, \sqrt{2}]$, has derivative $\phi'(t) = 2t \neq 0$ on this interval which is continuous.

We set

$$f(x) = \frac{e^{-x}}{2\sqrt{x}}, \quad g(t) = e^{-t^2},$$

and verify that

$$f\left(\phi(t)\right)\phi'(t) = f\left(t^2\right)2t = e^{-t^2} = g(t).$$

Therefore, by Theorem 19.22,

$$\int\limits_1^2 f(x)\,dx = \int\limits_{\phi^{-1}(1)}^{\phi^{-1}(2)} g(t)\,dt = \int\limits_1^{\sqrt{2}} e^{-t^2}\,dt,$$

which is precisely the transformed integral we used above.

**Example 19.4.** Calculate the integral

$$\int\limits_0^{\frac{a}{2}} \frac{dx}{\sqrt{a^2 - x^2}}.$$

*Solution.* The antiderivative is known and corresponds to the inverse sine function, so

$$\int \frac{dx}{\sqrt{a^2 - x^2}} = \arcsin\frac{x}{a} + c.$$

Applying the fundamental theorem of calculus, we get

$$\int\limits_0^{\frac{a}{2}} \frac{dx}{\sqrt{a^2 - x^2}} = \arcsin\frac{1}{2} - \arcsin 0 = \frac{\pi}{6}. \qquad \square$$

**Example 19.5.** Calculate the integral

$$\int\limits_{\frac{\pi}{3}}^{\frac{\pi}{2}} \frac{\cos x + 1}{3\sin x + \sin^3 x}dx.$$

*Solution.* The function to integrate has singularities at points $x = k\pi$ with $k = 0, 1, 2, \ldots$. The integrand is continuous on the interval $[\frac{\pi}{3}, \frac{\pi}{2}]$, so the antiderivative is

$$F(x) = \frac{1}{3}\ln(1 - \cos x) - \frac{1}{4}\ln(2 - \cos x) - \frac{1}{12}\ln(\cos x + 2) + c.$$

Therefore, we find

$$\int_{\frac{\pi}{3}}^{\frac{\pi}{2}} \frac{\cos x + 1}{3\sin x + \sin^3 x}\, dx = F\left(\frac{\pi}{2}\right) - F\left(\frac{\pi}{3}\right) \approx 0.177. \qquad \square$$

---

**Example 19.6** (Integral involving a quadratic under a square root). Calculate the integral

$$\int_a^{\frac{1}{3}} \frac{dx}{\sqrt{-2x^2 + x + 1}},$$

for $a \in (-\frac{1}{2}, \frac{1}{3})$.

---

*Solution.* The antiderivative is

$$F(x) = \begin{cases} \sqrt{2}\arctan \frac{1 - \sqrt{-2x^2 + x + 1}}{x\sqrt{2}}, & x \in (-\frac{1}{2}, 0) \cup (0, 1), \\ -\sqrt{2}\arctan \frac{1}{2\sqrt{2}}, & x = 0. \end{cases}$$

Therefore,

$$\int_a^{\frac{1}{3}} \frac{dx}{\sqrt{-2x^2 + x + 1}} = F\left(\frac{1}{3}\right) - F(a). \qquad \square$$

## 19.6 Numerical integration

In many cases, it is impossible or impractical to compute the exact value of a definite integral analytically. Numerical integration techniques provide efficient ways to approximate definite integrals to any desired accuracy. One of the simplest and most widely used methods is the *trapezoidal rule*, which approximates the area under a curve by dividing the interval into small subintervals and replacing the curve on each subinterval by a straight line segment (the trapezoid).

---

**Theorem 19.23** (Trapezoidal rule). *Let $f$ be a function continuous on $[a, b]$. Divide $[a, b]$ into $n$ equal subintervals of width $h = \frac{b-a}{n}$, with nodes $x_0 = a, x_1 = a + h, \ldots, x_n = b$. Then the definite integral of $f$ can be approximated by*

$$\int_a^b f(x)\,dx \approx T_n = \frac{h}{2}\left[f(x_0) + 2\sum_{i=1}^{n-1} f(x_i) + f(x_n)\right].$$

*Moreover, if $f''$ is continuous on $[a,b]$, the error satisfies*

$$\left|\int_a^b f(x)\,dx - T_n\right| \le \frac{(b-a)^3}{12n^2}\max_{x\in[a,b]}\left|f''(x)\right|.$$

*Proof.* We derive the trapezoidal rule by approximating $f(x)$ on each subinterval $[x_{i-1}, x_i]$ by the linear interpolant (see Theorem 13.3) that passes through $(x_{i-1}, f(x_{i-1}))$ and $(x_i, f(x_i))$.

On each subinterval,

$$\int_{x_{i-1}}^{x_i} f(x)\,dx \approx \frac{h}{2}[f(x_{i-1}) + f(x_i)].$$

Summing over all subintervals, the terms $f(x_1), f(x_2), \dots, f(x_{n-1})$ appear twice, except the endpoints

$$\int_a^b f(x)\,dx \approx \frac{h}{2}[f(x_0) + 2f(x_1) + 2f(x_2) + \cdots + 2f(x_{n-1}) + f(x_n)].$$

This yields the trapezoidal rule formula.

For the error, the local error on each subinterval is (by Taylor's theorem)

$$\int_{x_{i-1}}^{x_i} f(x)\,dx - \frac{h}{2}[f(x_{i-1}) + f(x_i)] = -\frac{h^3}{12}f''(\xi_i)$$

for some $\xi_i \in [x_{i-1}, x_i]$. Summing over all $n$ subintervals,

$$\text{Total error} = -\frac{h^3}{12}\sum_{i=1}^{n} f''(\xi_i) = -\frac{(b-a)h^2}{12}\frac{1}{n}\sum_{i=1}^{n} f''(\xi_i).$$

Thus,

$$\left|\int_a^b f(x)\,dx - T_n\right| \le \frac{(b-a)h^2}{12}\max_{x\in[a,b]}\left|f''(x)\right| = \frac{(b-a)^3}{12n^2}\max_{x\in[a,b]}\left|f''(x)\right|. \qquad \square$$

**Example 19.7** (Approximating $\int_0^1 e^x\,dx$ using the trapezoidal rule). Use the trapezoidal rule with $n = 4$ subintervals to approximate the value of $\int_0^1 e^x\,dx$.

*Solution.* First, divide $[0,1]$ into $n = 4$ equal subintervals of width $h = \frac{1-0}{4} = 0.25$. The nodes are:

$$x_0 = 0, \quad x_1 = 0.25, \quad x_2 = 0.5, \quad x_3 = 0.75, \quad x_4 = 1.$$

Evaluate the function at these points:

$$f(x_0) = e^0 = 1,$$
$$f(x_1) = e^{0.25} \approx 1.2840,$$
$$f(x_2) = e^{0.5} \approx 1.6487,$$
$$f(x_3) = e^{0.75} \approx 2.1170,$$
$$f(x_4) = e^1 \approx 2.7183.$$

Apply the trapezoidal rule

$$T_4 = \frac{h}{2}[f(x_0) + 2f(x_1) + 2f(x_2) + 2f(x_3) + f(x_4)],$$

$$T_4 = \frac{0.25}{2}[1 + 2(1.2840) + 2(1.6487) + 2(2.1170) + 2.7183]$$

$$= 0.125[1 + 2.5680 + 3.2974 + 4.2340 + 2.7183]$$

$$= 0.125 \times (1 + 2.5680 + 3.2974 + 4.2340 + 2.7183)$$

$$= 0.125 \times 13.8177 = 1.7272.$$

The exact value is $\int_0^1 e^x dx = e^1 - e^0 = 2.7183 - 1 = 1.7183$. So, the error is about $1.7272 - 1.7183 = 0.0089.$  □

**Example 19.8** (Approximating $\int_1^2 \frac{\cos x}{x} dx$ using the trapezoidal rule). We approximate

$$I = \int\limits_1^2 \frac{\cos x}{x} dx$$

by the composite trapezoidal rule with a small number of subintervals.
   Let $f(x) = \frac{\cos x}{x}$ on $[a, b] = [1, 2]$. Choose $n = 4$, so

$$h = \frac{b-a}{n} = \frac{2-1}{4} = 0.25, \quad x_i = a + ih = 1 + 0.25\,i \ (i = 0, \ldots, 4).$$

Function values (rounded to 6 decimals):

$$f(x_0) = \frac{\cos 1}{1} \approx 0.540302,$$
$$f(x_1) = \frac{\cos 1.25}{1.25} \approx 0.252258,$$
$$f(x_2) = \frac{\cos 1.5}{1.5} \approx 0.047158,$$

$$f(x_3) = \frac{\cos 1.75}{1.75} \approx -0.101855,$$

$$f(x_4) = \frac{\cos 2}{2} \approx -0.208073.$$

**Trapezoidal approximation.**

$$T_4 = \frac{h}{2}\left[f(x_0) + 2\sum_{i=1}^{3} f(x_i) + f(x_4)\right]$$

$$= \frac{0.25}{2}\left[0.540302 + 2(0.252258 + 0.047158 - 0.101855) + (-0.208073)\right].$$

Numerically,

$$T_4 \approx 0.0909188948.$$

**Rigorous error bound for the trapezoidal rule.**
   If $f''$ is continuous on $[a, b]$, then

$$\left|\int_a^b f(x)\,dx - T_n\right| \le \frac{(b-a)^3}{12\,n^2} \max_{x\in[a,b]} \left|f''(x)\right|.$$

For $f(x) = \frac{\cos x}{x}$ we compute

$$f''(x) = \frac{2\sin x}{x^2} + \frac{2\cos x}{x^3} - \frac{\cos x}{x}.$$

On $[1, 2]$ one may take the (sharp) endpoint bound $\max_{[1,2]} |f''(x)| \le |f''(1)| = 2\sin 1 + \cos 1 \approx 2.223244$. Hence, with $n = 4$,

$$\left|\int_1^2 f(x)\,dx - T_4\right| \le \frac{(2-1)^3}{12\cdot 4^2}(2.223244) = \frac{2.223244}{192} \approx 1.15794 \times 10^{-2}.$$

## 19.7  Stirling's formula

We calculate the indefinite integral

$$\int \sin^n x\,dx = -\frac{1}{n}\sin^{n-1} x \cos x + \frac{n-1}{n}\int \sin^{n-2} x\,dx.$$

Thus,

$$\int_0^{\pi/2} \sin^n x\,dx = \frac{n-1}{n}\int_0^{\pi/2} \sin^{n-2} x\,dx.$$

Continuing this process, we obtain

$$\int\limits_0^{\pi/2} \sin^{2m} x \, dx = \frac{2m-1}{2m} \cdot \frac{2m-3}{2m-2} \cdots \frac{1}{2} \cdot \frac{\pi}{2},$$

$$\int\limits_0^{\pi/2} \sin^{2m+1} x \, dx = \frac{2m}{2m+1} \cdot \frac{2m-2}{2m-1} \cdots \frac{2}{3}.$$

Dividing these relations yields

$$\frac{\pi}{2} = \frac{2 \cdot 2}{1 \cdot 3} \cdot \frac{4 \cdot 4}{3 \cdot 5} \cdots \frac{2m \cdot 2m}{(2m-1)(2m+1)} \cdot \frac{\int_0^{\pi/2} \sin^{2m} x \, dx}{\int_0^{\pi/2} \sin^{2m+1} x \, dx}.$$

The last fraction on the right-hand side converges to 1 as $m \to \infty$. Indeed, for $0 < x < \frac{\pi}{2}$ we have

$$0 < \int\limits_0^{\pi/2} \sin^{2m+1} x \, dx \le \int\limits_0^{\pi/2} \sin^{2m} x \, dx \le \int\limits_0^{\pi/2} \sin^{2m-1} x \, dx.$$

Dividing term-wise by $\int_0^{\pi/2} \sin^{2m+1} x \, dx$ and using

$$\int\limits_0^{\pi/2} \sin^n x \, dx = \frac{n-1}{n} \int\limits_0^{\pi/2} \sin^{n-2} x \, dx,$$

we get

$$1 \le \frac{\int_0^{\pi/2} \sin^{2m} x \, dx}{\int_0^{\pi/2} \sin^{2m+1} x \, dx} \le 1 + \frac{1}{2m}.$$

Hence, the desired limit holds. Finally, the equality

$$\frac{\pi}{2} = \lim_{m \to \infty} \frac{2}{1} \cdot \frac{2}{3} \cdots \frac{2m}{2m-1} \cdot \frac{2m}{2m+1} \quad \text{(Wallis product)}$$

holds.

Next, we rewrite this limit as

$$\frac{\pi}{2} = \lim_{m \to \infty} \frac{2^2 \cdot 4^2 \cdots (2m-2)^2}{3^2 \cdot 5^2 \cdots (2m-1)^2} \cdot 2m,$$

and thus,

$$\sqrt{\frac{\pi}{2}} = \lim_{m \to \infty} \frac{2^2 \cdot 4^2 \cdots (2m-2)^2}{(2m)!} \sqrt{2m} = \lim_{m \to \infty} \frac{(2^2 \cdot 1^2)(2^2 \cdot 2^2) \cdots (2^2 \cdot m^2)}{(2m)! \sqrt{2m}}.$$

Hence,

$$\sqrt{\pi} = \lim_{m \to \infty} \frac{(m!)^2 2^{2m}}{(2m)! \sqrt{m}} \quad \text{(Wallis formula)}. \tag{19.1}$$

Using the Wallis product, one can prove the Stirling inequality (see [3], p. 504):

$$\sqrt{2\pi n}\left(\frac{n}{e}\right)^n < n! < \sqrt{2\pi n}\left(\frac{n}{e}\right)^n\left(1 + \frac{1}{4n}\right). \tag{19.2}$$

## 19.8 Infinite sums and integrals

Next, we present a useful result that relates infinite sums and integrals via Riemann sums, which is helpful for computing limits of sequences.

**Proposition 19.3** (Riemann sums for continuous functions). *Let f be continuous on [a, b]. Then*

$$\lim_{n \to \infty} \frac{b-a}{n} \sum_{k=1}^{n} f\left(a + \frac{k(b-a)}{n}\right) = \int_a^b f(x)\, dx.$$

*In particular, if [a, b] = [0, 1], then*

$$\lim_{n \to \infty} \frac{1}{n} \sum_{k=1}^{n} f\left(\frac{k}{n}\right) = \int_0^1 f(x)\, dx.$$

*Proof.* Since $f$ is continuous on $[a, b]$, it is integrable. Divide the interval into $n$ equal parts of length $\frac{b-a}{n}$ and choose sample points $\xi_k = a + k\frac{b-a}{n}$, $k = 1, 2, \ldots, n$. The Riemann sum is

$$\sum_{k=1}^{n} f(\xi_k)(x_k - x_{k-1}) = \sum_{k=1}^{n} f\left(a + \frac{k(b-a)}{n}\right)\frac{b-a}{n}.$$

Taking the limit $n \to \infty$, we get the integral. $\qquad \square$

**Example 19.9.** Express the definite integral $\int_0^1 x^2\, dx$ as a limit of an infinite sum.

*Solution.* Since the function $x^2$ is continuous on $[0, 1]$, it is integrable and the proposition applies. Hence,

$$\int_0^1 x^2\, dx = \lim_{n \to \infty} \frac{1}{n} \sum_{k=1}^{n} \frac{k^2}{n^2}.$$

Also, using the formula for the sum of squares,

$$\frac{1}{n} \sum_{k=1}^{n} \frac{k^2}{n^2} = \frac{1}{n^3}(1^2 + 2^2 + \cdots + n^2) = \frac{n(n+1)(2n+1)}{6n^3}.$$

Therefore,

$$\int_0^1 x^2 \, dx = \lim_{n\to\infty} \frac{n(n+1)(2n+1)}{6n^3} = \frac{1}{3},$$

which can also be verified by direct integration. ☐

**Example 19.10.** Calculate the limit

$$\lim_{n\to\infty} \left( \frac{1}{n+1} + \frac{1}{n+2} + \cdots + \frac{1}{n+n} \right).$$

*Solution.* Rewrite the sum as

$$\lim_{n\to\infty} \frac{1}{n} \left( \frac{1}{1+\frac{1}{n}} + \frac{1}{1+\frac{2}{n}} + \cdots + \frac{1}{1+\frac{n}{n}} \right) = \lim_{n\to\infty} \frac{1}{n} \sum_{k=1}^{n} \frac{1}{1+\frac{k}{n}}.$$

Set $f(x) = \frac{1}{1+x}$. By the previous proposition,

$$\int_0^1 \frac{1}{1+x} \, dx = \lim_{n\to\infty} \frac{1}{n} \sum_{k=1}^{n} \frac{1}{1+\frac{k}{n}}.$$

Therefore,

$$\lim_{n\to\infty} \left( \frac{1}{n+1} + \frac{1}{n+2} + \cdots + \frac{1}{n+n} \right) = \int_0^1 \frac{1}{1+x} \, dx = \ln 2. \qquad ☐$$

**Example 19.11.** Calculate the limit

$$\lim_{n\to\infty} \frac{1}{n} \left( \sin \frac{t}{n} + \sin \frac{2t}{n} + \cdots + \sin \frac{(n-1)t}{n} \right).$$

*Solution.* Set $f(x) = \sin x$. Then

$$\int_0^t \sin x \, dx = \lim_{n\to\infty} \frac{t}{n} \sum_{k=1}^{n} \sin \frac{kt}{n} = 1 - \cos t.$$

Note that

$$\lim_{n\to\infty} \frac{1}{n} \left( \sin \frac{t}{n} + \sin \frac{2t}{n} + \cdots + \sin \frac{(n-1)t}{n} \right)$$

$$= \lim_{n\to\infty} \frac{1}{n} \sum_{k=1}^{n-1} \sin\frac{kt}{n}$$

$$= \lim_{n\to\infty} \frac{1}{n} \left( \sum_{k=1}^{n} \sin\frac{kt}{n} - \sin t \right)$$

$$= \frac{1}{t} \lim_{n\to\infty} \frac{t}{n} \left( \sum_{k=1}^{n} \sin\frac{kt}{n} - \sin t \right)$$

$$= \frac{1 - \cos t}{t} - \lim_{n\to\infty} \frac{\sin t}{n}$$

$$= \frac{1 - \cos t}{t}. \qquad\qquad \square$$

**Exercise 23.** Calculate the following limits:

$$\lim_{n\to\infty} \left( \frac{n}{n^2 + 1^2} + \frac{n}{n^2 + 2^2} + \cdots + \frac{n}{n^2 + n^2} \right),$$

$$\lim_{n\to\infty} \frac{1^p + 2^p + \cdots + n^p}{n^{p+1}}, \quad p > -1,$$

$$\lim_{n\to\infty} \left( \frac{1}{\sqrt{n^2 + 1^2}} + \frac{1}{\sqrt{n^2 + 2^2}} + \cdots + \frac{1}{\sqrt{n^2 + n^2}} \right),$$

$$\lim_{n\to\infty} \sum_{k=1}^{n} \frac{n}{n^2 + k^2 x^2}, \quad x \neq 0,$$

$$\lim_{n\to\infty} \frac{\sqrt{n+1} + \sqrt{n+2} + \cdots + \sqrt{2n-1}}{n^{3/2}}.$$

## 19.9 Python and AI

**AI Request**

Calculate the limit

$$\lim_{n\to\infty} \left( \frac{1}{n+1} + \frac{1}{n+2} + \cdots + \frac{1}{n+n} \right).$$

Write a Python code to compare the above result.

## 19.10 Exercises

### True or false

Decide whether each statement is **true** or **false**. Justify briefly.

1. If $f$ is integrable on $[a, b]$, then the function $F(x) = \int_a^x f(t)\, dt$ is continuous on $[a, b]$.

   **Answer:** _____

2. If $f$ is integrable on $[a, b]$ and continuous at $x_0 \in [a, b]$, then $F(x) = \int_a^x f(t)\, dt$ is differentiable at $x_0$ and $F'(x_0) = f(x_0)$.

   **Answer:** _____

3. The integral $\int_a^b f(t)\, dt$ can always be evaluated using an antiderivative of $f$.

   **Answer:** _____

4. If $f(x) = e^{-x^2}$, then $\int_0^\infty f(x)\, dx = \sqrt{\pi}$.

   **Answer:** _____

5. The substitution rule for definite integrals requires that the substitution function $\phi(t)$ be strictly monotonic.

   **Answer:** _____

## Multiple choice questions

Choose the correct option (A, B, C, or D).

1. Which theorem guarantees that $F(x) = \int_a^x f(t)\, dt$ is continuous if $f$ is integrable?

   A. Fundamental theorem of calculus    B. Mean value theorem

   C. Continuity of the indefinite integral    D. Riemann integrability criterion

   **Answer:** _____

2. If $G(x)$ is an antiderivative of $f(x)$, which formula expresses $\int_a^b f(x)\, dx$?

   A. $G(a) - G(b)$    B. $G(b) - G(a)$

   C. $G(b) + G(a)$    D. $G(a) + G(b)$

   **Answer:** _____

3. To evaluate $\int_0^1 x^2\, dx$ as a limit of sums, which expression should be used?

   A. $\lim_{n\to\infty} \frac{1}{n} \sum_{k=1}^n \frac{k^2}{n^2}$    B. $\lim_{n\to\infty} \frac{1}{n} \sum_{k=1}^n \frac{k}{n}$

   C. $\lim_{n\to\infty} \frac{1}{n^2} \sum_{k=1}^n k^2$    D. $\lim_{n\to\infty} \frac{1}{n^3} \sum_{k=1}^n k^3$

   **Answer:** _____

4. What is the value of $\int_0^\infty e^{-t^2}\, dt$?

   A. $\frac{\sqrt{\pi}}{2}$    B. $\sqrt{\pi}$

   C. $\frac{\pi}{2}$    D. $\pi$

   **Answer:** _____

5. Which substitution simplifies $\int_1^2 \frac{e^{-x}}{2\sqrt{x}}\, dx$?

   A. $t = \sqrt{x}$    B. $t = x^2$

   C. $t = e^{-x}$    D. $t = \ln x$

   **Answer:** _____

## Fill in the blanks

Complete each sentence with the most appropriate word or phrase.

1. The function $F(x) = \int_a^x f(t)\, dt$ is guaranteed to be _____ if $f$ is integrable on $[a, b]$.

   **Answer:** _____

2. The fundamental theorem of calculus connects _____ and _____.

   **Answer:** _____

3. The integral $\int_a^b f(x)\,dx$ can be computed using an antiderivative $G(x)$ via the formula $\int_a^b f(x)\,dx =$ _____.

Answer: _____

4. The error function $\mathrm{erf}(x)$ is defined as $\frac{2}{\sqrt{\pi}} \int_0^x$ _____ $dt$.

Answer: _____

5. To evaluate $\int_0^1 \frac{1}{1+x}\,dx$ as a limit of sums, we use the function $f(x) =$ _____.

Answer: _____

## Computational problems

Evaluate the following integrals or limits:

1. Compute $\int_0^{\frac{\pi}{2}} \sin^2 x\,dx$.

Answer: _____

2. Evaluate $\lim_{n\to\infty} \frac{1}{n} \sum_{k=1}^n \frac{1}{1+\frac{k}{n}}$.

Answer: _____

3. Use substitution to compute $\int_1^2 \frac{e^{-x}}{2\sqrt{x}}\,dx$.

Answer: _____

4. Find the value of $\int_0^1 x^2\,dx$ using the formula for the sum of squares.

Answer: _____

5. Compute $\int_0^\infty e^{-t^2}\,dt$.

Answer: _____

## Python-based exercises

### Exercise 1: Darboux sums for $f(x) = x^2$

Write a Python program to compute the lower and upper Darboux sums for the function $f(x) = x^2$ on the interval $[0, 1]$ using $n = 4, 8, 16$ subintervals. Compare the results as $n$ increases and observe how the sums converge to the integral.

### Exercise 2: Trapezoidal rule approximation

Implement the trapezoidal rule in Python to approximate the integral $\int_0^1 e^x\,dx$ using $n = 4, 8, 16$ subintervals. Compare the numerical results with the exact value of the integral $(e - 1)$.

### Exercise 3: Composite trapezoidal rule for $\int_1^2 \frac{\cos x}{x}\,dx$

Use Python to approximate the integral $\int_1^2 \frac{\cos x}{x}\,dx$ using the composite trapezoidal rule with $n = 4$ subintervals. Compute the function values at the nodes and calculate the approximation.

### Exercise 4: Riemann sum approximation

Write a Python program to approximate the integral $\int_0^1 x^2\, dx$ using Riemann sums. Use the formula

$$\int_0^1 x^2\, dx = \lim_{n\to\infty} \frac{1}{n} \sum_{k=1}^{n} \frac{k^2}{n^2}.$$

Compute the sum for $n = 10, 100, 1000$ and compare the results with the exact value $\left(\frac{1}{3}\right)$.

### Exercise 5: Limit of a sequence using Riemann sums

Consider the sequence

$$\lim_{n\to\infty} \left( \frac{1}{n+1} + \frac{1}{n+2} + \cdots + \frac{1}{n+n} \right).$$

Rewrite this as a Riemann sum and use Python to compute the limit numerically for large $n$. Compare your result with the theoretical value ($\ln 2$).

### Exercise 6: Numerical integration of sin x

Write a Python program to compute the integral $\int_0^t \sin x\, dx$ for $t = \pi/2$ using the trapezoidal rule with $n = 10, 50, 100$ subintervals. Compare the results with the exact value $(1 - \cos t)$.

### Exercise 7: Wallis formula approximation

The Wallis formula is given by

$$\sqrt{\pi} = \lim_{m\to\infty} \frac{(m!)^2 2^{2m}}{(2m)! \sqrt{m}}.$$

Write a Python program to approximate $\sqrt{\pi}$ using this formula for $m = 10, 100, 1000$. Compare the results with the actual value of $\sqrt{\pi}$.

### Exercise 8: Gaussian integral approximation

Approximate the Gaussian integral $\int_0^\infty e^{-t^2}\, dt$ using numerical integration in Python. Use the trapezoidal rule on a finite interval $[0, L]$ (e. g., $L = 10$) and refine the partition to estimate the integral. Compare the result with the theoretical value $\left(\frac{\sqrt{\pi}}{2}\right)$.

### Exercise 9: Error function approximation

The error function is defined as

$$\mathrm{erf}(x) = \frac{2}{\sqrt{\pi}} \int_0^x e^{-t^2}\, dt.$$

Write a Python program to approximate erf(1) using the trapezoidal rule with $n$ = 10, 50, 100 subintervals. Compare the result with the exact value.

### Exercise 10: Convergence of Darboux sums

Write a Python program to compute the lower and upper Darboux sums for a general function $f(x)$ on an interval $[a, b]$. Test the program for $f(x) = x^2$ on $[0, 1]$ and observe the convergence of the sums as the number of subintervals increases.

# 20 The logarithmic and exponential functions

## Contents

The logarithmic and exponential functions play a fundamental role in mathematics, modeling growth, decay, and many natural phenomena. Beginning with the concept of integer powers, we extend the idea of exponentiation to arbitrary real exponents using the natural logarithm, defined via integration. The logarithm function, strictly increasing and continuous, provides a bridge between multiplication and addition, while its inverse, the exponential function, generalizes powers to all real exponents. These functions possess elegant properties—such as $e^x$ being its own derivative and $\ln(xy) = \ln x + \ln y$—that underpin both theoretical and applied mathematics. In this chapter, we explore their definitions, properties, and connections, culminating in a unified framework for powers, roots, and logarithms. Computational and graphical examples, along with Python-based exercises, provide further insight into their behavior and applications.

## 20.1 Introduction

Logarithmic and exponential functions are two of the most fundamental building blocks in mathematics, with profound implications across pure and applied sciences. Historically, exponentiation was first understood for positive integers as repeated multiplication, but real progress came when the operation was extended to rational and then arbitrary real exponents. This leap requires a deep understanding of the structure of the real numbers and the notion of continuity.

The key tool for this extension is the natural logarithm, a function initially defined through the definite integral $\ln x = \int_1^x \frac{1}{t}\, dt$ for $x > 0$. This approach not only gives meaning to the logarithm for all positive real numbers, but also connects it fundamentally to the area under the hyperbola $y = 1/x$. The logarithm converts products into sums and powers into multiples, providing a powerful algebraic and analytic tool.

The exponential function, $e^x$, emerges as the unique function that is its own derivative and as the inverse of the natural logarithm. Its remarkable properties—such as $e^{x+y} = e^x e^y$, strict monotonicity, and continuity—form the basis for the general definition of $a^x$ for any $a > 0$ and $x \in \mathbb{R}$ via $a^x = e^{x \ln a}$. This unifies all cases: integer, rational, and real exponents, while remaining compatible with our original notion of repeated multiplication.

https://doi.org/10.1515/9783112228289-020

The logarithmic and exponential functions are central to solving equations involving growth and decay, modeling processes in physics, biology, finance, and information theory. Their properties also play a critical role in calculus, differential equations, and complex analysis. For example, the function $f(x) = e^x$ appears in solutions to many linear differential equations, while logarithms linearize exponential growth and allow us to solve equations involving unknown exponents.

In this chapter, we develop a rigorous foundation for logarithmic and exponential functions. We will start by recalling the algebraic definition of powers, then introduce the natural logarithm through integration, establish its key properties, and use it to define the exponential function and real powers. We will explore the domains, ranges, and differentiability of these functions, their inverse relationships, and the extension of exponentiation to all real exponents. Connections with rational powers, roots, and even complex exponents will be discussed.

Through a careful blend of theoretical exposition, illustrative examples, and computational exercises (including applications of Python for visualization and symbolic manipulation) you will acquire a deep and intuitive understanding of these functions. The chapter closes with a variety of exercises (theoretical, computational, and multiple choice) designed to reinforce the material and challenge your ability to apply the concepts to new problems.

The journey through logarithms and exponentials is not merely an abstract mathematical exercise: it provides a window into the very nature of numbers, growth, and change. Mastery of these concepts is essential for any further study in mathematics, science, and engineering.

## 20.2 Definitions

**Definition 20.1** (Integer powers of a real number). Let $x > 0$ and $n \in \mathbb{N} = \{1, 2, \dots\}$. The $n$-th power of $x$ is

$$x^n := \underbrace{x \cdot x \cdots x}_{n \text{ times}}.$$

For negative exponents (with $x \neq 0$) and $n \in \mathbb{N}$,

$$x^{-n} := \frac{1}{x^n}, \quad \text{and we set } x^0 := 1.$$

Finally, for any integer $n$ and any $x < 0$, the above agrees with

$$x^n := (-1)^n \, |x|^n.$$

In this section, we will generalize the notion of powers of positive real numbers so that it is defined for exponents in $\mathbb{R}$. To do this, we define a new function, the logarithm, and then its inverse, the exponential function. We will show that the exponential function is the generalization we need.

**Definition 20.2** (Natural logarithm). If $x > 0$, the natural logarithm is defined by

$$\ln x = \int_1^x \frac{1}{t} dt.$$

**Theorem 20.1** (Properties of the natural logarithm). *If $x, y \in \mathbb{R}^+$ and $a \in \mathbb{Z}$, then*
- $\ln(xy) = \ln x + \ln y$.
- $\ln(\frac{x}{y}) = \ln x - \ln y$.
- $\ln(x^a) = a \ln x$.

*Proof.* From the fundamental theorem of calculus, $\frac{d}{dx} \ln x = \frac{1}{x}$. Fix $y > 0$ and define $f(x) = \ln(xy)$. Then

$$f'(x) = \frac{1}{x} = \ln'(x),$$

so $f(x) - \ln x = c$ for some constant $c \in \mathbb{R}$. For $x = 1$,

$$\ln(1 \cdot y) = \ln(1) + c = c,$$

thus $c = \ln y$, and therefore

$$\ln(xy) = \ln x + \ln y.$$

For the second identity, set $x = \frac{1}{y}$ in the first identity:

$$\ln 1 = \ln \frac{1}{y} + \ln y.$$

Since $\ln 1 = 0$, it follows that

$$\ln \frac{1}{y} = -\ln y.$$

Hence, the second identity follows easily.

For the last property, assume initially $a \in \mathbb{N}$. Then

$$\ln x^a = \ln \underbrace{x \cdots x}_{a \text{ times}} = \underbrace{\ln x + \cdots + \ln x}_{a \text{ times}} = a \ln x.$$

If $-a \in \mathbb{N}$, then write

$$\ln x^a = \ln \frac{1}{x^{-a}},$$

and proceed as before using the second property of the logarithm. □

We will later generalize the third property to real $a$.

The function ln is continuous and one-to-one since it is strictly increasing. We will prove that its range is all $\mathbb{R}$.

Indeed, the area of the rectangle with base $[1, 2]$ and height $\frac{1}{2}$ is less than $\ln 2$ (see Figure 20.1). Thus, for every $n \in \mathbb{N}$,

$$\ln 2^n = n \ln 2 > \frac{n}{2}, \quad \text{and} \quad \ln 2^{-n} = -n \ln 2 < -\frac{n}{2}.$$

Therefore, by continuity, ln attains all intermediate values and

$$\lim_{x \to 0^+} \ln x = -\infty, \qquad \lim_{x \to \infty} \ln x = +\infty.$$

Hence, ln is a bijection onto $\mathbb{R}$ and has an inverse.

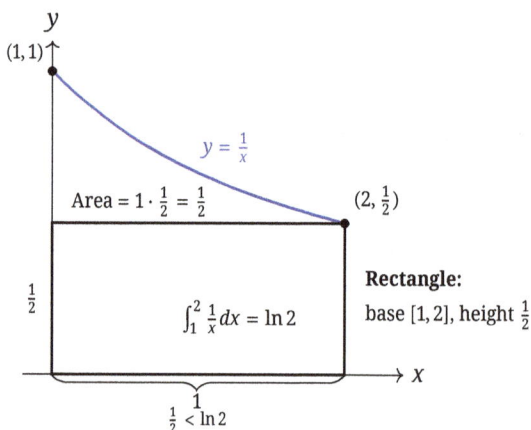

**Figure 20.1:** The area of the rectangle $[1, 2] \times [0, \frac{1}{2}]$ is less than the area under $y = 1/x$ (ln 2).

**Definition 20.3** (Exponential function). We define the exponential function as the inverse of ln, denoted by $\exp(x)$.

**Definition 20.4** (The number $e$ and exponential notation). We denote by $e$ the number $\exp(1)$. Also, define $e^x := \exp(x)$. We call $e$ the base and $x$ the exponent.

The notation $e^x$ suggests powers of real numbers, but this will be proven later to be consistent. For now, this notation simply denotes the inverse of the natural logarithm.

Note that $e^x : \mathbb{R} \to \mathbb{R}^+$ since it is the inverse of $\ln x : \mathbb{R}^+ \to \mathbb{R}$. Since $\ln x$ is continuous and differentiable with nonzero derivative, its inverse is also continuous and differentiable, see Theorem 10.2.

> **Theorem 20.2** (Derivative and integral of $e^x$; normalization of ln). *Let* $\ln : (0, \infty) \to \mathbb{R}$ *be the natural logarithm with* $(\ln x)' = 1/x$ *for* $x > 0$, *and let* $\exp = \ln^{-1}$. *Writing* $e^x := \exp(x)$, *the following hold:*
> (i)  $\frac{d}{dx} e^x = e^x$ *for all* $x \in \mathbb{R}$; *consequently* $\int e^x \, dx = e^x + C$.
> (ii)  $\ln 1 = 0$ *and* $e^0 = 1$.

*Proof.* We use the derivative-of-inverse formula (Theorem 10.2): if $f$ is differentiable with $f'(x) \neq 0$, then $(f^{-1})'(f(x)) = 1/f'(x)$. Apply this with $f = \ln$. For any $x \in \mathbb{R}$,

$$\frac{d}{dx} e^x = (\exp)'(x) = \frac{1}{(\ln)'(\exp(x))} = \frac{1}{1/\exp(x)} = \exp(x) = e^x,$$

which yields the antiderivative $\int e^x \, dx = e^x + C$.

By the defining property of ln, $\ln 1 = \int_1^1 \frac{dt}{t} = 0$. Since exp and ln are inverse functions, $e^0 = \exp(0)$ is the unique $y > 0$ with $\ln y = 0$; hence $y = 1$, so $e^0 = 1$. ◻

> **Definition 20.5** (Definition of $a^x$ for $a > 0$). If $a > 0$, for every real number $x$ define
> $$a^x := e^{x \ln a}.$$
> Also define $0^x = 0$ when $x > 0$.

Obviously, $a^x : \mathbb{R} \to \mathbb{R}^+$. Its derivative is

$$\frac{d}{dx} a^x = a^x \ln a,$$

which follows easily by differentiating $e^{x \ln a}$.

The property $\ln x^a = a \ln x$, previously proven for $a \in \mathbb{N}$, now holds for all $a \in \mathbb{R}$ since

$$\ln x^a = \ln e^{a \ln x} = a \ln x,$$

since $\ln x$, $e^x$ are inverse of each other. If $x > 0$, as $a \to 0$ we have $a^x = e^{x \ln a} \to 0$, so it is natural to define $0^x = 0$ for $x > 0$.

We will now prove that the exponential function is indeed the generalization of powers of positive numbers, which is not obvious. For example, it is obvious that $1^x = 1$. If $x$ is an integer, using the first definition of power,

$$1^x = \underbrace{1 \cdots 1}_{x \text{ times}},$$

which is obviously 1. If $x \notin \mathbb{Z}$, we cannot use this definition. Using the second definition via the logarithm, we have

$$1^x = e^{x \ln 1} = e^0 = 1,$$

showing that the two definitions coincide when the exponent is an integer and the base is 1.

> **Definition 20.6** (Logarithm to the base $a$). Since the function $a^x$ is one-to-one (strictly monotone when $a \neq 1$), its inverse exists and is denoted by $\log_a x$.

As the inverse of $a^x$, we have $\log_a x : \mathbb{R}^+ \to \mathbb{R}$ (for $a \neq 1$). We can compute $\log_a x$ in terms of the natural logarithm and prove that

$$\log_a x = \frac{\ln x}{\ln a}.$$

Indeed,

$$\log_a x = y \iff x = a^y,$$

and

$$\frac{\ln x}{\ln a} = z \iff x = e^{z \ln a} \iff x = a^z.$$

Thus,

$$a^y = a^z \implies y = z.$$

From this relation, we can compute the derivative of $\log_a x$ and prove that $\log_a x$ satisfies the same properties as $\ln x$.

Using the above, we can prove the properties of the exponential function: for any $a > 0$,

- $a^{x+y} = a^x a^y$. Indeed,

$$a^{x+y} = a^{\log_a a^x + \log_a a^y} = a^{\log_a(a^x a^y)} = a^x a^y.$$

- $(a^x)^y = a^{xy}$. If $b = a^x$, then

$$(a^x)^y = b^y = e^{y \ln b} = e^{y \ln a^x} = e^{xy \ln a} = a^{xy}.$$

- $(ab)^x = a^x b^x$ for $a, b \in \mathbb{R}^+, x \in \mathbb{R}$. Indeed,

$$(ab)^x = e^{x \ln(ab)} = e^{x \ln a + x \ln b} = e^{x \ln a} e^{x \ln b} = a^x b^x.$$

Note that $a^0 = e^{0 \cdot \ln a} = 1$ and $\log_a a = \frac{\ln a}{\ln a} = 1$. Also, from the first property,

$$a^{-x} = \frac{1}{a^x}$$

since

$$a^0 = a^{x-x} = a^x a^{-x} = 1.$$

The function $f(x) = x^t$ for $t \in \mathbb{R}$ is continuous on $\mathbb{R}^+$ since by definition $x^t = e^{t \ln x}$. Its derivative is

$$\frac{d}{dx} x^t = tx^{t-1},$$

which follows easily by differentiating $e^{t \ln x}$.

We are now ready to prove that the function $a^x$ truly behaves like the power function with real exponent. We first show that it coincides with the integer power definition. Using the first property, for $x \in \mathbb{N}$,

$$a^x := e^{x \ln a} = e^{\overbrace{\ln a + \cdots + \ln a}^{x \text{ times}}} = \underbrace{e^{\ln a} \cdots e^{\ln a}}_{x \text{ times}} = \underbrace{a \cdots a}_{x \text{ times}}.$$

The $n$-th root of a positive real number $a$, denoted $\sqrt[n]{a}$, where $n \in \mathbb{N}$, is defined as the unique positive root of

$$x^n = a.$$

Using the second property of the exponential function, we see that

$$a^{1/n} = e^{\frac{1}{n} \ln a}$$

is a root of the above equation since

$$\left(e^{\frac{1}{n} \ln a}\right)^n = e^{\ln a} = a.$$

Therefore, if $q = \frac{m}{n} \in \mathbb{Q}$, then the exponential function coincides with the rational power:

$$\sqrt[n]{x^m} = e^{q \ln x}.$$

Similarly for negative powers since $a^{-x} = \frac{1}{a^x}$.

Every real number $y$ can be approximated by a sequence of rational numbers:

$$q_k = \frac{m_k}{n_k} \to y \quad \text{as } k \to \infty.$$

We have proven that

$$\sqrt[n_k]{x^{m_k}} = e^{q_k \ln x}.$$

By continuity of the exponential function,

$$e^{q_k \ln x} \to e^{y \ln x} = x^y \quad \text{as } k \to \infty,$$

which implies

$$\sqrt[n_k]{x^{m_k}} \to e^{y \ln x} = x^y \quad \text{as } k \to \infty.$$

Note that this convergence relies on the continuity of the exponential function as the inverse of the natural logarithm.

In summary, the second definition (using logarithms) generalizes the first because it coincides with it when the exponent is an integer and is defined for non-integer exponents as well. However, if one wants to define powers of negative quantities with real exponents, it can only be done using the first definition for integer exponents.

In the complex numbers $\mathbb{C}$, one can write

$$(-a)^x = e^{ix\pi} a^x,$$

where $a \in \mathbb{R}^+$, $x \in \mathbb{R}$, and $e^{ix} = \cos x + i \sin x$ with $i^2 = -1$.

The construction of the logarithmic and exponential functions by a different approach can be found in [8].

## 20.3 Python and AI

**AI Request**

Show me all the possible constructions of the logarithmic and exponential functions. Write a Python code to compare all these constructions.

## 20.4 Exercises

### True or false

Decide whether each statement is **true** or **false**. Justify briefly.

1. The natural logarithm function is defined as $\ln x = \int_1^x \frac{1}{t}\, dt$.
   Answer: _____
2. The exponential function is the inverse of the logarithmic function.
   Answer: _____
3. The domain of $\ln x$ is $\mathbb{R}$.
   Answer: _____
4. For any real number $a$, $\ln e^a = a$.
   Answer: _____
5. The function $f(x) = a^x$ is differentiable and its derivative is $a^x \ln a$.
   Answer: _____

### Multiple choice questions

Choose the correct option (A, B, C, or D).

1. The derivative of $\ln(x^2)$ with respect to $x$ is
   A. $\frac{1}{x}$   B. $\frac{2}{x}$
   C. $\frac{x}{2}$   D. $2x$
   Answer: _____

2. The limit $\lim_{x \to \infty} \frac{\ln x}{x}$ equals

   A. $\infty$   B. 1

   C. 0   D. does not exist

   **Answer:** _____

3. The inverse function of $y = e^{2x}$ is

   A. $\ln(x)$   B. $\frac{\ln(x)}{2}$

   C. $2\ln(x)$   D. $e^{-2x}$

   **Answer:** _____

4. The integral $\int \frac{1}{x} \, dx$ equals

   A. $e^x + C$   B. $\frac{1}{x} + C$

   C. $\ln |x| + C$   D. $\frac{x^2}{2} + C$

   **Answer:** _____

5. The value $\ln 1$ equals

   A. 0   B. 1

   C. $e$   D. does not exist

   **Answer:** _____

## Matching exercise

Match each expression or function with its corresponding property or result.

| | |
|---|---|
| 1. $\ln(xy)$ | A. $\frac{1}{x}$ |
| 2. Derivative of $\ln x$ | B. $\ln x - \ln y$ |
| 3. $\ln \frac{x}{y}$ | C. $\ln x + \ln y$ |
| 4. Inverse of $e^x$ | D. $\frac{\ln x}{\ln a}$ |
| 5. $\log_a x$ | E. $\ln x$ |

**Answers:** 1._____   2._____   3._____   4._____   5._____

## Fill in the blanks

Fill in the blanks with the most appropriate word or expression.

1. The logarithmic function $\ln x$ is defined for _____ real numbers.
2. The function $e^x$ is _____ increasing.
3. The derivative of $e^x$ with respect to $x$ is _____.
4. $a^x = e$————————.
5. The inverse function of $a^x$ is _____.

## Theoretical and computational problems

1. Prove the identity $\ln(xy) = \ln x + \ln y$.
2. Compute the derivative of the function $f(x) = x^x$.
3. Show that $\lim_{x \to 0^+} \ln x = -\infty$.
4. Solve the equation $e^{2x} - 5e^x + 6 = 0$.
5. Prove that the exponential function $e^x$ is continuous for all $x \in \mathbb{R}$.

## Python-based exercises

Use Python (numpy, sympy, matplotlib) for the following:

1. Plot the natural logarithm function $y = \ln x$ on the interval $[0.1, 5]$.
2. Use sympy to compute the derivative of $e^{x^2}$.
3. Graphically illustrate that $e^x$ is the inverse of $\ln x$.
4. Numerically approximate $\ln 2$ using numerical integration.
5. Solve the equation $\ln x + x^2 = 4$ numerically.

# 21 Area of a plane region

## Contents

The concept of area is fundamental in geometry and calculus, measuring the extent of regions in the plane. While the area of simple shapes like rectangles or circles is familiar, more general plane regions require a rigorous analytical approach. Using definite integrals, we can precisely define and compute the area under curves, between functions, or even more complex sets. In this chapter, we introduce the formal definition of area for regions bounded by graphs, explain the method of setting up and evaluating the corresponding integrals, and present applications including the calculation of lengths, surface areas, and volumes. Through worked examples and exercises, you will learn how integration translates geometric problems into analytic solutions.

## 21.1 Introduction

The notion of area lies at the heart of both ancient geometry and modern mathematical analysis. The area of a plane region, at first glance, seems intuitive: it quantifies the "amount of space" a shape occupies in the plane. From the earliest days of mathematics, people sought ways to calculate the area of familiar figures—rectangles, triangles, circles, ellipses—and, gradually, more complicated shapes bounded by curves. Yet, while the concept is simple for polygons, it becomes far subtler when dealing with regions bounded by arbitrary or curved lines.

To rigorously define and compute area, we require the full power of the integral calculus developed in previous chapters. The definite integral, originally invented to formalize the computation of areas under curves, provides both a precise definition and powerful computational techniques. By partitioning the domain and summing up the contributions of infinitesimal slices, we capture the exact extent of even the most irregular regions.

In this chapter, we begin with the formal definition of the area of a region under a curve, connecting the geometric intuition with the analytical machinery of the Riemann integral. We show that if a function $f$ is integrable over $[a, b]$, then the area under its graph (above the $x$-axis) is given by $\int_a^b |f(x)|\, dx$. This formula allows us not only to calculate the area between a curve and the axis, but also to handle situations where the function may dip below the axis, by taking the absolute value.

https://doi.org/10.1515/9783112228289-021

We extend this approach to find the area between two curves, reducing the geometric problem to evaluating the integral of the absolute difference $|f(x) - g(x)|$ over the interval of interest. This powerful idea enables us to tackle classic problems: the area of a circle or ellipse, regions bounded by intersecting parabolas and lines, and even regions described by piecewise functions.

But area is just the beginning. We further explore related geometric quantities that arise naturally: the length of a curve (arc length), the surface area generated by rotating a curve about an axis, and the volume of the resulting solid. Each of these quantities is given by a specific type of definite integral, reflecting the underlying geometry.

Through carefully chosen examples, we demonstrate not only how to set up these integrals, but also how to interpret their geometric meaning. The calculation of the area of a circle, for instance, is re-expressed as an integral over a semicircular function; the area between a cubic and a linear function requires finding their points of intersection and analyzing which function lies above the other.

Applications of these concepts abound: in physics (computing mass and moments), engineering (determining material requirements), probability (finding cumulative distribution functions), and many other fields. Understanding how to translate a geometric problem into an integral is a critical skill, bridging the gap between the abstract world of analysis and concrete, real-world questions.

In this chapter, you will gain mastery over the process of translating a wide variety of geometric area and volume problems into precise analytic form, and develop the tools to solve them both by hand and with the aid of technology. The included exercises, ranging from basic calculations to more challenging conceptual and computational problems, will help you internalize both the theory and its practical applications.

## 21.2 Definitions

We start this section by associating the notion of area to a plane region. Informally, the area represents the measure of the extent of a set on the plane in square units. We will give a rigorous approach to define the area by using the concept of a plane region.

As a first step, we give a formal and precise definition of the measure of the set in question.

**Proposition 21.1** (Area under a curve and the integral). *If a function f is integrable on $[a, b]$, then the area of the set*

$$E(f) = \{(x,y) : 0 \le y \le |f(x)|, \ a \le x \le b\}$$

*is given by the integral of the absolute value of the function, that is,*

$$\varepsilon\big(E(f)\big) = \int_a^b |f(x)| \, dx.$$

Based on the above definition, it is natural to measure the area of the set limited by the graph of a function $f$, the $x$-axis, and the vertical lines $x = a$ and $x = b$.

Thus, if the function $f$ is positive, the area under its graph can be naturally interpreted as the area of the region bounded by the curve of the function and these lines. Conversely, if the function is negative, the area is defined as the area under the graph of $|f|$.

We also consider a similar definition for the area between two graphs.

**Proposition 21.2** (Area between two curves). *If $f$, $g$ are integrable functions on $[a, b]$, then the area of the set*

$$E(f, g) = \{(x, y) : g(x) \leq y \leq f(x), \ a \leq x \leq b\}$$

*is given by*

$$\varepsilon\big(E(f, g)\big) = \int_a^b |f(x) - g(x)| \, dx.$$

**Example 21.1** (Area of a circle). Calculate the area of a circle of radius $a$.

*Solution.* The equation of a circle does not define a function. However, by symmetry, we can calculate the area of a semicircle (the upper half of the circle). The semicircle is the graph of the function

$$f(x) = \sqrt{a^2 - x^2}, \quad x \in [-a, a].$$

We are going to compute

$$E = 2 \int_{-a}^{a} \sqrt{a^2 - x^2} \, dx.$$

**Step 1: Symmetry of the integrand**

Notice that the function $f(x) = \sqrt{a^2 - x^2}$ is even (since $f(-x) = f(x)$). Thus

$$\int_{-a}^{a} \sqrt{a^2 - x^2} \, dx = 2 \int_{0}^{a} \sqrt{a^2 - x^2} \, dx.$$

Therefore,

$$E = 2 \cdot 2 \int_{0}^{a} \sqrt{a^2 - x^2} \, dx = 4 \int_{0}^{a} \sqrt{a^2 - x^2} \, dx.$$

### Step 2: Substitution
Let us use the trigonometric substitution

$$x = a \sin \theta \quad \text{for } \theta \in [0, \pi/2].$$

Then

$$dx = a \cos \theta \, d\theta$$

and

$$\sqrt{a^2 - x^2} = \sqrt{a^2 - a^2 \sin^2 \theta} = a \cos \theta.$$

**Changing the limits:**
When $x = 0$, $\theta = 0$. When $x = a$, $\sin \theta = 1 \Rightarrow \theta = \pi/2$.

### Step 3: Substituting in the integral
Thus,

$$\int_0^a \sqrt{a^2 - x^2} \, dx = \int_0^{\pi/2} a \cos \theta \cdot a \cos \theta \, d\theta = a^2 \int_0^{\pi/2} \cos^2 \theta \, d\theta.$$

### Step 4: Evaluating the integral
Recall that

$$\cos^2 \theta = \frac{1 + \cos 2\theta}{2}.$$

Therefore,

$$a^2 \int_0^{\pi/2} \cos^2 \theta \, d\theta = a^2 \int_0^{\pi/2} \frac{1 + \cos 2\theta}{2} \, d\theta = \frac{a^2}{2} \int_0^{\pi/2} (1 + \cos 2\theta) \, d\theta.$$

Now, integrate term by term:

$$\int_0^{\pi/2} 1 \, d\theta = \frac{\pi}{2},$$

$$\int_0^{\pi/2} \cos 2\theta \, d\theta = \frac{\sin 2\theta}{2} \Big|_0^{\pi/2} = \frac{\sin \pi}{2} - \frac{\sin 0}{2} = 0.$$

So,

$$a^2 \int_0^{\pi/2} \cos^2\theta \, d\theta = \frac{a^2}{2} \cdot \frac{\pi}{2} = \frac{a^2\pi}{4}.$$

**Step 5: Substitute back to** $E$

Recall that

$$E = 4\int_0^a \sqrt{a^2 - x^2} \, dx = 4 \cdot \frac{a^2\pi}{4} = a^2\pi.$$

**Final answer**

$$\boxed{E = 2\int_{-a}^a \sqrt{a^2 - x^2} \, dx = a^2\pi}$$  ☐

**Example 21.2** (Area of an ellipse). Calculate the area of the ellipse with equation $\frac{x^2}{a^2} + \frac{y^2}{b^2} = 1$.

*Solution.* The upper half of the ellipse corresponds to the graph of the function

$$f(x) = \frac{b}{a}\sqrt{a^2 - x^2}.$$

Hence the area of the ellipse is

$$E = 4\int_0^a \frac{b}{a}\sqrt{a^2 - x^2} \, dx = \pi ab,$$

where we have used the substitution $x = a\sin t$ as before.  ☐

**Example 21.3** (Area between $f(x) = x + 1$ and $g(x) = x^2 - x$ on $[0, 2]$). Calculate the area of the set bounded by the graphs of the functions $f(x) = x + 1$ and $g(x) = x^2 - x$ for $x \in [0, 2]$.

*Solution.* The area is

$$E = \int_0^2 |f(x) - g(x)| \, dx.$$

To compute this integral, we find where the graphs intersect (i. e., solve $f(x) = g(x)$) to determine the sign of the difference. We compute

$$f(x) - g(x) = (x + 1) - (x^2 - x) = -x^2 + 2x + 1.$$

This quadratic has roots at $x = 1 \pm \sqrt{2}$. The only root in $[0, 2]$ is $x = 1 + \sqrt{2} \approx 2.414$, which is **greater than 2**, so on the entire interval $[0, 2]$, we have $f(x) - g(x) > 0$. Therefore, $|f(x) - g(x)| = -x^2 + 2x + 1$, and no splitting is needed.

Thus, the area is

$$E = \int_0^2 (-x^2 + 2x + 1)\, dx = \left[ -\frac{x^3}{3} + x^2 + x \right]_0^2 = -\frac{8}{3} + 4 + 2 = \frac{10}{3}. \qquad \square$$

**Example 21.4** (Area between $f(x) = |x|$ and $g(x) = [x]$ on $[-2, 2]$). Calculate the area of the set bounded by the graphs of $f(x) = |x|$ and $g(x) = [x]$ (the floor function) for $x \in [-2, 2]$.

*Solution.* We compute the area between the graphs:

$$E = \int_{-2}^2 (|x| - [x])\, dx.$$

Since both $|x|$ and $[x]$ are piecewise-defined, we split the integral at integer points: $x = -2, -1, 0, 1, 2$.

On each subinterval:
- For $x \in [-2, -1)$: $|x| = -x$, $[x] = -2$, so $|x| - [x] = -x + 2$.
- For $x \in [-1, 0)$: $|x| = -x$, $[x] = -1$, so $|x| - [x] = -x + 1$.
- For $x \in [0, 1)$: $|x| = x$, $[x] = 0$, so $|x| - [x] = x$.
- For $x \in [1, 2)$: $|x| = x$, $[x] = 1$, so $|x| - [x] = x - 1$.
- At $x = 2$: single point.

Now compute each integral:

$$\int_{-2}^{-1} (-x + 2)\, dx = \left[ -\frac{x^2}{2} + 2x \right]_{-2}^{-1} = \left( -\frac{1}{2} - 2 \right) - (-2 - 4) = -2.5 + 6 = 3.5,$$

$$\int_{-1}^{0} (-x + 1)\, dx = \left[ -\frac{x^2}{2} + x \right]_{-1}^{0} = 0 - \left( -\frac{1}{2} - 1 \right) = 0 - (-1.5) = 1.5,$$

$$\int_0^1 x\, dx = \left[ \frac{x^2}{2} \right]_0^1 = \frac{1}{2},$$

$$\int_1^2 (x - 1)\, dx = \left[ \frac{(x-1)^2}{2} \right]_1^2 = \frac{1}{2} - 0 = \frac{1}{2}.$$

Summing:

$$E = 3.5 + 1.5 + 0.5 + 0.5 = 6. \qquad \square$$

**Example 21.5** (Area between parabola and the x-axis). Calculate the area of the set bounded by the graph of the parabola $y = x^2 - 3x + 2$ and the x-axis.

*Solution.* First, find where the parabola intersects the x-axis by solving

$$x^2 - 3x + 2 = 0 \quad \Rightarrow \quad (x-1)(x-2) = 0 \quad \Rightarrow \quad x = 1, 2.$$

So the graph crosses the x-axis at $x = 1$ and $x = 2$.

The quadratic $x^2 - 3x + 2$ opens upwards, so it is negative between the roots. Therefore, for $x \in [1, 2]$, we have $y < 0$, and the area between the curve and the x-axis is given by

$$E = \int_1^2 |x^2 - 3x + 2| dx = \int_1^2 -(x^2 - 3x + 2) \, dx = \int_1^2 (-x^2 + 3x - 2) \, dx.$$

Compute the integral

$$\int_1^2 (-x^2 + 3x - 2) \, dx = \left[ -\frac{x^3}{3} + \frac{3x^2}{2} - 2x \right]_1^2.$$

At $x = 2$,

$$-\frac{8}{3} + \frac{12}{2} - 4 = -\frac{8}{3} + 6 - 4 = -\frac{8}{3} + 2 = -\frac{2}{3}.$$

At $x = 1$,

$$-\frac{1}{3} + \frac{3}{2} - 2 = -\frac{1}{3} - \frac{1}{2} = -\frac{5}{6}.$$

Now subtract

$$\left( -\frac{2}{3} \right) - \left( -\frac{5}{6} \right) = -\frac{4}{6} + \frac{5}{6} = \frac{1}{6}.$$

Thus, the area is

$$E = \frac{1}{6}. \qquad \qquad \square$$

**Example 21.6** (Area between $y = x^3$ and $y = 4x$ (Figure 21.1)). Calculate the area of the set bounded by the graphs of $y = x^3$ and $y = 4x$.

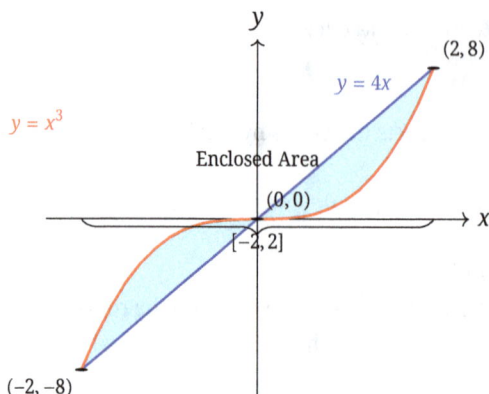

Figure 21.1: The finite area enclosed between $y = 4x$ and $y = x^3$ on $[-2, 2]$.

*Solution.* Find the points of intersection by solving

$$x^3 = 4x \quad \Rightarrow \quad x^3 - 4x = 0 \quad \Rightarrow \quad x(x^2 - 4) = 0 \quad \Rightarrow \quad x = -2, 0, 2.$$

So the curves intersect at $x = -2$, $x = 0$, and $x = 2$.

We analyze the sign of $x^3 - 4x = x(x - 2)(x + 2)$ on the intervals:

- On $[-2, 0]$: $x \le 0$, $x - 2 < 0$, $x + 2 \ge 0$ so $x^3 \ge 4x$.
- On $[0, 2]$: $x \ge 0$, $x - 2 \le 0$, $x + 2 > 0$ so $x^3 \le 4x$.

Therefore, the total area is

$$E = \int_{-2}^{0} (x^3 - 4x)\, dx + \int_{0}^{2} (4x - x^3)\, dx.$$

Compute each integral:
First

$$\int_{-2}^{0} (x^3 - 4x)\, dx = \left[ \frac{x^4}{4} - 2x^2 \right]_{-2}^{0} = 0 - \left( \frac{16}{4} - 2 \cdot 4 \right) = -(4 - 8) = 4.$$

Second

$$\int_{0}^{2} (4x - x^3)\, dx = \left[ 2x^2 - \frac{x^4}{4} \right]_{0}^{2} = \left( 8 - \frac{16}{4} \right) - 0 = 8 - 4 = 4.$$

Add

$$E = 4 + 4 = 8.$$

Thus, the total area bounded by the curves is 8. ◻

Now we present the formula for the length of the graph of a function over an interval.

**Theorem 21.1** (Length of a curve). *If $f$ is continuously differentiable on $[a, b]$, then the length of the graph of $f$ on $[a, b]$ is given by*

$$S = \int_a^b \sqrt{1 + |f'(x)|^2}\, dx.$$

**Example 21.7** (Length of the graph of $f(x) = x$ on $[0, 1]$). Calculate the length of the graph of $f(x) = x$ on $[0, 1]$.

*Solution.* Since $f'(x) = 1$, the length is

$$S = \int_0^1 \sqrt{1 + 1}\, dx = \int_0^1 \sqrt{2}\, dx = \sqrt{2}.$$

This can be confirmed using the Pythagorean theorem: the graph is the diagonal of a right triangle with legs of length 1, so $S = \sqrt{1^2 + 1^2} = \sqrt{2}$. □

**Example 21.8** (Length of the graph of $f(x) = x^2$ on $[0, 1]$). Calculate the length of the graph of $f(x) = x^2$ on $[0, 1]$.

*Solution.* First, compute the derivative

$$f'(x) = 2x.$$

The arc length is given by

$$S = \int_0^1 \sqrt{1 + [f'(x)]^2}\, dx = \int_0^1 \sqrt{1 + (2x)^2}\, dx = \int_0^1 \sqrt{1 + 4x^2}\, dx.$$

Let

$$x = \frac{1}{2} \sinh t \quad \Longrightarrow \quad dx = \frac{1}{2} \cosh t\, dt$$

and

$$\sqrt{1 + 4x^2} = \sqrt{1 + 4\left(\frac{1}{2} \sinh t\right)^2} = \sqrt{1 + \sinh^2 t} = \cosh t.$$

So the integral becomes

$$I = \int \sqrt{1 + 4x^2}\, dx = \int \cosh t \cdot \frac{1}{2} \cosh t\, dt = \frac{1}{2} \int \cosh^2 t\, dt.$$

Recall

$$\cosh^2 t = \frac{1}{2}(\cosh 2t + 1).$$

So,

$$I = \frac{1}{2} \int \left[\frac{1}{2} \cosh 2t + \frac{1}{2}\right] dt = \frac{1}{4} \int \cosh 2t\, dt + \frac{1}{4} \int dt$$

$$= \frac{1}{4} \cdot \frac{1}{2} \sinh 2t + \frac{1}{4} t + C = \frac{1}{8} \sinh 2t + \frac{1}{4} t + C.$$

**Back-substitute in $x$:**
Recall that $x = \frac{1}{2} \sinh t \implies t = \sinh^{-1}(2x)$, and $\sinh 2t = 2 \sinh t \cosh t = 4x\sqrt{1 + 4x^2}$.
So,

$$I = \frac{1}{8} \sinh 2t + \frac{1}{4} t + C = \frac{1}{8} \cdot 4x\sqrt{1 + 4x^2} + \frac{1}{4} \sinh^{-1}(2x) + C = \frac{1}{2} x\sqrt{1 + 4x^2} + \frac{1}{4} \sinh^{-1}(2x) + C.$$

Now evaluate from $x = 0$ to $x = 1$:

$$S = \left[\frac{1}{2} x\sqrt{1 + 4x^2} + \frac{1}{4} \sinh^{-1}(2x)\right]_0^1.$$

At $x = 1$: $\sqrt{1 + 4(1)^2} = \sqrt{5}$, $\sinh^{-1}(2 \cdot 1) = \sinh^{-1}(2)$.
At $x = 0$: $\sqrt{1} = 1$, $\sinh^{-1}(0) = 0$.
So,

$$S = \frac{1}{2}(1)\sqrt{5} + \frac{1}{4} \sinh^{-1}(2) - [0 + 0]$$

$$= \frac{\sqrt{5}}{2} + \frac{1}{4} \sinh^{-1}(2).$$

$$\boxed{S = \frac{\sqrt{5}}{2} + \frac{1}{4} \sinh^{-1}(2)}$$

Alternatively, you may write $\sinh^{-1}(2) = \ln(2 + \sqrt{5})$. $\square$

---

**Theorem 21.2** (Volume of solid of revolution). *If $f$, $g$ are continuous functions on $[a, b]$ such that either $f(x) \geq 0, g(x) \geq 0$ or $f(x) \leq 0, g(x) \leq 0$ for all $x \in [a, b]$, then the volume of the solid of revolution generated by rotating the region between the graphs about the x-axis is*

$$V = \pi \int_a^b \left| f^2(x) - g^2(x) \right| dx.$$

If the functions satisfy the above conditions and are integrable, the volume is well-defined.

**Example 21.9** (Volume of a disk). Calculate the volume of the solid obtained by rotating the graph of $f(x) = 1$ (the line $y = 1$) and $g(x) = 0$ (the x-axis) about the x-axis for x in $[0, 1]$.

*Solution.* Here $f(x) = 1$ and $g(x) = 0$ on $[0, 1]$. The formula gives

$$V = \pi \int_0^1 \left| f^2(x) - g^2(x) \right| dx = \pi \int_0^1 (1^2 - 0^2) \, dx = \pi \int_0^1 1 \, dx = \pi.$$

This matches the volume of a cylinder of radius 1 and height 1: $V = \pi r^2 h = \pi$.  ☐

**Theorem 21.3** (Surface area of a solid of revolution). *If f is continuously differentiable on $[a, b]$ and non-negative, then the surface area of the solid generated by revolving the graph of f about the x-axis is*

$$E = 2\pi \int_a^b |f(x)| \sqrt{1 + |f'(x)|^2} \, dx.$$

**Example 21.10** (Surface area of a paraboloid). Calculate the surface area of the solid formed by revolving the graph of $f(x) = x^2$ about the x-axis for $x \in [0, 1]$.

*Solution.* To evaluate the surface area of the solid formed by revolving $f(x) = x^2$ about the x-axis for $x \in [0, 1]$, we use the formula

$$S = 2\pi \int_0^1 x^2 \sqrt{1 + (2x)^2} \, dx.$$

Let

$$x = \frac{1}{2} \sinh t \quad \Longrightarrow \quad dx = \frac{1}{2} \cosh t \, dt$$

and

$$\sqrt{1 + 4x^2} = \sqrt{1 + 4\left(\frac{1}{2} \sinh t\right)^2} = \sqrt{1 + \sinh^2 t} = \cosh t.$$

So,

$$I = \int x^2 \sqrt{1 + 4x^2} \, dx$$

$$= \int \left(\frac{1}{2} \sinh t\right)^2 \cdot \cosh t \cdot \frac{1}{2} \cosh t \, dt$$

$$= \int \frac{1}{4} \sinh^2 t \cdot \frac{1}{2} \cosh^2 t \, dt$$

$$= \frac{1}{8} \int \sinh^2 t \cosh^2 t \, dt.$$

We use the identity

$$\sinh^2 t \cosh^2 t = \frac{1}{4} \sinh^2(2t).$$

Therefore,

$$I = \frac{1}{8} \cdot \frac{1}{4} \int \sinh^2(2t) \, dt = \frac{1}{32} \int \sinh^2(2t) \, dt.$$

Recall

$$\sinh^2(2t) = \frac{1}{2}(\cosh(4t) - 1).$$

So,

$$I = \frac{1}{32} \int \frac{1}{2}(\cosh(4t) - 1) \, dt = \frac{1}{64} \int (\cosh(4t) - 1) \, dt.$$

Integrating:

$$\int \cosh(4t) \, dt = \frac{1}{4} \sinh(4t).$$

So,

$$I = \frac{1}{64}\left(\frac{1}{4} \sinh(4t) - t\right) + C = \frac{1}{256} \sinh(4t) - \frac{1}{64}t + C.$$

Back-substitute in $x$:
Recall $x = \frac{1}{2} \sinh t \implies t = \sinh^{-1}(2x)$.
Compute $S$ on $[x = 0, x = 1]$:
For $x = 0$: $t_0 = \sinh^{-1}(0) = 0$.
For $x = 1$: $t_1 = \sinh^{-1}(2)$.
So,

$$S = 2\pi[I]_{x=0}^{x=1} = 2\pi(I(x = 1) - I(x = 0)).$$

Plug in the bounds:

$$S = 2\pi\left\{\left[\frac{1}{256} \sinh(4t_1) - \frac{1}{64}t_1\right] - \left[\frac{1}{256} \sinh(0) - \frac{1}{64} \cdot 0\right]\right\}$$

$$= 2\pi\left(\frac{1}{256} \sinh(4t_1) - \frac{1}{64}t_1\right),$$

where $t_1 = \sinh^{-1}(2)$.

**Final answer:**

$$S = 2\pi\left(\frac{1}{256}\sinh(4\sinh^{-1}(2)) - \frac{1}{64}\sinh^{-1}(2)\right)$$  □

## 21.3 Python and AI

**AI Request**

Calculate the area of the set bounded by the graph of the parabola $y = x^2 - 3x + 2$ and the $x$-axis and give me the Latex code of the solution step-by-step. Write also a Python code for this computation and plot the area.

## 21.4 Exercises

### True or false

Decide whether each statement is **true** or **false**. Justify briefly.

1. The area under the curve $y = f(x)$ for $a \le x \le b$ is always $\int_a^b f(x)\,dx$.
   **Answer:** _____

2. The area between two curves $f$ and $g$ is $\int_a^b |f(x) - g(x)|\,dx$.
   **Answer:** _____

3. The length of the curve $y = f(x)$ from $a$ to $b$ is given by $\int_a^b \sqrt{1 + f'(x)^2}\,dx$.
   **Answer:** _____

4. The volume generated by rotating $y = f(x)$ around the $x$-axis is $\pi \int_a^b f(x)^2\,dx$.
   **Answer:** _____

5. The surface area generated by rotating $y = f(x)$ around the $x$-axis is $2\pi \int_a^b f(x) \times \sqrt{1 + f'(x)^2}\,dx$.
   **Answer:** _____

### Multiple choice questions

Choose the correct option (A, B, C, or D).

1. The area of a circle with radius $r$ is
   A. $\pi r$    B. $2\pi r$
   C. $\pi r^2$    D. $4\pi r^2$
   **Answer:** _____

2. The area bounded by $y = x$ and $y = x^2$ for $0 \le x \le 1$ is
   A. $\frac{1}{2}$    B. $\frac{1}{3}$
   C. $\frac{1}{6}$    D. $\frac{2}{3}$
   **Answer:** _____

3. The length of the curve $y = x^2$ from 0 to 1 is

A. $\frac{\sqrt{5}}{2}$    B. $\frac{\sqrt{3}}{2}$

C. $\frac{5}{2}$    D. $\frac{3}{2}$

**Answer:** _____

4. The surface area generated by rotating $y = x$ for $0 \le x \le 1$ about the $x$-axis is

A. $\pi \sqrt{2}$    B. $2\pi \sqrt{2}$

C. $\frac{\pi}{2} \sqrt{2}$    D. $4\pi \sqrt{2}$

**Answer:** _____

5. The volume generated by rotating $y = \sqrt{x}$ for $0 \le x \le 1$ around the $x$-axis is

A. $\frac{\pi}{2}$    B. $\pi$

C. $2\pi$    D. $\frac{2\pi}{3}$

**Answer:** _____

## Matching exercise

Match each integral or expression with the corresponding description.

1. $\int_a^b |f(x)|\, dx$            A. Volume of revolution

2. $\int_a^b \sqrt{1 + (f'(x))^2}\, dx$      B. Surface area of revolution

3. $\pi \int_a^b |f^2(x) - g^2(x)|\, dx$      C. Area between two curves

4. $\int_a^b |f(x) - g(x)|\, dx$         D. Arc length

5. $2\pi \int_a^b |f(x)| \sqrt{1 + (f'(x))^2}\, dx$    E. Area under the curve

**Answers:** 1._____    2._____    3._____    4._____    5._____

## Fill in the blanks

Complete the following sentences:

1. The area under the curve $f(x)$ is given by the integral of _____.
2. The area between two curves $f$ and $g$ is calculated using the integral of _____.
3. The arc length of the function $f(x)$ is computed by integrating _____.
4. The volume of a solid generated by rotating about the $x$-axis is given by integrating

    _____.

5. The surface area generated by rotating $y = f(x)$ about the $x$-axis is computed by integrating _____.

## Theoretical and computational problems

Compute the area of each region. Set up the appropriate definite integral(s), determine convergence if relevant, and evaluate exactly.

1. Find the area of the region bounded by $y = x^2$ and $y = 2x$.
2. Compute the area enclosed by the polar curve $r = 2\sin\theta$ for $\theta \in [0, \pi]$.
3. Find the area of the region bounded by $y = \sqrt{x}$, $y = \frac{x}{2}$, and the vertical lines $x = 0$ and $x = 4$.
4. Derive the area of the ellipse $\frac{x^2}{a^2} + \frac{y^2}{b^2} = 1$ using an integral.

5. Determine for which values of $p > 0$ the area

$$\int_1^\infty \frac{1}{x^p}\, dx$$

is finite, and compute it when it converges.

## Python-based exercises

Use Python (numpy, sympy, matplotlib) to perform the following tasks. In each case compare the numerical result to the known analytic value and provide a brief comment on accuracy or convergence.

1. Numerically evaluate the area between $y = x^2$ and $y = 2x$ on $[0, 2]$; plot the two curves, shade the region, and compare with the exact value.
2. Estimate the area of the unit circle $x^2 + y^2 \le 1$ via Monte Carlo simulation and compare with $\pi$; include a plot illustrating the random sampling.
3. Compute numerically the area enclosed by the polar curve $r = 1+\cos\theta$ for $\theta \in [0, 2\pi]$ using the polar area formula. Plot the curve in Cartesian coordinates and report the estimated area versus the theoretical value.
4. Use sympy to derive symbolically the formula for the area of an ellipse and then numerically verify it for $a = 3$, $b = 2$; compare the result to $\pi ab$.
5. Given noisy sample data of $f(x) = \sin x$ on $[0, \pi]$ (e. g., sample at 100 equally spaced points with small random perturbation), approximate the area under the curve using the trapezoidal rule and Simpson's rule. Compare both with the exact value 2 and discuss the effect of noise and method.

# 22 Improper integrals

## Contents

In calculus, many important problems involve integrating functions over infinite intervals or where the integrand becomes unbounded within the interval. Such integrals, called improper integrals, arise naturally in analysis, probability, and physics. In this chapter, we extend the notion of the definite integral to include cases where one or both endpoints are infinite, or where the function has singularities. We develop criteria for the convergence of improper integrals and study their properties, including comparison and limit comparison tests. Special attention is given to integrals with oscillatory behavior and to important examples like the Gaussian integral. Through theory, examples, and exercises, you will gain practical tools for determining convergence and evaluating improper integrals.

## 22.1 Introduction

In many branches of mathematics, science, and engineering, we encounter integrals that go beyond the scope of ordinary (proper) definite integrals. These are integrals where either the limits of integration are infinite, or the integrand becomes unbounded within the interval of integration. Such cases give rise to the concept of **improper integrals**.

Improper integrals are crucial in the study of infinite series, probability theory, differential equations, and mathematical physics. For example, calculating the total mass of an infinitely long rod, evaluating the probability distributions with infinite support, or analyzing wave functions in quantum mechanics often leads to improper integrals. Their study allows us to make sense of quantities that, at first glance, may appear to be infinite or undefined.

This chapter introduces improper integrals in a rigorous way. We first describe the two main types: integrals over unbounded intervals and integrals of unbounded functions. For each type, we define convergence through limits and give precise criteria for when such integrals have finite values. We then develop a set of tools to analyze convergence, including the comparison test, the limit comparison test, and criteria based on asymptotic behavior near infinity or near singularities.

Special cases such as integrals involving oscillatory functions are treated using tests like Dirichlet's test, which ensures convergence in situations where simple estimates are

https://doi.org/10.1515/9783112228289-022

insufficient. We also discuss absolute and conditional convergence, emphasizing their distinction and importance.

Numerous examples illustrate the main ideas, from classic integrals like $\int_1^\infty \frac{dx}{x^p}$ and the Gaussian integral $\int_{-\infty}^\infty e^{-x^2}\,dx$, to more subtle cases involving singularities and oscillations. Worked exercises show how to determine convergence, estimate values, and apply the general theorems in practice.

By the end of this chapter, you will have developed a solid understanding of improper integrals, the conditions under which they converge, and the techniques required for their evaluation. Mastery of this material is fundamental for advanced calculus and many applications across the mathematical sciences.

## 22.1.1  Integrals on an unbounded interval

**Definition 22.1** (Improper integral on $[a, +\infty)$ (Figure 22.1)). Let $f : [a, +\infty) \to \mathbb{R}$ be a function such that, for every $t \in (a, +\infty)$, the function $f$ is integrable on the interval $[a, t]$. Define the function

$$F : (a, +\infty) \to \mathbb{R}, \quad F(t) = \int_a^t f(x)\,dx.$$

If there exists a number $\lambda \in \mathbb{R}$ such that

$$\lim_{t \to +\infty} F(t) = \lim_{t \to +\infty} \int_a^t f(x)\,dx = \lambda,$$

then we say that the function $f$ is **integrable on** $[a, +\infty)$, and the number $\lambda$ is called the **improper integral of** $f$ over $[a, +\infty)$, denoted by

$$\int_a^{+\infty} f(x)\,dx.$$

If this limit does not exist, the improper integral is said to **diverge**.

**Definition 22.2** (Improper integral on $(-\infty, +\infty)$). Let $f : \mathbb{R} \to \mathbb{R}$ be a function such that, for every $m < k$, the function $f$ is integrable on the interval $[m, k]$. The **improper integral of** $f$ **over** $(-\infty, +\infty)$ is defined as the iterated limit

$$\int_{-\infty}^{\infty} f(x)\,dx := \lim_{m \to -\infty} \lim_{k \to \infty} \int_m^k f(x)\,dx = \lim_{k \to \infty} \lim_{m \to -\infty} \int_m^k f(x)\,dx = \lambda \in \mathbb{R},$$

provided that both limits exist and are equal. If the limit does not exist or is infinite, the integral **diverges**.

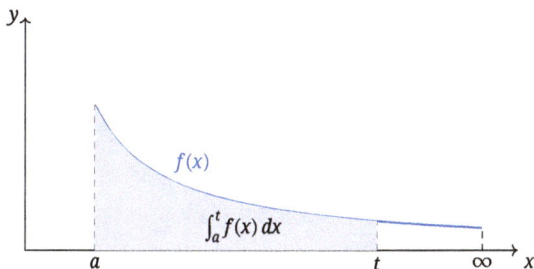

Figure 22.1: Improper integral $\int_a^{+\infty} f(x)\,dx$ as the area under a decreasing function on $[a, +\infty)$.

**Example 22.1.** Consider the improper integral

$$\int_{-\infty}^{\infty} x\,dx.$$

For $m < k$, we have

$$\int_m^k x\,dx = \frac{k^2}{2} - \frac{m^2}{2}.$$

Taking the limit as $k \to +\infty$, the integral diverges. However,

$$\lim_{k \to \infty} \int_{-k}^k x\,dx = 0,$$

which is the Cauchy principal value of the integral, but the improper integral itself does not converge.

**Example 22.2** (Improper integrals of $1/(1 + x^2)$). Calculate the improper integrals

$$\int_a^{+\infty} \frac{dx}{1+x^2}, \quad \int_{-\infty}^0 \frac{dx}{1+x^2}.$$

*Solution.* We begin by noting that the function $f(x) = \frac{1}{1+x^2}$ is continuous and hence integrable on every finite interval.

First, consider the integral

$$\int_0^{+\infty} \frac{dx}{1+x^2} = \lim_{t \to +\infty} \int_0^t \frac{dx}{1+x^2}.$$

We compute

$$\int_0^t \frac{dx}{1+x^2} = \arctan x\big|_0^t = \arctan t - \arctan 0 = \arctan t.$$

Taking the limit as $t \to +\infty$,

$$\lim_{t \to +\infty} \arctan t = \frac{\pi}{2}.$$

Thus,

$$\int_0^{+\infty} \frac{dx}{1+x^2} = \frac{\pi}{2}.$$

Now, for the integral from $a$ to $+\infty$, assuming $a \in \mathbb{R}$, we write

$$\int_a^{+\infty} \frac{dx}{1+x^2} = \int_a^0 \frac{dx}{1+x^2} + \int_0^{+\infty} \frac{dx}{1+x^2},$$

but more directly

$$\int_a^{+\infty} \frac{dx}{1+x^2} = \lim_{t \to +\infty} \int_a^t \frac{dx}{1+x^2} = \lim_{t \to +\infty} (\arctan t - \arctan a) = \frac{\pi}{2} - \arctan a.$$

For the second integral,

$$\int_{-\infty}^0 \frac{dx}{1+x^2} = \lim_{s \to -\infty} \int_s^0 \frac{dx}{1+x^2} = \lim_{s \to -\infty} (\arctan 0 - \arctan s) = 0 - \left( -\frac{\pi}{2} \right) = \frac{\pi}{2}.$$

Therefore

$$\int_a^{+\infty} \frac{dx}{1+x^2} = \frac{\pi}{2} - \arctan a, \qquad \int_{-\infty}^0 \frac{dx}{1+x^2} = \frac{\pi}{2}. \qquad \square$$

**Example 22.3** (Convergence of $\int_a^{+\infty} \frac{dx}{x^m}$ for $a > 0$). Calculate the improper integral

$$\int_a^{+\infty} \frac{dx}{x^m}, \quad a > 0.$$

*Solution.* Let $f(x) = \frac{1}{x^m}$ for $x \geq a > 0$. We analyze the convergence of the improper integral

$$\int_a^{+\infty} \frac{dx}{x^m} = \lim_{t \to +\infty} \int_a^t \frac{dx}{x^m}.$$

We consider three cases:

- **Case** $m = 1$: Then

$$\int_a^t \frac{dx}{x} = \ln t - \ln a = \ln\left(\frac{t}{a}\right).$$

As $t \to +\infty$, $\ln(t/a) \to +\infty$, so the integral diverges.
- **Case** $m > 1$: We compute

$$\int_a^t \frac{dx}{x^m} = \left[\frac{x^{1-m}}{1-m}\right]_a^t = \frac{t^{1-m}}{1-m} - \frac{a^{1-m}}{1-m}.$$

Since $m > 1$, we have $1 - m < 0$, so $t^{1-m} \to 0$ as $t \to +\infty$. Thus,

$$\int_a^{+\infty} \frac{dx}{x^m} = \lim_{t \to +\infty}\left(\frac{t^{1-m}}{1-m} - \frac{a^{1-m}}{1-m}\right) = 0 - \frac{a^{1-m}}{1-m} = \frac{a^{1-m}}{m-1},$$

which is a finite real number. Hence, the integral converges.
- **Case** $m < 1$: Then $1 - m > 0$, so $t^{1-m} \to +\infty$ as $t \to +\infty$. Since $1 - m > 0$,

$$\int_a^t \frac{dx}{x^m} = \frac{t^{1-m} - a^{1-m}}{1-m} \to +\infty,$$

so the integral diverges.

Therefore, we conclude

$$\int_a^{+\infty} \frac{dx}{x^m} = \begin{cases} \frac{a^{1-m}}{m-1}, & m > 1, \\ +\infty, & m \le 1. \end{cases} \qquad \square$$

The following theorem provides a useful criterion to determine the convergence of improper integrals.

**Theorem 22.1** (Convergence of improper integrals for positive functions). *Let $f(x)$ be a positive and integrable function on $[a, x]$ for every $x \ge a$. If there exists a constant $M$ such that*

$$\int_a^x f(t)\, dt \le M,$$

*for all $x \ge a$, then the improper integral*

$$\int_a^\infty f(x)\, dx$$

*converges.*

*Proof.* Define $F(x) = \int_a^x f(t)\,dt$. The function $F$ is monotone increasing (since $f(x) \geq 0$) and bounded above by $M$, so the limit

$$\lim_{x \to \infty} F(x)$$

exists and is finite. Therefore, by definition,

$$\int_a^\infty f(x)\,dx = \lim_{x \to \infty} \int_a^x f(t)\,dt$$

converges. □

**Theorem 22.2** (Comparison test for improper integrals). *Let $f$ and $g$ be integrable functions on $[a, x]$ for all $x \geq a$ such that*

$$0 \leq f(x) \leq g(x).$$

*If the improper integral $\int_a^\infty g(x)\,dx$ converges, then the improper integral $\int_a^\infty f(x)\,dx$ also converges, and moreover,*

$$\int_a^\infty f(x)\,dx \leq \int_a^\infty g(x)\,dx.$$

*Proof.* Define

$$F(x) = \int_a^x f(t)\,dt \quad \text{and} \quad G(x) = \int_a^x g(t)\,dt.$$

Both are increasing functions (since $f, g \geq 0$) and satisfy $0 \leq F(x) \leq G(x)$ for all $x \geq a$. Since $\int_a^\infty g(x)\,dx$ converges, the limit $\lim_{x \to \infty} G(x)$ exists and is finite. Therefore, $G(x)$ is bounded above, and so is $F(x)$. By the monotone convergence theorem for functions, $F(x)$ converges to a finite limit as $x \to \infty$, which means

$$\int_a^\infty f(x)\,dx$$

converges. Furthermore, since $F(x) \leq G(x)$ for all $x$, taking limits gives

$$\int_a^\infty f(x)\,dx \leq \int_a^\infty g(x)\,dx.$$ □

**Definition 22.3** (Absolute convergence of improper integrals). The improper integral $\int_a^\infty f(x)\,dx$ is said to converge absolutely if the improper integral

$$\int_a^\infty |f(x)|\,dx$$

converges.

**Theorem 22.3** (Absolute convergence implies convergence). *Absolute convergence of $\int_a^\infty f(x)\,dx$ implies convergence of $\int_a^\infty f(x)\,dx$.*

*Proof.* Observe that

$$0 \le |f(x)| - f(x) \le 2|f(x)|.$$

Since $\int_a^\infty |f(x)|\,dx$ converges by assumption, the comparison test implies that $\int_a^\infty (|f(x)| - f(x))\,dx$ also converges.

Now, note that

$$f(x) = |f(x)| - (|f(x)| - f(x)).$$

Therefore,

$$\int_a^\infty f(x)\,dx = \int_a^\infty |f(x)|\,dx - \int_a^\infty (|f(x)| - f(x))\,dx$$

is the difference of two convergent improper integrals, hence it converges. $\square$

**Theorem 22.4** (Limit comparison test for improper integrals). *Let $f(x)$ and $g(x)$ be integrable functions on $[a, \infty)$ with $g(x) \ge 0$, and suppose that*

$$\lim_{x \to \infty} \frac{f(x)}{g(x)} = k \ne 0, \pm\infty.$$

*Then the improper integrals*

$$F = \int_a^\infty f(x)\,dx \quad and \quad G = \int_a^\infty g(x)\,dx$$

*either both converge or both diverge.*
*If $k = 0$ and $G$ converges, then $F$ converges absolutely.*
*If $k = \pm\infty$ and $G$ diverges, then $F$ also diverges.*

*Proof.* Assume $k > 0$. For every $\varepsilon > 0$, there exists $M > 0$ such that for all $x > M$,

$$(k - \varepsilon)g(x) < f(x) < (k + \varepsilon)g(x).$$

Since $g(x) \geq 0$ and the inequalities hold for $x > M$, we can apply the comparison test to the tails of the integrals. If $G$ converges, then so does $(k+\varepsilon)g(x)$, and thus $f(x)$ is bounded above by an integrable function, implying $F$ converges. Conversely, if $G$ diverges, then $(k - \varepsilon)g(x)$ also diverges (since $k - \varepsilon > 0$ for small $\varepsilon$), so $f(x) > (k - \varepsilon)g(x)$ implies $F$ diverges.

If $k < 0$, apply the same reasoning to $-f(x)$, since $\lim_{x \to \infty} \frac{-f(x)}{g(x)} = -k > 0$, and convergence of $\int f$ is equivalent to convergence of $\int(-f)$.

If $k = 0$, then for every $\varepsilon > 0$, there exists $M > 0$ such that for $x > M$,

$$|f(x)| < \varepsilon g(x).$$

If $G = \int_a^\infty g(x)\, dx$ converges, then so does $\int_M^\infty \varepsilon g(x)\, dx$, and by the comparison test, $\int_M^\infty |f(x)|\, dx$ converges. The integral from $a$ to $M$ is finite (since $f$ is integrable there), so $\int_a^\infty |f(x)|\, dx$ converges, i. e., $F$ converges absolutely.

If $k = +\infty$, then for any large $M > 0$, there exists $N > 0$ such that for $x > N$,

$$f(x) > Mg(x).$$

If $G$ diverges, then $\int_N^\infty Mg(x)\, dx$ also diverges (since $M > 0$), so $f(x) > Mg(x)$ implies $\int_N^\infty f(x)\, dx$ diverges, hence $F$ diverges. The case $k = -\infty$ is similar by considering $-f(x)$.

□

**Corollary 22.1** (Convergence of $\int_a^\infty f(x)\, dx$ based on asymptotic behavior). *Let $f(x)$ be integrable on $[a, \infty)$, and suppose*

$$\lim_{x \to \infty} x^p f(x) = k \neq \pm\infty,$$

*with*

$$p > 1.$$

*Then the integral $\int_a^\infty f(x)\, dx$ converges absolutely.*
*If*

$$\lim_{x \to \infty} x^p f(x) = k \neq 0, \pm\infty,$$

*with*

$$p \leq 1,$$

*then the integral diverges.*

*Proof.* Apply Theorem 22.4 with the comparison function $g(x) = \frac{1}{x^p}$. Then

$$\lim_{x \to \infty} \frac{f(x)}{g(x)} = \lim_{x \to \infty} x^p f(x) = k.$$

We know from Example 22.3 that the integral

$$G = \int_a^\infty \frac{1}{x^p}\, dx$$

converges if and only if $p > 1$.

- If $p > 1$ and $k \neq \pm\infty$, then by the limit comparison test, $\int_a^\infty f(x)\, dx$ converges absolutely (since $|f(x)| \sim |k|/x^p$, and $\int |k|/x^p\, dx$ converges).
- If $p \leq 1$ and $k \neq 0, \pm\infty$, then $G$ diverges and the limit is nonzero, so the limit comparison test implies that $\int_a^\infty f(x)\, dx$ also diverges.

This completes the proof. $\qquad\qquad\qquad\qquad\qquad\qquad\qquad\qquad$ □

## 22.2 Improper integrals with finite limits

**Definition 22.4** (Improper integral on a finite interval (Figures 22.2, 22.3)). Let $f : [a, b) \to \mathbb{R}$ be a function such that, for every $t \in (a, b)$, the function $f$ is integrable on the interval $[a, t]$, but not necessarily on $[a, b)$. If the limit

$$\lim_{t \to b^-} \int_a^t f(x)\, dx = \lambda \in \mathbb{R}$$

exists, then we say that $f$ is **improperly integrable on** $[a, b]$, and the improper integral

$$\int_a^b f(x)\, dx$$

**converges** to $\lambda$. If this limit does not exist, the improper integral is said to **diverge**.

**Remark 2** (Substitution for endpoint singularities). Suppose the function $f$ has a singularity at the left endpoint $a$, i.e., $f(a^+) = \infty$. Then, using the substitution $x = a + \frac{1}{t}$, the improper integral becomes

$$\int_{a+\varepsilon}^b f(x)\, dx = \int_{\frac{1}{b-a}}^{\frac{1}{\varepsilon}} \frac{1}{t^2} f\!\left(a + \frac{1}{t}\right) dt,$$

and taking the limit as $\varepsilon \to 0$, we obtain

$$\int_a^b f(x)\, dx = \int_{\frac{1}{b-a}}^{+\infty} \frac{1}{t^2} f\!\left(a + \frac{1}{t}\right) dt.$$

Similarly, if the function $f$ has a singularity at the right endpoint $b$, i.e., $f(b^-) = \infty$, then using the substitution $x = b - \frac{1}{t}$, we get

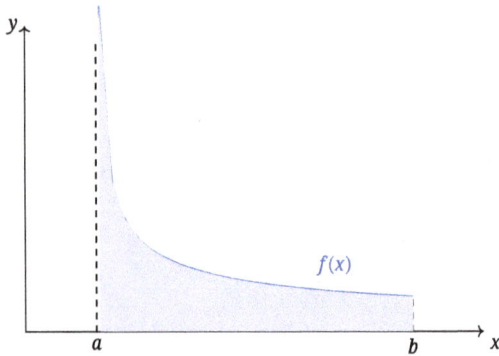

Figure 22.2: Improper integral on $[a, b]$ due to singularity at $a$. The graph of $f(x)$ diverges as $x \to a^+$.

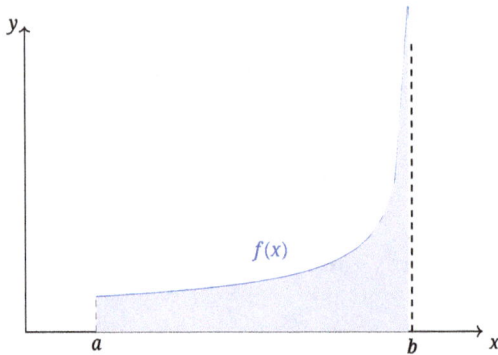

Figure 22.3: Improper integral on $[a, b]$ due to singularity at $b$. The graph of $f(x)$ diverges as $x \to b^-$.

$$\int_a^b f(x)\, dx = \int_{\frac{1}{b-a}}^{+\infty} \frac{1}{t^2} f\left(b - \frac{1}{t}\right) dt.$$

These transformations allow us to reduce the problem of integration near a singularity to an integral over an unbounded domain, which can be more tractable for analysis or numerical computation.

The following corollaries provide comparison tests near such finite endpoints.

**Corollary 22.2** (Comparison test for improper integrals at an endpoint). *Let $f$ and $g$ be integrable on $[a +$ $\varepsilon, b]$ for every $\varepsilon > 0$ or on $[a, b - \varepsilon]$ for every $\varepsilon > 0$, with $0 \leq f(x) \leq g(x)$. If $\int_a^b g(x)\, dx$ converges (as an improper integral), then so does $\int_a^b f(x)\, dx$.*

**Corollary 22.3** (Limit comparison test near a singularity). *Let $f(x)$ and $g(x)$ be integrable on $(a, b]$ with $g(x) \geq 0$, and suppose*

$$\lim_{x \to a^+} \frac{f(x)}{g(x)} = k \neq 0, \pm\infty.$$

*Then the improper integrals*

$$F = \int_a^b f(x)\, dx, \quad G = \int_a^b g(x)\, dx$$

*either both converge or both diverge. If $k = 0$ and $G$ converges, then $F$ converges absolutely. If $k = \pm\infty$ and $G$ diverges, then $F$ diverges.*

**Corollary 22.4** (Convergence based on power behavior near a singularity). *Let $f$ be integrable on $(a, b]$ and suppose*

$$\lim_{x \to a^+} (x - a)^q f(x) = k \neq \pm\infty, \quad q \in (0, 1).$$

*Then $\int_a^b f(x)\, dx$ converges absolutely. If*

$$\lim_{x \to a^+} (x - a)^q f(x) = k \neq 0, \pm\infty, \quad q \geq 1,$$

*then the integral diverges.*

*Proof.* Apply the limit comparison test (Corollary 22.3) with the comparison function $g(x) = \frac{1}{(x-a)^q}$. Then

$$\lim_{x \to a^+} \frac{f(x)}{g(x)} = \lim_{x \to a^+} (x - a)^q f(x) = k.$$

Now consider the improper integral

$$G = \int_a^b \frac{dx}{(x - a)^q}.$$

This is a standard improper integral of Type II. For $\varepsilon > 0$,

$$\int_{a+\varepsilon}^b \frac{dx}{(x - a)^q} = \left[ \frac{(x - a)^{1-q}}{1 - q} \right]_{a+\varepsilon}^b = \frac{(b - a)^{1-q}}{1 - q} - \frac{\varepsilon^{1-q}}{1 - q}.$$

- If $q < 1$, then $1 - q > 0$, so $\varepsilon^{1-q} \to 0$ as $\varepsilon \to 0^+$, and the integral converges to $\frac{(b-a)^{1-q}}{1-q}$.
- If $q \geq 1$, then $1 - q \leq 0$, so $\varepsilon^{1-q} \to +\infty$ as $\varepsilon \to 0^+$, and the integral diverges.

Therefore:

- If $q \in (0,1)$ and $k \neq \pm\infty$, then $G$ converges, so by the limit comparison test, $\int_a^b f(x)\,dx$ converges absolutely.
- If $q \geq 1$ and $k \neq 0, \pm\infty$, then $G$ diverges and the limit is nonzero, so $F$ also diverges.

This completes the proof. □

The next theorem is a classical test for convergence of integrals involving oscillatory functions.

**Theorem 22.5** (Dirichlet test for improper integrals). *Let $g$ be an integrable and monotone function on $[a, \infty)$. Then the improper integral*

$$I = \int_a^\infty g(x) \sin x \, dx$$

*converges. Moreover, if the series*

$$\sum_{k=m}^\infty g(k\pi), \quad m\pi \geq a,$$

*converges absolutely, then $I$ converges absolutely.*

*Proof.* We aim to show convergence of

$$I = \int_a^\infty g(x) \sin x \, dx,$$

where $g$ is monotone and integrable on $[a, \infty)$.

Let $m, n \in \mathbb{N}$ such that

$$a \leq m\pi < n\pi < b < (n+1)\pi.$$

We split the integral

$$\int_a^b g(x) \sin x \, dx = \int_a^{m\pi} g(x) \sin x \, dx + \int_{m\pi}^{n\pi} g(x) \sin x \, dx + \int_{n\pi}^b g(x) \sin x \, dx.$$

As $b \to \infty$, the third integral vanishes in the limit (since it's bounded by $\sup_{[n\pi,(n+1)\pi]} |g(x)| \cdot \pi$, and $g(x) \to 0$ if $g$ is monotone and integrable—a standard result).

The first integral is over a finite interval and hence exists.

Now consider the middle term

$$\int_{m\pi}^{n\pi} g(x) \sin x \, dx = \sum_{k=m}^{n-1} \int_{k\pi}^{(k+1)\pi} g(x) \sin x \, dx.$$

Using the substitution $x = t + k\pi$, so $\sin x = \sin(t + k\pi) = (-1)^k \sin t$, we get

$$\int_{k\pi}^{(k+1)\pi} g(x) \sin x \, dx = (-1)^k \int_0^\pi g(t + k\pi) \sin t \, dt.$$

Since $\sin t \geq 0$ on $[0, \pi]$, and $g$ is monotone, we can apply the mean value theorem for integrals: there exists $\xi_k \in (0, \pi)$ such that

$$\int_0^\pi g(t + k\pi) \sin t \, dt = g(k\pi + \xi_k) \int_0^\pi \sin t \, dt = 2g(k\pi + \xi_k).$$

Thus,

$$\int_{k\pi}^{(k+1)\pi} g(x) \sin x \, dx = (-1)^k \cdot 2g(k\pi + \xi_k).$$

Therefore,

$$\sum_{k=m}^{n-1} \int_{k\pi}^{(k+1)\pi} g(x) \sin x \, dx = 2 \sum_{k=m}^{n-1} (-1)^k g(k\pi + \xi_k).$$

Since $g$ is monotone and tends to zero (as it is monotone and integrable on $[a, \infty)$), the sequence $g(k\pi + \xi_k)$ is monotone decreasing to zero. Hence, by the alternating series test, the series

$$\sum_{k=m}^{\infty} (-1)^k g(k\pi + \xi_k)$$

converges. Therefore, the integral $I$ converges.

Now, for **absolute convergence**, suppose that

$$\sum_{k=m}^{\infty} |g(k\pi)| < \infty.$$

We consider

$$\int_a^\infty |g(x) \sin x| \, dx.$$

Note that

$$|g(x) \sin x| \geq |g(x)| \cdot |\sin x|,$$

but we need a lower bound. However, since $|\sin x|$ is periodic and bounded below on average, and $g$ is monotone, one can show that if $\int_a^\infty |g(x)\sin x|\,dx$ converges, then so does $\sum |g(k\pi)|$. Conversely, if $\sum |g(k\pi)|$ converges, then $g(x) \to 0$ fast enough that the integral

$$\int_a^\infty |g(x)\sin x|\,dx$$

converges due to the uniform bound over intervals and summability.

Thus, under the assumption of absolute convergence of $\sum g(k\pi)$, the integral $I$ converges absolutely.

This completes the proof. □

### 22.2.1 Examples

**Example 22.4** (Improper integral involving $\frac{1}{\sqrt{1-x^2}}$). Consider the improper integral

$$\int_0^1 \frac{dx}{\sqrt{1-x^2}}.$$

The function $f(x) = \frac{1}{\sqrt{1-x^2}}$ is integrable on $[0,t]$ for every $t \in (0,1)$, but has a singularity at $x = 1$. We compute

$$\int_0^t \frac{dx}{\sqrt{1-x^2}} = \arcsin x\big|_0^t = \arcsin t.$$

Thus,

$$\lim_{t\to 1^-}\int_0^t \frac{dx}{\sqrt{1-x^2}} = \lim_{t\to 1^-}\arcsin t = \frac{\pi}{2}.$$

Therefore, the improper integral converges and

$$\int_0^1 \frac{dx}{\sqrt{1-x^2}} = \frac{\pi}{2}.$$

**Example 22.5** (Gaussian integral $\int_{-\infty}^\infty e^{-x^2}\,dx$). Evaluate the improper integral

$$I = \int_{-\infty}^{+\infty} e^{-x^2}\,dx,$$

which converges.

We split

$$I = \int\limits_{-\infty}^{0} e^{-x^2}\, dx + \int\limits_{0}^{+\infty} e^{-x^2}\, dx = 2 \int\limits_{0}^{+\infty} e^{-x^2}\, dx.$$

Further split

$$\int\limits_{0}^{\infty} e^{-x^2}\, dx = \int\limits_{0}^{1} e^{-x^2}\, dx + \int\limits_{1}^{\infty} e^{-x^2}\, dx.$$

On $[0, 1]$, $e^{-x^2}$ is continuous, hence integrable. For $x \geq 1$, we have

$$0 \leq e^{-x^2} \leq e^{-x},$$

and since $\int_1^{\infty} e^{-x}\, dx = 1 < \infty$, the comparison test implies $\int_1^{\infty} e^{-x^2}\, dx$ converges. Hence, $I$ converges. It is known that $I = \sqrt{\pi}$ (a classic result proven via polar coordinates).

**Example 22.6** (Convergence of $\int_1^{\infty} \frac{dx}{1+x^2}$). The integral

$$\int\limits_{1}^{\infty} \frac{dx}{1 + x^2}$$

converges.

Since for $x \geq 1$,

$$\frac{1}{1 + x^2} \leq \frac{1}{x^2},$$

and $\int_1^{\infty} \frac{1}{x^2}\, dx = 1 < \infty$, the comparison test implies convergence. Indeed,

$$\int\limits_{1}^{t} \frac{1}{x^2}\, dx = \left[ -\frac{1}{x} \right]_1^{t} = 1 - \frac{1}{t} \to 1,$$

so

$$\int\limits_{1}^{\infty} \frac{1}{x^2}\, dx = 1.$$

**Example 22.7** (Divergence of $\int_1^{\infty} \frac{1}{\sqrt{x}}\, dx$). The integral

$$\int\limits_{1}^{\infty} \frac{1}{\sqrt{x}}\, dx$$

diverges.

For $x \geq 1$,

$$\frac{1}{x} \leq \frac{1}{\sqrt{x}},$$

and $\int_1^\infty \frac{1}{x} \, dx = \infty$. By the comparison test, it follows that

$$\int_1^\infty \frac{1}{\sqrt{x}} \, dx = \lim_{t \to \infty} \int_1^t x^{-1/2} \, dx = \lim_{t \to \infty} 2\sqrt{t} - 2 = \infty.$$

**Example 22.8** (Limit comparison test examples). The integrals

$$\int_1^\infty \frac{dx}{x^2}, \quad \int_1^\infty \frac{dx}{1+x^2}$$

both converge.

Let $f(x) = \frac{1}{x^2}, g(x) = \frac{1}{1+x^2}$. Then

$$\lim_{x \to \infty} \frac{f(x)}{g(x)} = \lim_{x \to \infty} \frac{1+x^2}{x^2} = 1 \neq 0, \infty.$$

By the limit comparison test, both integrals share the same behavior. Since $\int_1^\infty \frac{dx}{x^2} = 1$, both converge.
Explicitly:

$$\int_1^\infty \frac{dx}{1+x^2} = \arctan x \big|_1^\infty = \frac{\pi}{2} - \frac{\pi}{4} = \frac{\pi}{4}.$$

**Example 22.9** (Divergence of $\int_6^\infty \frac{1}{\sqrt{x-5}} \, dx$). The integral

$$\int_6^\infty \frac{1}{\sqrt{x-5}} \, dx$$

diverges.

Let $u = x - 5$, then

$$\int_6^\infty \frac{dx}{\sqrt{x-5}} = \int_1^\infty u^{-1/2} \, du = \lim_{t \to \infty} 2\sqrt{t} - 2 = \infty.$$

**Example 22.10** (Convergence of $\int_1^\infty \frac{dx}{x\sqrt{x^2+1}}$). Evaluate the integral

$$\int_1^\infty \frac{dx}{x\sqrt{x^2+1}}.$$

Consider

$$x^2 \cdot \frac{1}{x\sqrt{x^2+1}} = \frac{x}{\sqrt{x^2+1}} \to 1 \quad \text{as } x \to \infty.$$

So $f(x) \sim \frac{1}{x^2}$, and since $p = 2 > 1$, by the $x^{-p}$ test, the integral converges.

**Example 22.11** (Divergence of $\int_0^\infty \frac{x^2}{\sqrt{x^5+1}}\, dx$). The integral

$$\int_0^\infty \frac{x^2}{\sqrt{x^5+1}}\, dx$$

diverges.

As $x \to \infty$,

$$\frac{x^2}{\sqrt{x^5+1}} \sim \frac{x^2}{x^{5/2}} = x^{-1/2},$$

and $\int_1^\infty x^{-1/2}\, dx$ diverges ($p = 1/2 \le 1$). Hence, the tail diverges.

At $x = 0$, the integrand $\to 0$, so no issue. Therefore, the integral diverges.

**Example 22.12** (Convergence of $\int_0^1 \frac{\ln x}{\sqrt{x}}\, dx$). The integral

$$\int_0^1 \frac{\ln x}{\sqrt{x}}\, dx$$

converges.

Near $x = 0^+$, we analyze

$$\lim_{x \to 0^+} x^{3/4} \cdot \frac{\ln x}{x^{1/2}} = \lim_{x \to 0^+} x^{1/4} \ln x = 0,$$

since $x^a \ln x \to 0$ for any $a > 0$.

So $f(x) = \frac{\ln x}{\sqrt{x}} = o(x^{-3/4})$, and since $q = 3/4 < 1$, by Corollary 22.4, the integral converges.

**Example 22.13** (Convergence of $\int_0^\infty \frac{\sin x}{x}\, dx$). The integral

$$\int_0^\infty \frac{\sin x}{x}\, dx$$

converges conditionally.

As $x \to 0$, $\frac{\sin x}{x} \to 1$, so the integral over $[0, \pi]$ is proper.

For $[\pi, \infty)$, apply Dirichlet's test (Theorem 22.5): $g(x) = \frac{1}{x}$ is positive, decreasing, and tends to 0. So

$$\int_\pi^\infty \frac{\sin x}{x}\, dx$$

converges.

But $\int_0^\infty |\frac{\sin x}{x}| dx$ diverges, so convergence is conditional.

**Example 22.14** (Convergence of $\int_0^\infty \sin x^2\, dx$). The integral

$$\int_0^\infty \sin x^2\, dx$$

converges conditionally.

Substitute $x = \sqrt{t}$, so $dx = \frac{1}{2} t^{-1/2}\, dt$,

$$\int_{\sqrt{\pi}}^\infty \sin x^2\, dx = \int_\pi^\infty \sin t \cdot \frac{1}{2\sqrt{t}}\, dt = \frac{1}{2} \int_\pi^\infty \frac{\sin t}{\sqrt{t}}\, dt.$$

Now apply Dirichlet's test: $g(t) = \frac{1}{\sqrt{t}}$ is decreasing to 0, so the integral converges.

Since $\int_\pi^\infty |\frac{\sin t}{\sqrt{t}}| dt$ diverges, convergence is conditional.

**Example 22.15** (Convergence of $\int_{-1}^7 \frac{dx}{\sqrt[3]{x+1}}$). Determine whether the integral

$$\int_{-1}^7 \frac{dx}{\sqrt[3]{x+1}}$$

converges and, if so, compute its value.

*Solution.* Singularity at $x = -1$. Let $u = x + 1$, so

$$\int_{-1}^7 (x+1)^{-1/3}\, dx = \int_0^8 u^{-1/3}\, du = \left[\frac{u^{2/3}}{2/3}\right]_0^8 = \frac{3}{2} \cdot 8^{2/3} = \frac{3}{2} \cdot 4 = 6.$$

Since $p = 1/3 < 1$, the improper integral converges. Value is 6. □

**Example 22.16** (Beta function integral). For $m, n \in \mathbb{N}$, evaluate

$$\int_0^1 x^{m-1}(1-x)^{n-1}\, dx,$$

and prove the Beta function property.

*Solution.* The integral in question is a well-known representation of the **Beta function**, denoted as $B(m, n)$. The Beta function is defined as

$$B(m, n) = \int_0^1 x^{m-1}(1-x)^{n-1}\, dx,$$

where $m, n > 0$. For $m, n \in \mathbb{N}$, this is a proper integral because the integrand $x^{m-1}(1 - x)^{n-1}$ is continuous on $[0, 1]$.

### Step 1: Evaluating the integral for natural numbers

For $m, n \in \mathbb{N}$, the result of the integral is given by

$$B(m, n) = \frac{(m - 1)!(n - 1)!}{(m + n - 1)!}.$$

### Proof by induction:

We proceed by induction on $m$ and $n$.

1. **Base case ($m = 1$):** When $m = 1$, the integral becomes

$$B(1, n) = \int_0^1 (1 - x)^{n-1} \, dx.$$

This is a standard power rule integral

$$\int_0^1 (1 - x)^{n-1} \, dx = \left[ -\frac{(1 - x)^n}{n} \right]_0^1 = \frac{1}{n}.$$

Thus, $B(1, n) = \frac{1}{n}$, which matches the formula

$$B(1, n) = \frac{(1 - 1)!(n - 1)!}{(1 + n - 1)!} = \frac{0!(n - 1)!}{n!} = \frac{1}{n}.$$

2. **Inductive step:** Assume the formula holds for $B(m, n)$, i. e.,

$$B(m, n) = \frac{(m - 1)!(n - 1)!}{(m + n - 1)!}.$$

We need to show that it holds for $B(m + 1, n)$. Using integration by parts, let

$$u = x^m, \quad dv = (1 - x)^{n-1} \, dx.$$

Then

$$du = mx^{m-1} \, dx, \quad v = -\frac{(1 - x)^n}{n}.$$

Applying integration by parts,

$$\int_0^1 x^m (1 - x)^{n-1} \, dx = \left[ -\frac{x^m (1 - x)^n}{n} \right]_0^1 + \int_0^1 \frac{mx^{m-1}(1 - x)^n}{n} \, dx.$$

The boundary term vanishes at both $x = 0$ and $x = 1$. Thus

$$B(m + 1, n) = \frac{m}{n} \int_0^1 x^{m-1}(1 - x)^n \, dx.$$

Recognizing the remaining integral as $B(m, n + 1)$, we have

$$B(m + 1, n) = \frac{m}{n} B(m, n + 1).$$

Using the inductive hypothesis for $B(m, n + 1)$, we get

$$B(m + 1, n) = \frac{m}{n} \cdot \frac{(m - 1)!n!}{(m + n)!} = \frac{m!(n - 1)!}{(m + n)!}.$$

This completes the induction.

Thus, for $m, n \in \mathbb{N}$, we have

$$\int_0^1 x^{m-1}(1 - x)^{n-1} \, dx = \frac{(m - 1)!(n - 1)!}{(m + n - 1)!}.$$

**Step 2: Connection to the Gamma function**
The Beta function is closely related to the Gamma function, defined as

$$\Gamma(z) = \int_0^\infty t^{z-1} e^{-t} \, dt, \quad z > 0.$$

The relationship between the Beta and Gamma functions is

$$B(m, n) = \frac{\Gamma(m)\Gamma(n)}{\Gamma(m + n)}.$$

**Proof:** □

**Exercise 24.** Determine whether the following integrals converge or diverge:

$$\int_1^\infty \frac{x \, dx}{3x^4 + 5x^2 + 1}, \quad \int_2^\infty \frac{x^2 - 1}{\sqrt{x^6 + 16}} \, dx, \quad \int_1^\infty \frac{\ln x}{x + a} \, dx, \quad a > 0,$$

$$\int_0^\infty \frac{1 - \cos x}{x^2} \, dx, \quad \int_{-\infty}^{-1} \frac{e^x}{x} \, dx, \quad \int_{-\infty}^\infty \frac{x^3 + x^2}{x^6 + 1} \, dx.$$

## 22.3 Python and AI

___

**AI Request**

Determine whether the following integrals converge or diverge:

$$\int_1^\infty \frac{x\,dx}{3x^4 + 5x^2 + 1}, \quad \int_2^\infty \frac{x^2 - 1}{\sqrt{x^6 + 16}}\,dx, \quad \int_1^\infty \frac{\ln x}{x + a}\,dx, \quad a > 0,$$

$$\int_0^\infty \frac{1 - \cos x}{x^2}\,dx, \quad \int_{-\infty}^{-1} \frac{e^x}{x}\,dx, \quad \int_{-\infty}^\infty \frac{x^3 + x^2}{x^6 + 1}\,dx.$$

Write Python codes to compare the above results.

___

## 22.4 Exercises

### True or false

Decide whether each statement is **true** or **false**. Justify briefly.

1.  The improper integral $\int_1^\infty \frac{dx}{x}$ converges.

    **Answer:** _____

2.  If $\int_a^\infty |f(x)|\,dx$ converges, then $\int_a^\infty f(x)\,dx$ converges.

    **Answer:** _____

3.  The integral $\int_0^1 \frac{dx}{x}$ is an improper integral due to the integrand being unbounded.

    **Answer:** _____

4.  If $f(x) \le g(x)$ for $x \ge a$, and $\int_a^\infty g(x)\,dx$ diverges, then $\int_a^\infty f(x)\,dx$ diverges.

    **Answer:** _____

5.  The integral $\int_{-\infty}^\infty x\,dx$ converges to zero.

    **Answer:** _____

### Multiple choice questions

Choose the correct option (A, B, C, or D).

1.  The integral $\int_1^\infty \frac{dx}{x^2}$ equals

    A. 0    B. 1

    C. $\infty$    D. does not exist

    **Answer:** _____

2.  The integral $\int_0^1 \frac{dx}{\sqrt{x}}$ is

    A. convergent           B. divergent

    C. absolutely convergent   D. conditionally convergent

    **Answer:** _____

3.  By the comparison test, if $0 \le f(x) \le \frac{1}{x^p}$ for $x \ge 1$, the integral $\int_1^\infty f(x)\,dx$ converges if

A. $p > 1$   B. $p = 1$
C. $p < 1$   D. Always
**Answer:** _____

4.   The integral $\int_0^\infty e^{-x^2}\, dx$ converges to
A. 0   B. 1
C. $\frac{\sqrt{\pi}}{2}$   D. $\sqrt{\pi}$
**Answer:** _____

5.   Dirichlet's test guarantees convergence of integrals involving
A. monotone and bounded functions   B. oscillatory functions like $\sin x$
C. exponential functions   D. rational functions
**Answer:** _____

## Matching exercise

Match each integral with the correct convergence behavior.
1. $\int_1^\infty \frac{dx}{x}$   A. Absolutely convergent
2. $\int_1^\infty \frac{dx}{x^2}$   B. Divergent
3. $\int_0^1 \frac{dx}{\sqrt{x}}$   C. Conditionally convergent
4. $\int_0^\infty \frac{\sin x}{x}\, dx$   D. Convergent
5. $\int_0^\infty e^{-x}\, dx$   E. Absolutely divergent
**Answers: 1.** _____   2. _____   3. _____   4. _____   5. _____

## Fill in the blanks

Complete the following sentences:
1.   An integral whose limits of integration are infinite or whose integrand is unbounded is called an _____.
2.   The integral $\int_a^\infty f(x)\, dx$ converges absolutely if _____ converges.
3.   By Dirichlet's test, the integral $\int_a^\infty g(x) \sin x\, dx$ converges if $g(x)$ is _____.
4.   If $f(x) \sim \frac{1}{x^p}$ as $x \to \infty$, the integral $\int_1^\infty f(x)\, dx$ converges if _____.
5.   The Gaussian integral $\int_{-\infty}^\infty e^{-x^2}\, dx$ converges to _____.

## Theoretical and computational problems

Solve each integral, determining convergence or divergence.
1.   $\int_1^\infty \frac{\ln x}{x^2}\, dx$
2.   $\int_0^1 \frac{dx}{\sqrt[3]{x}}$
3.   $\int_1^\infty \frac{dx}{x(x^2+1)}$
4.   $\int_0^\infty x e^{-x}\, dx$
5.   $\int_{-\infty}^\infty \frac{dx}{x^2+4x+5}$

### Python-based exercises

Use Python (numpy, sympy, matplotlib) to perform the following:

1. Numerically evaluate the integral $\int_1^\infty \frac{dx}{x^2}$ and compare with its exact value.

2. Plot the function $f(x) = e^{-x^2}$ for $x \in [-3,3]$ and verify numerically the value of the integral $\int_{-\infty}^\infty e^{-x^2}\, dx$.

3. Compute numerically the integral $\int_1^\infty \frac{\sin x}{x}\, dx$ and discuss its convergence.

4. Use sympy to verify if the integral $\int_0^1 \frac{dx}{\sqrt{x}}$ converges or diverges.

5. Implement and demonstrate Dirichlet's test for the integral $\int_1^\infty \frac{\sin x}{x^2}\, dx$.

# Bibliography

[1]   J. Bak, D. J. Newman, *Complex Analysis*, Springer, 1997.

[2]   L. Brand, *Advanced Calculus*, Dover, 2006.

[3]   R. Courant, F. John, *Introduction to Calculus and Analysis I*, Springer, 1989.

[4]   P. Davis, *Interpolation and Approximation*, Dover, 1975.

[5]   H. Hamilton, The partial decomposition of a rational function, *Mathematics Magazine* 45(3), 117–119, 1972.

[6]   J. Havil, *Gamma: Exploring Euler's Constant*, Princeton University Press, 2003.

[7]   W. J. Kaczor, M. T. Nowak, *Problems in Mathematical Analysis I, II and III*, AMS, 2000.

[8]   E. Landau, *Differential and Integral Calculus*, Chelsea, 1951.

[9]   E. Mendelson, *3000 Solved Problems in Calculus*, McGraw Hill, 1988.

[10]  M. Protter, C. Morrey, *A First Course in Real Analysis*, second edition, Springer, 1997.

[11]  W. Rudin, *Principles of Mathematical Analysis*, McGraw-Hill, 1976.

[12]  K. Stromberg, *Introduction to Classical Real Analysis*, AMS, 2015.

[13]  A. Zygmunt, *Trigonometric Series*, Cambridge University Press, 2002.

https://doi.org/10.1515/9783112228289-023

# Index

https://doi.org/10.1515/9783112228289-024

www.ingramcontent.com/pod-product-compliance
Lightning Source LLC
Chambersburg PA
CBHW080130220326
41598CB00032B/5014